Citrus Health Management

edited by

L. W. Timmer
Larry W. Duncan
University of Florida
Citrus Research and Education Center, Lake Alfred

APS PRESS
The American Phytopathological Society

Front cover photograph and back cover photographs (upper left and lower left) reproduced, by permission, from the Florida Department of Citrus

Reference in this publication to a trademark, proprietary product, or company name by personnel of the U.S. Department of Agriculture or anyone else is intended for explicit description only and does not imply approval or recommendation to the exclusion of others that may be suitable.

Library of Congress Catalog Card Number: 98-83296
International Standard Book Number: 0-89054-227-9

Printed in the United States of America on acid-free paper

The American Phytopathological Society
3340 Pilot Knob Road
St. Paul, Minnesota 55121-2097, USA

Preface

The principal emphasis of *Citrus Health Management* is the economical production and maintenance of healthy citrus groves in the broadest sense. We have considered all aspects of citrus crop management, from the production of healthy nursery trees to the maintenance of fruit quality after harvest. Since production cost is also a vital element in maintaining citrus health, we have included a chapter on economic decision making for growers.

The *Compendium of Citrus Diseases* (American Phytopathological Society, 1988) provides a pictorial guide to citrus diseases and related problems, with detailed information on each. In *Citrus Health Management,* we have attempted to provide means of addressing these problems as well as citrus pests and horticultural problems. We have emphasized procedures for diagnosis and for detection and enumeration of pathogens and pests. Where economic thresholds have been established and new equipment and methods are available, we have highlighted them to put health management on a sounder, more quantitative basis. These techniques are not as advanced as some that have been developed for more intensively studied crops, but the application of these methods should greatly facilitate and improve citrus production.

Improved diagnostic procedures, such as dot blot assays and polymerase chain reaction, have made the detection of viruses and other systemic pathogens much easier and more rapid. Quantitation of pests and pathogens is essential for control, and thus methods for assaying rust mite and nematode populations, *Phytophthora* propagule density in the soil, and nutrient levels in citrus trees are described. Weather-based disease forecasts are beginning to be used in citrus management, and a predictive model for postbloom fruit drop is presented. These procedures and techniques are highlighted in boxes throughout the book.

Citrus is produced for the fresh fruit market or for processing into juice. The emphasis on production for the fresh market or for processing differs in various areas, and production practices must be adjusted accordingly. In fruit destined for processing, external quality is not important, and there is little need to manage pests and diseases which cause only external blemishes. Biological control is much more easily implemented in fruit grown for processing, since the level of damage which is tolerable is much greater than for fresh fruit. In various chapters the different needs of citrus grown for the fresh market and citrus grown for processing are addressed, and adjustments in management practices are described accordingly.

Most of the authors are from Florida, but the scope of the book goes far beyond there. Authors from areas with Mediterranean climates have been included to broaden the treatment of certain subjects. In all cases, the authors have considerable international experience and have approached their subjects with a global perspective. Problems and practices in citrus production vary considerably in different growing regions, from arid and semiarid areas with alkaline soils producing primarily fresh-market fruit to humid areas with acid soils producing more processing fruit, and there are great differences between near-temperate production areas, such as Japan, and the humid tropics of Asia, Africa, and South America. The authors have described adjustments in production practices needed under these different circumstances. Although the emphasis is on Florida, producers worldwide will find *Citrus Health Management* an informative resource in addressing crop production problems.

We thank all of the authors, who contributed their time and expertise to make this a high-quality publication. In addition, others provided slides, illustrations, and information which have added greatly to the book. The assistance of Diana Drouillard in the preparation of many of the figures is greatly appreciated. We are grateful to Barbara Thompson and the secretarial staff at the Citrus Research and Education Center for word processing services and for handling much correspondence in the preparation of this book.

We hope readers find this publication a useful tool in the production of high-quality citrus crops.

L. W. Timmer
Larry W. Duncan

Contents

CITRUS HEALTH MANAGEMENT

PART ONE

Introduction

CHAPTER 1

Larry W. Duncan
University of Florida
Citrus Research and Education Center, Lake Alfred

The Basis of Citrus Health Management

In its broadest sense, the term *citrus health management* means more than managing the health of trees in an orchard. It also means protecting the interests of citriculture into the foreseeable future. Understanding the influence of management practices on the profitability of citrus industries and the well-being of the environment on which those industries depend are vital concerns of health management. One of the goals of this book is to reinforce an understanding of the interrelationships between the various and diverse aspects of citriculture.

Just as a plant, a pathogen, and the environment interact to give rise to disease, as represented by the familiar disease triangle, so too profitability, the environment, and management practices determine the sustainability of agricultural endeavors such as citriculture (Fig. 1.1). Profitability is the basic incentive for growers to produce citrus. Profits also regulate the types of management practices growers employ, which in turn affect the orchard environment. Environmental quality regulates the resources available to growers for health management, which in turn affects orchard profitability. Management practices are the only means by which growers can influence profitability and are more or less sustainable according to their effect on profits and the environment.

Most growers are familiar with examples of management practices which, though once effective, could not be sustained because of negative consequences for the environment, lack of profitability, or even loss of efficacy of the practices themselves. In citriculture, as in many agricultural endeavors, restrictions on the use of many fertilizer nutrients and pesticides are being imposed to reduce groundwater pollution. The ability to market a commodity can be seriously damaged by public reports of chemical residues in fruit. Competition between growers and some urban communities for diminishing supplies of land and water has increased the cost of these commodities to levels at which citriculture is less profitable than other enterprises. We have learned through long experience with secondary pests and diseases that one problem can be quickly replaced by another when naturally occurring biological control agents are suppressed by pesticides. Misuse of pesticides and overreliance on them have also resulted in the development of pest and pathogen resistance to a broad range of chemicals. For all of these reasons, it has become necessary to view health management practices with a careful eye toward their economic and environmental consequences today and in the future. To do so requires growers to deal with a complex array of problems that were unrecognized by previous generations.

This book provides a holistic treatment of citrus health management, recognizing the interdependence of markets, the environment, and management practices. Many of the topics involve a set of limiting factors that cannot successfully operate independently of one another. Chapters dealing with land and water management, horticultural requirements of bearing and nonbearing trees, and economic principles outline the basic requirements of productive orchards and the most sustainable practices to provide those requirements, as we currently understand them. The successful application of the principles of pest and disease management, dealt with in subsequent chapters, is predicated on sound orchard management practices. As with any broad treatment of a subject, readers will often require more detailed information on specific regions and topics. Selected references for additional information are listed at the end of the book.

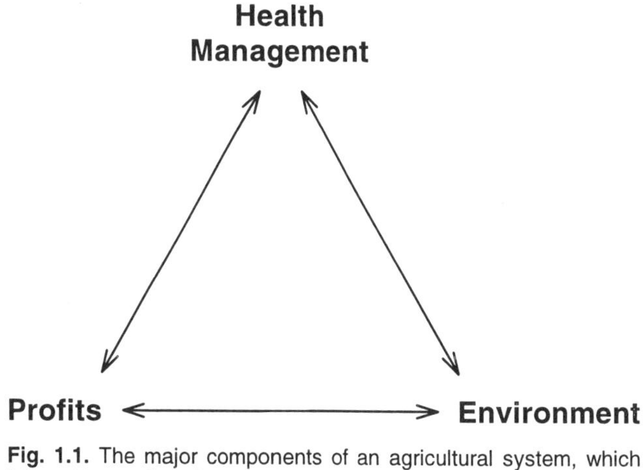

Fig. 1.1. The major components of an agricultural system, which interact to regulate the sustainability of the system.

Information Requirements and Health Management Concepts

Managing the health of citrus trees requires foresight and a great deal of information about the physical and biotic factors which continually interact to influence the development of trees and fruit and, ultimately, the longevity of the orchard. Throughout this book we have attempted to highlight current methods of obtaining and interpreting such information. Integrating information into a rational management program has always been complicated by the range of climates, soils, scion-rootstock combinations, ages of trees, resident pests and pathogens, adjacent habitats, past management histories, etc., making each orchard somewhat different from all others. Nevertheless, the importance of such information to management decisions has increased in recent times as a result of technical innovations and the competition for resources in growing human populations. For example, modern transit systems and new trade agreements have increased regional and international commerce. To remain competitive, most growers must increase production efficiency, primarily by reducing management costs, and many are learning to make management decisions based on anticipated market demand for fruit of various types and qualities. Moreover, significant economic cost is associated with the increased scrutiny given modern agricultural practices to ensure that they are safe for workers, consumers, and the environment. Again we see that growers today must process a great deal of information to understand with reasonable certainty how their management decisions influence not only the economic well-being of their enterprise but also, increasingly, the natural environment.

Is there a systematic framework with which to sort through this complexity, to identify the practices necessary to sustain production and profitability, while minimizing unwanted effects on the environment? In recent years, several related approaches have been proposed to achieve these ends in agriculture.

The concept of *sustainable agriculture* emphasizes practices that will sustain food production at necessary levels well into the foreseeable future. Implicit in sustainable practices is the need to conserve the resources necessary to produce future crops, by reducing waste and pollution.

Sustainable agriculture, BMP, and IPM are management concepts firmly based on principles of economics and ecology.

Integrated pest management, or IPM, is a philosophy of plant protection that is compatible with sustainable agriculture. In essence, IPM constitutes a rational integration of management tactics with the aim of maximizing profitability by optimizing the relationship between fruit yield and quality and the direct management costs and indirect environmental costs involved in its production. In its ideal form, IPM is based on physical and biological interactions known to occur between crops and populations of pests, disease organisms, and beneficial organisms.

The concept of *best management practices,* or BMP, refers to horticultural practices that are most compatible with sustainability. BMP encompasses IPM practices and all other operations designed to optimize fruit production with respect to the costs of managing an orchard.

IPM and BMP systems are generally developed in terms of current economic conditions. However, as we shall see, it is possible to consider the indirect costs of agricultural practices to other segments of society and to future generations in order to gain insights necessary to extend the time frame in which citrus industries can be profitably maintained.

As we might expect from Figure 1.1, sustainable agriculture, BMP, and IPM are management concepts firmly based on principles of economics and ecology. In order to sustain the production of citrus in a region, a certain environmental balance must be maintained to preserve the resources needed by the trees. Just as important, however, is that the practices necessary to maintain a productive balance in the ecosystem must be profitable to growers, or they will not be utilized. To better understand principles of effective and sustainable citrus health management it is important to review some of the ecological, economic, and IPM principles upon which they are based. A more detailed treatment of the following topics can be found in references at the end of the book, from which much of this material was drawn.

Ecological Foundations of Sustainable Citriculture

Ecology is the scientific study of the interrelationships between organisms and their environment. More than any other discipline, ecology considers organisms in the broadest sense, from molecular and cell biology to the biology of organisms and the behavior of populations. The need for a broad perspective in the discipline arises from an understanding that the basis for the evolution of populations is the genetic variability of individual organisms. Ecological research also tends to be quantitative in nature, because mathematical and statistical procedures are necessary to detect real trends and relationships among the highly variable data collected from studies in the field. Mathematical models are commonly derived to simplify and make comprehensible some of the fundamental principles governing interactions between organisms and their environment. The following is an overview of ecological concepts important to pest management, beginning with population models to explore the questions and implications they pose about plant-pest interactions.

The earliest models of population growth were logistic models, which were used extensively to study insect populations for several decades beginning in the 1920s. Variations on logistic models have been developed in all plant protection disciplines. Logistic models are based on the formula for exponential growth, $N_t = N_0 e^{rt}$, where N_t = population density at a specified time t, N_0 = initial population density, r = intrinsic rate of increase (births minus deaths), and e is the base of the natural logarithms. With the inclusion of a term to reduce the rate of growth as a population increases, these models are able to represent competition for resources, a key factor regulating the size of populations. The simplest example of a logistic equation is $N_t = K/\{1 + [(k - N_0)/N_0]e^{-rt}\}$, or $dN/dt = rN(K - N/K)$, where K = the carrying capacity of the system for the organism in question. This basic model has been used by entomologists to study a wide array of questions dealing with the population dynamics of insects and mites (Fig. 1.2). It is currently used, with a term to correct for tree canopy temperature, as a predictive model by commercial laboratories to schedule sprays to control the citrus rust mite (Plate 58).

Similarly, plant pathologists have recognized that disease progression tends to be of two forms. Diseases of foliage and fruit that are dispersed by wind, rain, or insect vectors (citrus canker, postbloom fruit drop, tristeza, etc.) are generally *polycyclic,* in that plants are infected by successive generations of disease organisms during a growing season. Polycyclic diseases increase in a manner comparable to compound interest in a bank (Fig. 1.3) until the number of available uninfected plants becomes limiting (i.e., logistic growth). The progress of most diseases caused by soilborne pathogens (such as citrus foot rot, citrus blight, and nematodes) is sometimes described as *monocyclic,* because disease propagules spread more slowly through soil, limiting the number of plants infected by successive generations of organisms during a season. The rate of increase of monocyclic diseases often resembles simple bank interest, with the final number of diseased plants being basically in linear proportion to the initial number (Fig. 1.3).

Nematologists, too, have used a variation of the logistic formula to describe citrus nematode population growth from the beginning to the end of a growing season (Fig. 1.4A). The model $P_f = rKP_i[(r-1)P_i + K]^{-1}$ relates final population density P_f to the initial density P_i rather than to time. Unlike the *multiple-point* models described above, which can predict population density at different points in time, the net reproduction function shown in Figure 1.4A is a *single-point model,*

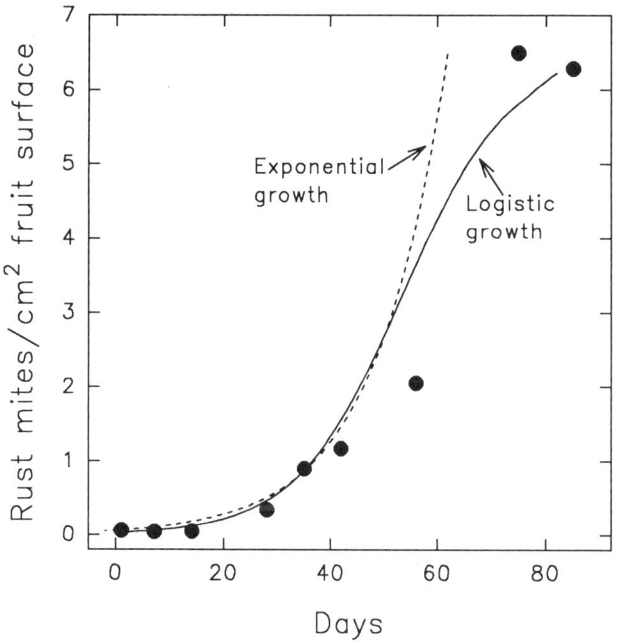

Fig. 1.2. The logistic growth curve as a mathematical representation of the population development of the citrus rust mite. The dotted line represents exponential (unrestrained) growth. (Data from Steven Rogers, Ecostat, Inc.)

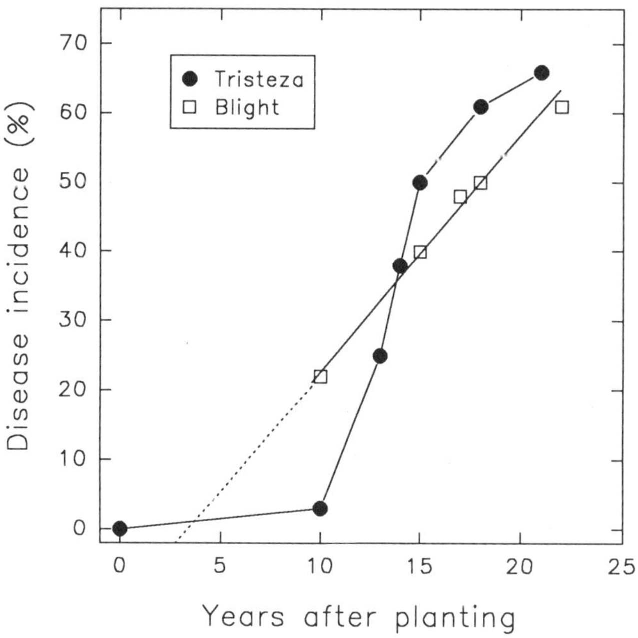

Fig. 1.3. Mathematical representation of the development of epidemics of citrus blight and tristeza according to models of monocyclic and polycyclic growth, respectively. (Redrawn from D. O. Chellemi, R. M. Sonoda, R. R. Pelosi, and M. Cohen, 1991, Temporal and spatial comparisons between epidemics of citrus blight and citrus tristeza virus, pages 289–296 in: Proc. Conf. Int. Org. Citrus Virol., 11th, R. H. Brlansky, R. F. Lee, and L. W. Timmer, eds., IOCV, Riverside, Calif.)

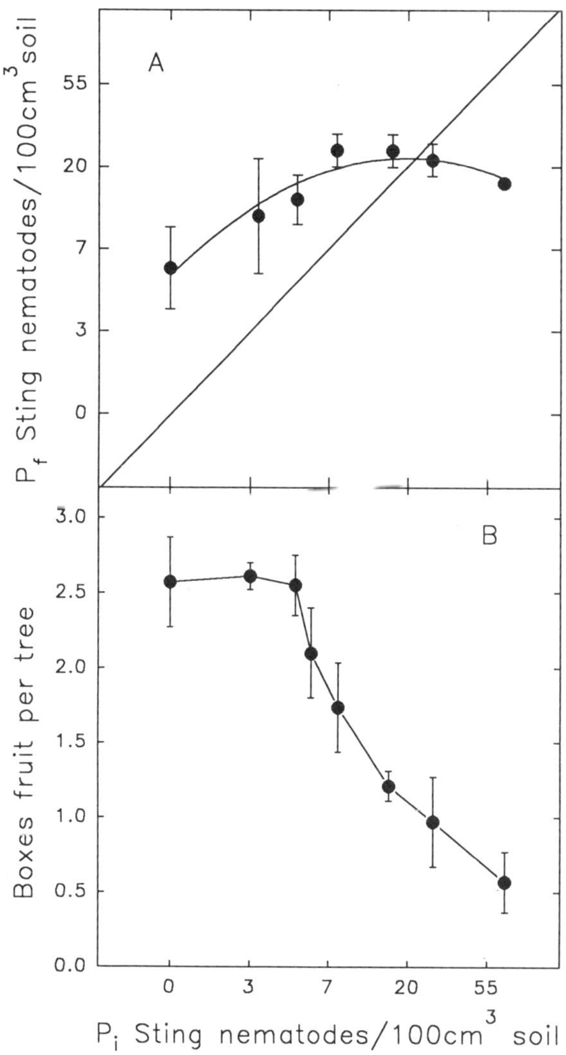

Fig. 1.4. Nematode reproduction and damage functions depicting the relationship between the sting nematode population density at the time of flowering of citrus trees (P_i) and the population density at fruit harvest (P_f) (**A**) and the relationship between P_i and fruit yield (**B**). The straight line in **A** represents a maintenance level at which no net change occurs in the population.

because only the population density at the end of the season is estimated. Single-point of models have proved useful for organisms that are difficult and costly to sample repeatedly. Frequently, the relationships between P_i and P_f and between P_i and the subsequent crop yield (Fig. 1.4B) are determined for use in cropping systems analysis (see below).

The models described so far are simple in two respects. They do not contain terms to describe interactions of pests with the physical environment or with other organisms, and they are *deterministic,* in that they do not provide information about the probability of obtaining the predicted result. As computers became widely available, computational ability increased enough to permit the development of *stochastic* models, which predict the odds that a particular event will occur—obviously an important consideration in management decisions. However, logistic models have yielded abundant information about pest management.

For a given set of environmental conditions, the two parameters of the logistic model are the rate at which a population grows (r) and the maximum population density that those conditions will support (K). The independent and dependent variables are the initial population density and the final population density, respectively. We might ask whether growers can more effectively reduce the spread of an epidemic by using a nonpersistent biocide or sanitation to reduce the existing pest population (N_0) in a grove or by cultural practices such as changing irrigation systems to reduce the rate at which the pest population increases. There are many implications of such practices, but from the standpoint of population dynamics we can say that reducing r is the most effective way to reduce polycyclic epidemics of pests and diseases, such as spider mites and most foliar diseases of citrus (Fig. 1.5A). This is because N_0 has relatively little effect on N_t when growth is exponential. In contrast, when growth is a linear product of r and N_0, as in monocyclic epidemics due to nematodes and soilborne fungi, reducing N_0 has the same effect as reducing r by a comparable amount (Fig. 1.5B). Consequently, preplant fallow and soil fumigation are sometimes recommended to manage certain soilborne pests and pathogens, whereas sanitation to remove inoculum or existing pests is of less importance in controlling rapidly increasing aboveground organisms.

The study of biological control has also been strongly influenced by deterministic models. Logistic models of populations of pests and their predators have been used to predict that cyclical fluctuations of these populations should occur out of phase with one another (Fig. 1.6). Predators increase by feeding on pests, until the pest population declines sufficiently that the growth of the predator population is slowed and eventually reversed. The reduction of the predator population permits increased growth of the pest population, which begins the cycle anew. These *predator-prey* models are among the tools used to investigate characteristics of good predators—those that can maintain themselves sufficiently to continually suppress a pest, keeping the pest population density below a damaging level.

Entomologists pay close attention to these characteristics when selecting potential biological control agents for introduction into new areas. Thus, a specific wasp parasitoid was chosen for release in Florida to manage a recently introduced citrus leafminer because 1) the wasp population increases rapidly, with little lag time, when the leafminer population increases, 2) the wasp is reasonably host-specific, 3) it has a high search rate and a good ability to locate and aggregate in areas of high leafminer density, and 4) its parasitization rate is dependent on pest density, so that at low leafminer density some prey escape. All of these characteristics permit the wasp popu-

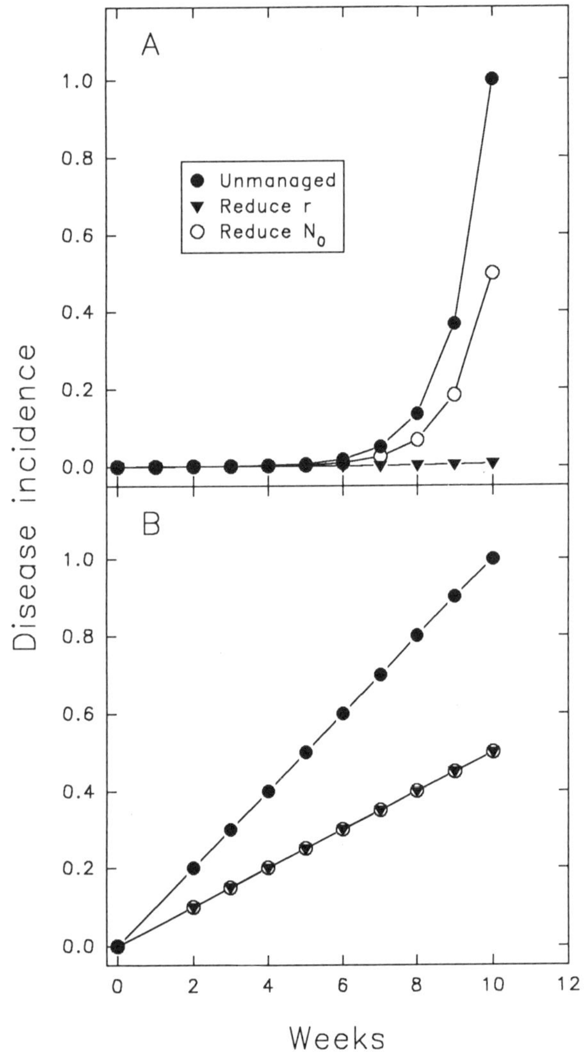

Fig. 1.5. Relative effects of pest management tactics that reduce inoculum density and tactics that reduce the rate of population increase by equivalent amounts in polycyclic epidemics (**A**) and monocyclic epidemics (**B**).

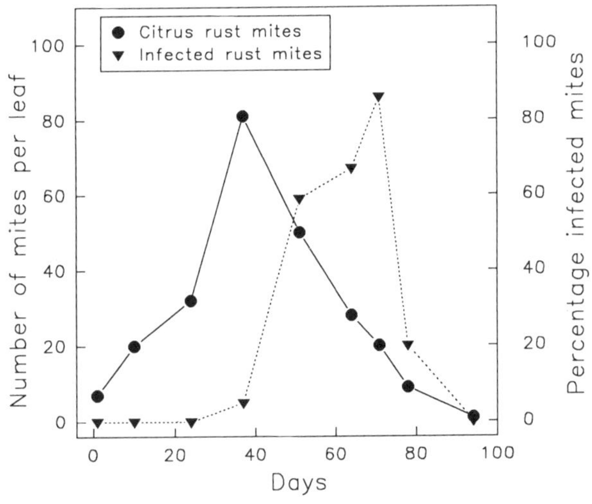

Fig. 1.6. Density-dependent population development of the citrus rust mite and *Hirsutella thompsonii,* a pathogen of the mite. (Data from Clayton McCoy, University of Florida)

lation to increase rapidly enough to prevent damage to citrus when the leafminer population increases, but without causing the extinction of the prey, which would exhaust the ability of the wasp to maintain itself. Many other potential agents were omitted from consideration for this program (thereby saving time and resources) because they lack one or more of these characteristics.

Many other questions about pest management are routinely addressed by modeling, but perhaps none is more important to understand than the relationship between the diversity of species and the stability of an ecosystem. In early laboratory experiments to validate predator-prey models, it was shown that simple systems are inherently unstable. For example, predators and prey become extinct if prey refuges are not present. This observation was supported by numerous other lines of field research. It is well understood that *secondary pests* are created by the use of pesticides that kill nontargeted biological control organisms, which had been responsible for maintaining those insects at nondamaging levels. *Island biogeography* is an approach to the study of how isolated habitats become colonized. In a classic experiment, methyl bromide was used to eradicate the insects on small coastal islands. Immigrant species of insects eventually recolonized the islands, forming communities of nearly the same diversity as those that had existed previously (Box 1.1). The level of community diversity was directly related to the degree of topographical diversity of the various islands.

Ecological succession refers to the colonization of simple habitats by increasing numbers of species to produce increasingly complex environments. Species involved in habitat succession have been characterized in terms of the logistic parameters to describe their competitive strategies. Species referred to as *r-strategists* generally invest resources in many small offspring, most of which die, rather than fewer well-defended large individuals. Their life spans tend to be short, and they disperse well. In brief, they tend to be opportunists capable of rapid colonization when opportunities arise. *K-strategists* are large species relative to others of their kind, and they tend to produce fewer, longer-lived offspring. They invest proportionally more energy in defense mechanisms than do *r*-strategists. Overall, *K*-strategists require greater habitat stability, because they respond too slowly to brief episodes of favorable conditions. If we examine successional changes in the composition of animal and plant communities by traveling from a beach to dunes, fields, and finally forests some distance from the beach, we will find that 1) the communities become much more diverse with distance from the beach, and 2) the composition of species in the communities changes to include proportionately more *K*-strategists in the later successional stages. These trends are well documented and give evidence of a direct effect of species diversity on community stability.

If species diversity promotes ecosystem stability, the implication for pest management is that diverse systems such as natural forests support a large variety of primary producers (Box 1.2) and also primary consumers (pests) and secondary consumers (predators). Therefore, plants in forests experience relatively few epidemics. In contrast, monoculture of genetically identical agronomic crops subject to frequent major disturbances of their habitats during planting and cultivation support far fewer animal and plant species, thereby decreasing the odds that pests or disease organisms will be adequately regulated by natural enemies. Such agricultural systems require a great deal of intervention to manage pests and diseases. Citrus orchards are intermediate in community diversity, compared to completely natural habitats and fields of agronomic crops. Theoretically, then, citrus orchards are likely to be intermediate in the number of potential pests and pathogens whose populations are maintained at nondamaging levels by natural forces and those that require grower intervention.

We can appreciate the dynamics of species complexity and ecosystem stability by observing wild citrus trees growing in the understory of a natural forest. Whiteflies, aphids, rust mites, scale insects, and greasy spot and melanose fungi can be isolated from these trees, but significant plant damage by any of these organisms is rarely evident, and sooty mold is virtually absent. Similarly, *Thielaviopsis basicola* and *Phytophthora nicotianae* cause severe root and stem rot epidemics in citrus nurseries where plants are grown in sterilized soil, although *T. basicola* is not a problem in citrus in the field, and *P. nicotianae* is much less a problem in the field than in nurseries. Citrus nematode egg masses contain twice as many eggs in sterilized soil as in the field. The closer one observes the biology of pests and pathogens, the more apparent is the effect of community diversity on disease and pest epidemiology.

Understanding the principles by which pest populations are regulated in nature is clearly necessary in developing rational and sustainable management systems. However, it is also important to recognize certain caveats about the value of ecological theories. First, principles revealed by research involve broad generalizations to which there are many exceptions. A disease that appears to progress in a monocyclic fashion under some conditions may behave in a polycyclic fashion under other conditions. The diversity of plant and insect species in citrus groves can be increased by mowing rather than disking row middles and by reducing pesticide usage. Such practices

Box 1.1

How Is Species Diversity Measured?

During successional change, the physical conditions of a habitat tend to become more complex, the diversity of species increases, and the species in the community shift toward those with characteristics of *K*-strategists rather than *r*-strategists. In order to investigate the relationships between species diversity and ecosystem attributes (e.g., physical complexity, community stability), it is necessary to have a measurable index of diversity. Of several such indices, the most widely used is the Shannon-Weaver index, $H' = -\sum p_i \log_e p_i$, where H' = diversity, and p_i is the estimated proportion of the ith species in the community.

To understand the properties of this formula, assume that we sample the foliage of a citrus tree and recover 2,000 rust mites and 10,000 spider mites. The estimate obtained from the Shannon-Weaver formula is $H' = 0.45$. If instead we had found 5,000 rust mites and 7,000 spider mites, our estimate would be $H' = 0.68$. Finally if we had recovered 2,500 rust mites, 7,000 spider mites, and 2,500 aphids, the estimate would be $H' = 0.97$. The total number of insects recovered is the same in each sample. Comparison of the first two examples shows that as the *evenness* among species increases, so does the diversity index. Intuitively, this is acceptable, because if a community is heavily dominated by one species it seems less diverse than if the numbers of individuals of all species are more or less equal (more even). Comparison of the last two examples shows that as the number of species in a community increases, so does our measurement of its diversity.

will increase biological control of some organisms, but they may increase the incidence of other pests and diseases. Unquestionably, reducing pesticide usage increases the level of natural control of large numbers of potentially harmful species of insects and mites. Where citrus leafminers have been introduced into citrus-growing regions worldwide, they initially caused serious economic damage by feeding on developing leaves. In all regions, however, parasites of leafminers responded within one or two years by moving from adjacent weeds and regulating the leafminer population to keep it below a damaging level. Nevertheless, it is difficult to profitably increase plant diversity in groves to levels anywhere approaching those in completely natural environments. Consequently, the effects on pests and diseases are difficult to predict. Growers in central Florida recently began mowing row middles rather than using tillage or herbicides to eradicate weeds. In one study, populations of citrus rust mites, citrus red mites, and Texas citrus mites were found to be lower in orchards whose middles were weedy rather than bare. However, it is unknown whether changes in microclimate or natural enemies were responsible for mite suppression. Moreover, sting nematodes, which reproduce well on many of the weeds in row middles, can increase to levels at which they cause severe damage to young trees.

Second, research is a continuing process. Data require interpretation, and new findings alter previous ideas. This means that ecological theories are frequently modified and sometimes discarded. It was recently proposed that geographically isolated populations of species can be considered regionally as fluid *metapopulations.* The concept of a metapopulation provides a basis for understanding that local extinction of prey and their predators does not automatically ensure that immigration of new pest or disease organisms will result in epidemics as predicted by simple population models. Rather, the degree to which epidemics occur depends partly on the rate at which biological control agents recolonize the area. Similarly, the desirable characteristics of biological control organisms (described above) supported by early population models are more descriptive of insect parasitoids than predators. Thus, parasitoids are often selected over predators in the search for exotic biological control agents for release in new areas. However, empirical research and revised models have revealed that omnivorous predators can be highly effective at pest suppression and can maintain themselves by switching to another prey species if a preferred species becomes extinct locally. Findings such as these may broaden the scope of what characteristics should be considered in the search for potentially useful biocontrol agents.

Finally, growers must optimize efforts to achieve pest control by natural processes and practices requiring human intervention. Agricultural systems are inherently less stable than natural systems, because the purpose of agriculture is to increase the abundance of selected plant species. By definition, it is necessary to reduce species diversity in order to produce citrus in quantities sufficient to sustain human needs. The challenge to growers is to find the right combination of practices to maximize the natural processes of pest control in systems that produce adequate fruit yield.

Box 1.2

The Flow of Energy Through a Habitat

The energy required by living organisms is captured from sunlight and stored in chemical form. As chemical energy passes through the food chain, some is used for metabolic processes, and the remainder is lost as heat. Organisms are sometimes classified on the basis of their nutrition, to indicate their relative positions, or *trophic levels,* in this energy exchange. Plants constitute the major group of organisms known as primary producers, for their role in converting sunlight and CO_2 into carbohydrate. Herbivores (including plant pests) and plant parasites are primary consumers, since they receive the first exchange of chemical energy between groups of organisms. Secondary consumers include biological control agents and other organisms that feed on primary consumers. *Hyperparasites,* organisms that parasitize biological control agents, and carnivores represent tertiary consumers, and so on. Omnivorous animals and detritivores can function at all consumer levels, depending on the source of their nutrition.

Trophic relationships provide a useful paradigm for the study of the flow of energy and matter through communities. While energy flows in one direction, with most of it being lost as heat, chemical exchange is cyclic, the most obvious example being CO_2 exchange between animals and plants. Quantitative indices of the trophic structure of communities are also used (much like diversity indices) to study ecosystem characteristics that pertain to agricultural sustainability. Moreover, the characteristics of organisms at various trophic levels help explain important habitat characteristics, such as stability. For example, forests are plant communities in which the primary production of energy is much less available to other organisms than it is in grasslands. This is because, unlike grasses, trees allocate a great deal of their production to lignin, which becomes a nutrient for only a few organisms when the trees die. As a consequence, much of the primary production in forests becomes available as food slowly and regularly, stabilizing the habitat. By comparison, the primary production in grasslands can be immediately consumed, but it varies greatly, depending on seasonal changes in the environment.

Economic Foundations of Sustainable Citriculture

Growers' Decisions

The concept of an *economic threshold* is central to IPM. It implies a rational approach to pest management based on sufficient information to maximize profitability. In its simplest form, the economic threshold is the pest density at which the cost of management equals the economic benefit that will be obtained from it. Implementation of management practices when the population density of a pest is below the economic threshold results in economic loss, wasted resources, and unnecessary risk to the environment.

Conceptually, we might consider crop loss to be a linear function of pest density and express cost equaling benefit as $MPDV = C$, where M = the proportional mortality inflicted by a pest management tactic, P = the population density of the pest, D = the proportion of crop value lost as a result of damage due to that pest, V = the value of the crop, and C = the cost of control. The formula can be rearranged as $P = C/MDV$ to solve for the economic threshold. Thus, if it were determined solve for the economic threshold. Thus, if it were determined that a pest reduced the yield of citrus by 0.05 tons per acre for each individual pest per lens field, the crop value was anticipated to be $133 per ton, and a pesticide spray was available that reduced the population density by 90% at a cost of $50 per acre, the economic threshold would be a pest population density of between eight and nine individuals per lens field.

In practice, it is usually desirable to elaborate on this simple definition, for several reasons. First, there is a threshold population density: at levels below this threshold, the pest population does not cause significant damage to the crop. Above this threshold, crop damage is seldom related to pest density in a linear manner, because each successive pest encounters less undamaged plant tissue available for it to damage. Therefore successive pests tend to cause proportionately less damage than the individuals preceding them. Graphically, curves representing crop loss as a function of pest density often resemble sigmoid or quadratic decay functions (Fig. 1.4B).

Second, by varying management tactics, it is often possible to obtain different levels of control. The difference between the cost of pest control and the value of the increase in yield varies with the level of control, and it is preferable to maximize this difference. The maximal difference is known as the *optimizing threshold,* and it describes the conditions that provide maximum profitability. Note that new information (a management efficacy function) is needed in order to estimate an optimizing threshold.

There are mathematical methods for estimating optimizing thresholds, but a simple example can be presented to illustrate this concept as well as that of risk assessment. We noted previously that stochastic predictions are preferable to deterministic ones because they give growers added information about the riskiness of decisions. Let us assume that from a large number of experiments we have characterized the probabilities of citrus packout losses associated with a given population of the citrus rust mite. Table 1.1 shows hypothetical odds of sustaining various levels of loss at this pest density and a method by which the average expected loss can be calculated. The uncertainty of the level of damage is due to such factors as errors in the estimate of rust mite density, the variability of weather conditions, or differences in levels of biological control from grove to grove, as well as other factors of which we may be unaware. In this example, on average a grower will lose $25.65 per acre without management and will profit from a management practice if it costs less than $25.65 per acre to control the mites so as to keep damages below the economic injury level. At a cost of $10.00 per acre the grower will profit 96% of the time. This approach provides growers with a very real basis for decisions about whether to implement a practice.

Moreover, this approach helps to optimize returns and costs of several possible management options. As shown in Table 1.2, a grower who manages rust mites at this particular density with a method that provides 95% control will on average sustain the least possible loss of profit due to the presence of the pest.

A third problem with our simple definition of the economic threshold involves pests and diseases that do not directly attack the fruit. It is difficult to estimate an economic threshold for these pests until we can predict the progression or *etiology* of plant responses that begin with damage to vegetative tissues and end with effects on fruit yield. For example, when a high-density population of a soilborne fungus or nematode is reduced with pesticides, the initial response of a citrus tree is to increase root and foliar growth rather that fruit growth. Similar effects occur in response to management of spider mites and foliar diseases. Thus, estimating the yield response to pest management requires an ability to predict plant behavior over several growing seasons. The delayed response in fruit yield to management of pests that do not directly attack the fruit is currently an important challenge to our ability to adequately use the concept of an economic threshold in perennial cropping systems such as citriculture.

Implementation of management practices when the population density of a pest is below the economic threshold results in economic loss, wasted resources, and unnecessary risk to the environment.

A final reason that more complex definitions of the economic threshold are desirable is that the management tactic used for a problem at hand influences subsequent management choices. This means that a series of pest management decisions that result in maximum profit over several years rarely includes

Table 1.1. Hypothetical probabilities of incurring various economic losses due to citrus rust mites at a given density[a]

Damage (per acre)	Probability	Expected loss (per acre)
$ 0	0.01	$ 0.00
5	0.01	0.05
10	0.02	0.20
15	0.06	0.90
20	0.15	3.00
25	0.40	10.00
30	0.20	6.00
35	0.10	3.50
40	0.05	2.00
Total	1.00	$25.65[b]

[a] Uncertainty is due to sampling error and variable conditions that influence damage by mites.

[b] Mean expected loss.

Table 1.2. Hypothetical probabilities of incurring various economic losses due to citrus rust mites at a specific density with various control strategies

Damage (per acre)	0% control	90% control	95% control	100% control
$ 0	0.01	0.30	0.60	1.00
5	0.01	0.08	0.15	0.00
10	0.02	0.12	0.10	0.00
15	0.06	0.13	0.05	0.00
20	0.15	0.12	0.04	0.00
25	0.40	0.11	0.03	0.00
30	0.20	0.09	0.02	0.00
35	0.10	0.04	0.01	0.00
40	0.05	0.01	0.00	0.00
Expected loss	$25.65	$13.20	$ 5.00	$ 0.00
Control costs	0.00	5.00	10.00	20.00
Total costs	$25.65	$18.20	$15.00	$20.00

the decision that would lead to maximum profit from the crop at hand. Methods by which to predict how today's decision will influence future decisions are discussed in terms of cropping systems in the next section. Thus, while the concept of the economic threshold is central to IPM, it should evolve along with our ability to refine its use.

Societal Decisions

Increased competition due to recent international trade agreements and other market forces has resulted in reduced fruit prices in many markets (see Chapter 15). Similarly, the increasing number of state and federal environmental regulations add significant costs to citrus-growing operations. Consequently, most growers are acutely aware that they need to become increasingly efficient to remain competitive. Because these changes to which growers must adapt are the result of social policies, it is useful to consider the economics of pest management at levels beyond which growers have direct control. We shall examine just two questions we might pose as examples of the importance of societal decisions in the economics of IPM. First, in the interest of increasing the efficiency of pest management, regulations can favor increasing yields through new management inputs, or they can favor maintaining current yields with less input. What are the relative advantages of these options? Second, are the costs of regulations aimed at environmental protection borne mainly by growers or by society?

When attempting to develop new, more profitable programs for growers, researchers and pest management experts may have the option of either providing new inputs to reduce pest populations or developing new practices to avoid the inappropriate use of current management practices. The difference in the expected economic benefits of the two approaches is difficult to predict and may sometimes be surprising because of marketplace dynamics.

Suppose citrus nematodes were widely distributed in an industry that produced an average of 450 boxes of oranges per acre at a value of $6.00 per box. Suppose too that a large percentage of growers routinely applied nematicides to manage the nematode population. If a scientist discovered a new, more effective nematicide that could increase yield by an average of 10% at an additional cost of $50 per acre, we might assume that the benefit to the industry would be substantial. A quick calculation suggests an average additional profit of $220 per acre if growers adopted the new nematicide. However, fruit prices would decline as a result of the increased supply of oranges. Worse, increased production in the absence of increased demand almost always causes prices to decline proportionally faster than the rise in supply. Even if the price of fruit declined only 10%, to $5.40 per box, in response to a 10% increase in fruit yield, growers would lose an average of $77 per acre by increasing their yields with the new management practice. Moreover, they would be increasing the risk to workers and to the environment.

An alternative strategy would be to attempt to improve the precision and frequency with which growers monitor nematode populations. By doing so, it should be possible to reduce nematicide use and restrict it to groves in which the likelihood of profitable yield response is high. Such an approach would have no effect on industry-wide fruit production, but it would increase grower efficiency and profitability (depending on sampling costs), by reducing nematicide use. In contrast to the first strategy, increased monitoring would reduce the risk to workers and the environment posed by the use of nematicides. This example demonstrates a clear need for integrating economic theory with the biological and technical information necessary to improve our current pest management practices.

In order to answer the question of whether growers or society pay most of the cost of government regulations, we must define some economic terminology.[1] *Consumer surplus* is the difference between the social benefit (in monetary terms) of a product and the amount consumers actually pay for the product. It is a hypothetical value derived by integrating areas in portions of the supply-demand curve for a particular product. *Producer surplus* is the profit to growers in a given supply-demand situation. *Net social benefit* is the sum of consumer surplus and producer surplus. Now consider the fact that citrus production frequently results in *hidden costs* to other industries or to society. An example of a hidden cost is the expense to communities for installing water filtration devices to prevent agrichemicals from reaching groundwater sources. It is important to recognize hidden costs, because they reduce net social benefit by increasing costs to producers in other industries or to society in the form of consumer expenses or taxes. They amount to a hidden subsidy to the industry that creates the cost. When an industry imposes hidden costs, supply and demand in that industry are not in market equilibrium, and the ratio of costs to benefits increases for consumers.

Society could address hidden costs, as in this example, by assessing citrus growers for water filtration devices or by regulating the use of pesticides to reduce the risk to groundwater. The cost of the regulation would change the supply-demand relationship that formerly existed, and therefore new estimates of consumer and producer surpluses and net social benefit could be derived. Different economic outcomes for growers and society will result from the manner in which hidden costs are addressed. If growers are assessed the costs of water filtration devices, we can say that prices will rise while the production of citrus and the net social benefit from citriculture will decline. Producer surplus will also decline somewhat, but frequently most of the reduction in net social benefit results from a reduction in the consumer surplus. Whether this occurs and the degree to which it occurs depend on how steeply the demand curve rises in relation to the supply curve. However, such analyses illustrate that regulations imposed on growers by society are never borne solely by growers in the form of reduced profits. These costs are frequently paid primarily by consumers as an expense for protecting the environment.

Sustainable Management Systems

The basis of IPM and BMP is the rational use of information about the current status of an orchard. These management paradigms require growers to monitor the condition of a grove and make decisions that optimize management practices in terms of profitability and environmental quality. The types of measuring and sampling procedures highlighted in insets throughout this book are the means by which growers obtain the necessary information. Publications, tables, models, expert systems (computer software that guides growers to decisions), and sometimes common sense are needed to maximize

[1] For a more detailed treatment of these concepts and other topics in this section, see J. C. Headley, 1975, The economics of pest management, pages 75–99 in: *Introduction to Insect Pest Management,* R. L. Metcalf and W. H. Luckmann, eds., John Wiley and Sons, New York.

the value of the information in the decision-making process. In this section we consider some aspects of decision making and some of the tactics currently available for IPM in citrus.

Sampling and Monitoring

The techniques by which populations of pests and disease organisms are measured are usually designed to provide the necessary level of precision at the least cost. Good estimates of population density can be more or less accurate, depending on 1) the economic risk if a wrong decision is made, 2) the cost of sampling, and 3) the frequency with which pest populations are present at densities near the economic threshold.

For example, soil sampling to detect and estimate the numbers of plant-parasitic nematodes is more intensive (and expensive) when the main concern is burrowing nematodes rather than citrus nematodes. If a grower wishes to purchase land for a grove, the economic consequences of an infestation by burrowing nematodes are more severe than if citrus nematodes are present. Estimates of citrus nematode population densities, derived by recommended methods, tend to be only about 50% accurate when the population density is fairly high and even less accurate at lower densities. This means that if a sample estimate is 2,000 nematodes per 100 cubic centimeters of soil, the actual number may be in the range of 1,000–3,000 nematodes. Such seemingly imprecise estimates are nonetheless valuable to growers in Florida, because most samples will indicate that no citrus nematodes are present in the grove, and therefore the decision not to manage is made. Of samples containing citrus nematodes, most will produce estimates well below 1,000 nematodes per 100 cubic centimeters of soil, again below an economic threshold. About half of the few remaining samples will produce estimates well above 3,000 nematodes, indicating a need for nematode management. Only a few samples will contain nematodes at densities for which it is difficult to conclude with much certainty whether they exceed or fall below the economic threshold. In these few cases the grower's decision is more difficult, but in most cases the options are straightforward.

The technical aspects of many sampling programs are derived from the formula for calculating the confidence interval about a mean:

$$\frac{\bar{x} \pm zs}{\sqrt{n}}$$

where n is the number of samples, z is the standard normal variate, and $\bar{x}$ is the mean and s is the standard deviation of the sample. A confidence interval is the range within which the estimate of the mean would be likely to fall if samples were repeatedly taken and analyzed. It is expressed in terms of probabilities, usually 95%. In the example of citrus nematodes given above, we could say that 95% of samples taken from an orchard on a given day would result in an estimate of 1,000 to 3,000 nematodes per 100 cubic centimeters of soil. By providing an estimate of the standard deviation and specifying a desired confidence interval half-length (d) as a proportion of the mean, we can rearrange the formula as

$$n = \frac{z^2 s^2}{d^2 \bar{x}^2} .$$

This new formula permits us to estimate the number of samples needed to achieve a particular level of precision and accuracy for a given population density.

In order to achieve the accuracy and reliability (repeatability) required from samples, we should recognize the factors that most affect our sample size formula. The most important factors are those that affect the variability associated with taking repeated samples. More samples are needed when variability is high than when it is low. Often sample variability is very high for pests, because they tend to aggregate in clumps within a grove (Fig. 1.7). The more highly aggregated a pest, the more likely it is that samples will contain many counts of 0 and a few high counts when a clump of pests is encountered. In this situation, a large number of samples is required for an accurate estimate of pest population density. When pests are less aggregated and more uniformly distributed, the variability of counts is reduced, and fewer samples are needed. As the population density increases, pests occupy more space, and the degree of aggregation declines.

Growers can take advantage of these relationships by restricting sampling to locations where pests are most likely to be present and, if possible, by sampling at times when the pest population density is high. Citrus nematodes are best sampled during late autumn and mid-spring, whereas detection of *P. nicotianae* in citrus roots is best accomplished in summer (Fig. 1.8). These and other soilborne pests occur in greatest numbers well beneath the tree canopy, where the fibrous root density is highest. Immature leaves are sampled for aphids, and newly mature leaves for spider mites, because this is where most individuals aggregate. Scab is most readily detected on off-bloom fruit. Young or declining trees should be examined for post-bloom fruit drop. Similarly, it is wise to stratify large areas into smaller sampling areas based on topography and variations in tree quality, soil texture, or other factors known to affect pest population densities. Understanding basic principles such as these helps foster an appreciation of the need to collect the requisite number of samples at the appropriate time and place.

Many growers of agronomic crops are familiar with the term *precision agriculture,* which will assume increasing importance in citriculture as researchers find ways of using the new technology. It refers to the ability to perform routine operations

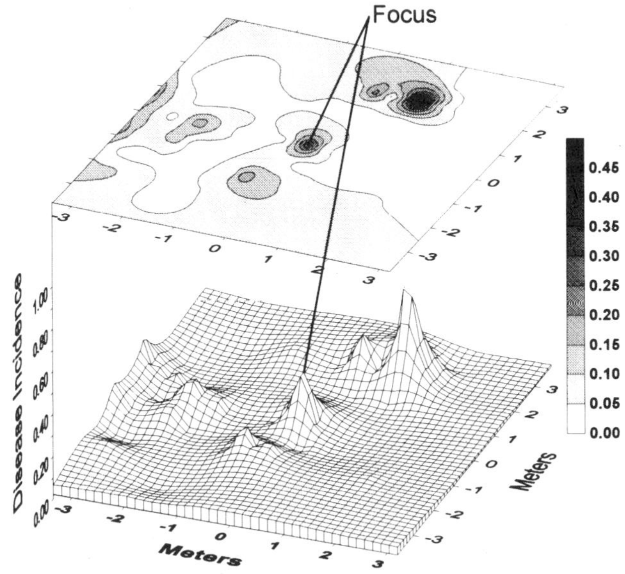

Fig. 1.7. Incidence of citrus canker in an Argentinean citrus nursery, illustrating aggregation typical of disease and pest organisms. (Reprinted, by permission, from T. R. Gottwald, L. W. Timmer, and R. G. McGuire, 1989, Analysis of disease progress of citrus canker in nurseries in Argentina, Phytopathology 79:1276–1283)

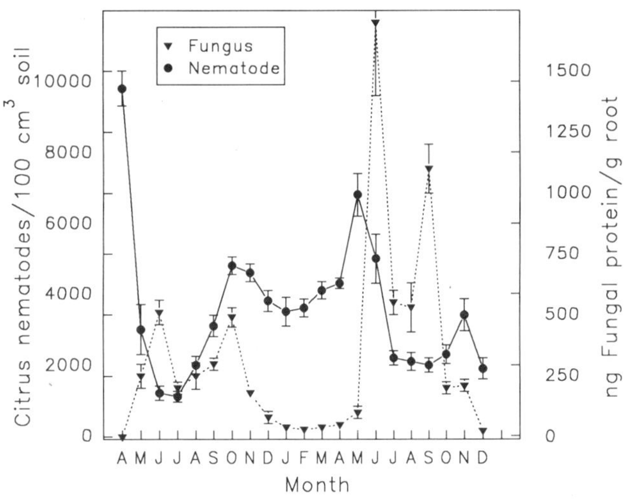

Fig. 1.8. Seasonal population development of the citrus nematode, *Tylenchulus semipenetrans,* and the fungal pathogen *Phytophthora nicotianae* over two years in a Florida citrus grove. (Redrawn from L. W. Duncan, J. H. Graham, and L. W. Timmer, 1993, Seasonal patterns associated with *Tylenchulus semipenetrans* and *Phytophthora parasitica* in the citrus rhizosphere, Phytopathology 83:573–581)

in a variable manner, so that greater precision is achieved in supplying the correct amount of fertilizer or pesticide, for example, to various parts of a field. Farm machinery is being developed to operate under the guidance of computer systems and global positioning devices, which use satellite signals to determine the precise location of the equipment as it moves through a field. Prior sampling using a global positioning system to map soil types, pest densities, etc., is necessary for programming the operation of the machinery. A promising aspect of precision agriculture is the possibility of using information about the influence of environmental variables on pest densities, plant nutrient availability, etc. For example, mapping the water table and soil textures in a grove just once could permit predictions about populations of *P. nicotianae* or nematodes or the availability of plant nutrients, which could be used thereafter to determine variable rate of pesticide and fertilizer application.

Crop Loss Assessment

Most parasites and arthropods for which citrus is a host are of no economic importance. Managing these organisms wastes money and resources, upsets natural biological control, and may threaten the environment. Therefore, various means are used to assess crop losses due to pests and pathogens, in order to identify truly noxious species and their economic threshold population densities. Figure 1.4B illustrates the relationship between citrus fruit yield and sting nematode population density during one season in Florida. The data were fitted by a *damage function,* in this case a model that describes the amount of loss due to a single nematode, the minimum yield that would be expected regardless of how many nematodes were present, and the tolerance threshold population density, below which losses do not occur.

The information in Figure 1.4B indicates that sting nematodes are associated with yield losses, and root damage by the nematode suggests that they are the cause of the losses. However, these data are insufficient for estimating an economic threshold for a particular management practice. As we noted previously regarding economic thresholds, it is difficult to predict the effect on fruit yield resulting from the management of pests that damage roots or foliage. An accurate assessment of the economic threshold for sting nematodes would require an integration of such data over several years, comparing yields in experimental plots in which nematodes were managed with yields in plots where they were not managed. Moreover, this relationship between yield and population density would undoubtedly appear somewhat different if it were measured in other groves or in different years, because of the influence of different environmental conditions, including soil, rainfall, and temperature. Repeated measurement of the relationship could provide insight into environmental factors that influence crop loss due to sting nematodes, and they could also provide an estimate of the probabilities of specific levels of crop loss.

For pests and diseases of citrus fruit, crop loss assessment and the determination of economic thresholds is more straightforward, by the method described above or similar methods. Many damage functions include environmental variables. For example, loss of marketable fruit due to rust mites is predicted from the mite population density and temperature, because of its importance in the growth of mite populations (see Chapter 11). Early-season monitoring of scale insects in California is used to predict year-end yield and thus aids in management decisions. A model predicting the development of *Colletotrichum acutatum* and subsequent fruit loss due to postbloom fruit drop takes into account fungal inoculum density and rainfall necessary for fungal dispersal and infection (see Chapter 10). The successful utilization of these types of models to optimize control costs and fruit returns argues strongly for the development of such models for a number of additional pests and pathogens, particularly those that respond rapidly to changes in the environment.

Cropping Systems

We suggested in our discussion of economic threshold that maximum profitability in the long term can be achieved only by considering the economic threshold as a dynamic rather than a static problem. Current management practices influence the pest density later in the year and in subsequent years. If we consider each time period (t) for which decisions are made, we see that the number of possible outcomes for n possible decisions is n^t. A simple example illustrates the complexity of forecasting optimum management when each decision alters the nature of subsequent decisions. Let us assume that in a grove infested by sting nematodes, we can choose not to manage nematodes or to reduce the population density with either of two nematicides. Product 1 costs $150 per acre and reduces the population density by 50%, while Product 2 costs $230 per acre and achieves a 99% reduction. There are 100 young trees per acre, and the expected yield in the absence of nematodes is according to Figure 1.4B. After production costs (excluding nematode management), each box of fruit is expected to produce a profit of $3.00. If we consider nematode management annually, over a two-year period there are $3^2 = 9$ possible management outcomes. For simplicity, we will assume that the damage function in Figure 1.4B predicts fruit yield based on nematode density at flowering. If the initial population density in the first year is nine nematodes per 100 cubic centimeters of soil, the first nematicide would reduce it to 4.5 nematodes, a level at which no yield loss is expected. Therefore, the profit would be expected to be roughly 2.5 boxes of fruit × 100 trees × $3 – $150 (the cost of the nematicide) = $600. The second nematicide would reduce the nematode population to less than

0.1 nematode per 100 cubic centimeters of soil, again below the damage threshold, but at a cost of $230, resulting in a profit of $520 per acre. If no nematode management is exercised, the predicted fruit yield is 144 boxes per acre, for a profit of $432. On the basis of these predictions, the use of the first product appears to be the most profitable choice. However, from Figure 1.4A we see that the population density following the use of Product 1 is likely to increase to approximately 15 nematodes per 100 cubic centimeters of soil by the beginning of the second year, compared to only four or five nematodes if Product 2 had been used. A second nematicide treatment in the second year would be profitable if Product 1 was used in the first year, whereas no treatment would be needed in the second year if Product 2 had been used in the first year. In this case, to achieve similar cumulative fruit yields, the two-year nematicide costs if Product 1 was chosen initially would total either $300 or $380 (depending on which product was selected in the second year), compared to only $230 for Product 2. Moreover, using Product 2 in the first year would reduce environmental risks by decreasing the number of nematicide applications. In this example, the predicted optimum treatments clearly differ according to the time scale we consider. Depending on conditions, if we considered the 27 (or 3^3) options over a three-year period, we may find that we should select a system that does not incorporate the practices we view as optimum for two years.

This method of analysis is sometimes referred to as a *cropping systems* approach, because it was developed to maximize profits in annual crops for which management options over several years (crop, cultivar, resistance, fallow, pest management, etc.) are often more numerous than in grove management. Of course, our ability to accurately predict these outcomes is severely limited by many factors that affect pests, yields, and profits but are not considered in the simulations. The degree to which these factors are likely to affect our predictions limits the time frame worth considering. Nevertheless, if the management systems and outcomes predicted to be most profitable over more than one management period appear reasonable, it is useful to consider their implications compared to the best short-term options.

Production Systems

Throughout this book, special reference is made to the variety of production systems involved in citriculture. Economic, environmental, and health management concerns differ for trees in nurseries and nonproducing and producing trees in the field. Young, actively growing trees do not respond to pests and diseases in the same way as mature trees, which have proportionately less young tissue. Organisms that feed on new leaves or roots of young trees, such as citrus leaf miners, aphids, whiteflies, and sting nematodes, may require management to prevent reductions in tree growth, whereas they may be of very little concern in mature trees. Similarly, fruit destined for the fresh market requires so much more protection from even cosmetic damage than processing fruit that this marketing decision radically effects the appropriate pest management strategy. Growers who understand the nature of these differences can benefit by making the appropriate management decisions in one of the few areas of operation that permits significant cost savings.

Pest Management Tactics

IPM decisions require not only information obtained from pest and disease monitoring but also information about available management tactics. If several tactics are available for dealing with a problem, they will differ in safety, efficacy, and cost. We have already seen how these variables can affect our estimate of an economic threshold. Often, considerable foresight may be required in order to use an optimum tactic. For example, if rootstock resistance is available and desirable, the time to know whether a soilborne pest is present in a grove is before rather than after planting.

Most management tactics can be classified in one of the following categories: sanitation, host plant resistance, cultural management, biological control, and chemical control. To a great degree, the order in which the categories are listed represents their relative importance in IPM. Sanitary measures and host plant resistance are highly desirable because they can prevent the presence or population growth of pests and diseases, obviating the need to consider more expensive or dangerous alternatives. Both management methods are highly effective and profitable means of dealing with many pests and diseases of citrus (see Chapters 9–14).

For some problems, cultural methods that conform to BMP may be needed to create the conditions under which other management tactics can be effective, and without which those tactics might not be effective. Just as providing optimum nitrogen is inadequate if other nutrients are limiting, so too is management of many chronic pest and disease problems in trees growing under suboptimum conditions.

Citriculture has a long, distinguished record of successful releases of exotic biocontrol agents to bring pests under long-term natural control, and some commercial products are available for periodic inundation with biocontrol agents (see Chapter 11). Often, the successful activity of these agents is predicated on the responsible use of chemical pesticides to avoid upsetting the balance of predators and prey. The use of pesticides is the most problematic of the tactics available to growers. Although pesticides are often necessary and in most cases reasonably safe, they are the most expensive tactic and have serious consequences when used improperly. Growers who employ the appropriate tactics and decision aids to minimize pesticide usage benefit from increased profits and reduced risk to the citriculture environment.

PART TWO

Orchard Establishment

Larry K. Jackson
University of Florida
Citrus Research and Education Center, Lake Alfred

CHAPTER 2

Citrus Cultivation

Distribution of Citrus and Requirements for Cultivation

The principal species of the genus *Citrus* include sweet orange (*C. sinensis*), grapefruit (*C. paradisi*), mandarin (*C. reticulata*), lemon (*C. limon*), and lime (*C. aurantifolia*). Some of these species comprise many cultivated varieties (cultivars), which generally differ in fruit size, shape, seed content, quality, and season of maturity more than they vary in cultural requirements.

Citrus is grown throughout the world in most areas where minimum temperatures stay above lethal limits (usually around 20–24°F). In general, this area forms a broad belt between 40° north and 40° south latitude (Plate 1). While many factors influence the world distribution of citrus, including soils and climatic factors, it is the minimum temperature which inevitably determines where citrus can be profitably grown. With this fact in mind, many areas between the 40° latitudes can be eliminated as potential sites for citrus cultivation. Where temperatures are favorable, other climatic factors can come into play and may adversely affect citrus production. Rainfall patterns, excessive wind, and cloud cover, with the accompanying reduction in solar radiation, are examples. Soil conditions may also be limiting, because of such factors as insufficient rooting depth, poor water-holding capability, flooding, or the presence of toxic materials.

The areas suitable for citrus can be subdivided into regions by climate (Plate 1). Most citrus is produced in subtropical areas located in two bands, between 35° and 15° north latitude and between 35° and 15° south latitude. In most areas within this zone, winter temperatures are sufficiently moderate to induce an inactive, dormant quiescence in citrus trees followed by synchronized flowering and new growth in the spring. These regions are optimal for citrus because of this growth pattern, and because conditions usually favor good coloration of fruit, good internal quality, high productivity, and vigorous tree growth.

Oranges and mandarins produced in the equatorial region situated between 15° north latitude and 15° south latitude are usually lower in quality and are consumed locally. Grapefruit and limes, on the other hand, will produce satisfactorily. Since the temperatures are quite high all year, citrus grown in this region produces multiple crops in most years, as flowering is triggered by release from drought and other environmental stress factors. The fruit (especially oranges and mandarins) is usually poorly colored, and internal quality is compromised. Altitude compensates for latitude in some cases, and some good-quality fruit is produced in high-elevation equatorial regions where soil and other climatic factors allow.

The greatest volume of citrus fruit produced in the world (and, arguably, the best in internal quality) is grown in subtropical areas near the latitudinal extremes. São Paulo, Brazil, and Florida are typical of such locations, and the citrus industries there have experienced phenomenal growth. World citrus production figures reflect the effect of climate, as shown in Table 2.1. Substantial plantings have recently been made in Mexico and China as a result of new government policies in those countries. Since those trees are quite young, the size of the industries in those countries is not accurately reflected in Table 2.1.

Citrus production in Florida and California accounts for most of the U.S. citrus industry. There are minor production areas in Texas and Arizona, and a smattering of citrus is grown along the Gulf Coast in Alabama, Mississippi, and Louisiana. There were 1,103,600 acres of bearing citrus of all varieties in the United States in 1996, with over 773,500 acres in Florida and just under 270,000 in California. Total U.S. production for the 1995–1996 season was just over 16 million tons. Florida produced 11.8 million tons, and California 3.6 million.

The greatest volume of citrus fruit produced in the world (and, arguably, the best in internal quality) is grown in subtropical areas.

Citrus tree acreage in Florida, in general, is fairly young at this time, as much of it was planted following several severe freezes in the 1980s. While acreage is not expected to increase, production is likely to increase dramatically as the young trees

Table 2.1. Production of the principal types of citrus fruit by country for the crop years 1992–1993 through 1995–1996 (thousands of metric tons)[a–c]

	1992–1993	1993–1994	1994–1995	1995–1996
Oranges				
Brazil	14,117	13,710	16,520	16,450
United States[d]	10,071	9,462	10,641	10,747
Mexico	2,700	3,174	3,500	2,600
Spain	2,989	2,509	2,644	2,440
Italy	2,111	2,100	1,710	1,770
China	1,070	1,575	1,633	1,725
Egypt	1,771	1,324	1,513	1,360
Turkey	820	840	920	880
Morocco	874	916	657	870
Greece	872	854	865	850
South Africa[e]	664	739	770	850
Argentina	660	746	712	640
Australia	572	651	416	543
Israel	377	365	405	460
Cuba	425	350	350	380
Cyprus	160	160	160	252
Gaza[f]	87	87	87	87
Total	40,340	39,562	43,503	42,904
Grapefruit				
United States	2,532	2,414	2,642	2,502
Israel	383	344	415	395
Cuba	307	232	230	250
Argentina	177	195	208	190
South Africa	106	145	154	172
Mexico	118	112	136	120
Cyprus	110	112	95	112
Total	3,733	3,554	3,880	3,741

	1992–1993	1993–1994	1994–1995	1995–1996
Lemons				
United States	844	893	831	896
Argentina	590	653	741	700
Italy	752	743	565	680
Spain	737	611	571	443
Turkey	420	440	470	440
Greece	119	137	140	135
Total	3,462	3,477	3,318	3,294
Tangerines				
China	3,990	4,265	4,423	4,667
Japan[g]	2,019	1,751	1,539	1,696
Spain	1,521	1,631	1,751	1,566
South Korea	719	619	549	615
Brazil[h]	553	620	560	535
Italy	500	509	468	510
United States[i]	352	425	378	412
Turkey	390	405	430	410
Morocco[j]	317	373	304	360
Argentina	345	394	344	325
Egypt	340	205	250	300
Mexico	185	165	180	145
Total	11,231	11,362	11,176	11,541

[a] Source: U.S. Department of Agriculture, Foreign Agricultural Service.
[b] Crop year is the harvest and marketing period, which usually begins in the fall and extends through the spring, or roughly from October through June in the Northern Hemisphere and from April through December in the Southern Hemisphere. In the Southern Hemisphere, the harvest is almost entirely during the second calendar year of the period. The lemon harvest usually begins earlier and often extends throughout the year.
[c] One metric ton is equivalent to 2,204.6 pounds.
[d] Including Temple oranges.
[e] Including a small quantity of tangerines.
[f] Including tangerines, the production of which is small.
[g] Mainly satsumas (also called mandarin or *unshu mikan*), but also including mandarin hybrids.
[h] State of São Paulo only, which apparently accounts for over half of Brazilian production. About 120,000 metric tons of tangerines that are processed are included with oranges.
[i] Including tangelos, which account for about half of the total production.
[j] Clementine only.

mature. California acreage and production have been stable for several years, and little change is anticipated.

Major growth in acreage and production elsewhere in the world is most likely to occur in Mexico and China, as previously stated. Such growth is anticipated as a result of rapid economic expansion in these countries but could be reduced should economic growth falter.

The Citrus Plant

Citrus is a long-lived perennial evergreen tree (Box 2.1). The economic life expectancy is on the order of 50 years, and individual specimens in good situations often live 100 years or more. The trees are planted in orchards in regular rows at varying distances. Between-row spacing is commonly 20–25 feet (or less in some areas), which allows adequate room for equipment necessary for caretaking and harvesting. In-row spacing varies greatly; most plantings are established with trees 10–20 feet apart. Larger trees, such as grapefruit, are planted further apart than orange and tangerine trees.

Citrus trees were originally grown as seedlings. Such plants have numerous inherent disadvantages. Chief among these are delayed fruit production, a vigorous and upright growth habit, thorniness of the branches, and a general lack of adaptation to soil conditions, including (but not limited to) drought and disease. These problems, among others, have resulted in the production of citrus on grafted trees. The rootstock (the lower portion of the tree) is selected for desirable characters that are imparted to the scion (the upper portion of the tree). Strong root systems which are well adapted to the planting site, resistance to and tolerance for pests and diseases, and tolerance for various environmental conditions are other desirable attributes.

Rootstock characters are many and varied, so the choice of a rootstock is, at best, difficult. Every rootstock has certain limitations, so the decision is often made to minimize potential problems. Some decisions can be made by eliminating rootstocks that are poorly adapted to the area to be planted. Rootstocks that are susceptible to cold should not be used in areas where cold hazard is high. Likewise, the presence of certain diseases or pests may preclude the use of certain susceptible rootstocks. The process of elimination is then used by most growers to develop a short list of candidates to consider. Detailed information on rootstocks is available elsewhere in this book.

Selection of a scion variety is also critical, since the citrus tree is long-lived. Choices which seem sound at the time of planting are sometimes less desirable when the trees reach maturity and peak productivity. Many growers spread the risks

Box 2.1

Botany of Citrus

Citrus and its close relatives are in the subfamily Aurantioideae of the very diverse family Rutaceae. This subfamily is further subdivided by taxonomists into tribes and subtribes. The tribe of interest to us is Citreae, and the subtribe Citrinae contains the citrus fruit trees. Within this subtribe are groupings based largely on fruit characters. The true citrus fruit trees are classified in six genera, three of which include all commercially grown citrus fruit.

The three genera of importance are *Poncirus* (trifoliate orange), *Fortunella* (kumquat), and *Citrus* (oranges, mandarins, grapefruit, and the other types of true citrus fruit). The three genera have certain characteristics in common:

1. The plants are trees or shrubs, usually thorny, with predominantly white flowers.
2. The leaves are compound but are usually reduced to a single leaflet, except in *Poncirus,* which has three leaflets.
3. The leaf petiole is often winged, and the width of the petiole is a useful character for identification.
4. The fruit is a specialized berry known as a hesperidium. It has a leathery rind containing oil glands. The rind consists of a colored exterior (the exocarp) and a white, spongy interior tissue (the mesocarp) known as the albedo. Beneath the rind, the interior of the fruit (the endocarp) is segmented and packed full of juice vesicles. Seeds, if any, are produced within the segments.
5. Seeds may be produced by normal sexual fertilization or, in some cases, contain adventive embryos arising from nucellar tissues. A nucellar embryo is asexual and genetically identical to the mother plant, while a sexual embryo is the product of self- or cross-pollination.
6. The species of the three important true citrus genera (and some other related genera) can hybridize freely. This characteristic has been exploited widely by plant breeders. Similarly, there is a wide range of graft compatibility between the genera, which allows a great variety of combinations of rootstocks and scions and allows the use of interstocks and trees with multiple scions.

The genus *Poncirus* is monotypic, containing a single species, *P. trifoliata.* It is deciduous, has leaves composed of three leaflets (as the species name implies), and produces an acrid, pubescent fruit, which is inedible. It makes a very fine rootstock for many citrus types under certain conditions. *Poncirus* hybridizes freely with *Citrus,* and plant breeders have produced many progeny, most of which have found use as rootstocks. Citranges are hybrids of trifoliate orange and sweet orange; the widely used rootstocks Carrizo and Troyer are examples. Citrumelos are hybrids of trifoliate orange and grapefruit; Swingle, a popular rootstock, is one such cross. Many bigeneric and even trigeneric hybrids have been produced, and time and research will ultimately reveal their true value.

The genus *Fortunella* contains four species, differing primarily in fruit characters. The kumquats are attractive ornamental plants with edible fruit, which is often used for decoration or in making marmalade. They have been hybridized with relatives, but not much of value has been forthcoming. The best-known of the hybrids are limequats, crosses between lime and kumquat, in which the cold-hardiness of kumquat is combined with the acidity of lime to produce a cold-hardy, acid fruit.

The genus *Citrus* comprises several species of great economic importance. The taxonomy of these species is complicated and has been the subject of debate for many years. For this discussion, we will consider eight species, which represent the important commercial varieties of the genus:

C. sinensis	orange
C. paradisi	grapefruit
C. reticulata	mandarin
C. limon	lemon
C. aurantifolia	lime
C. medica	citron
C. aurantium	sour orange
C. grandis	pummelo

These species have many characters in common, some of the more important of which are the following:

1. All are small to medium-sized evergreen trees. They usually have thorns, and seedlings, especially, are thorned.
2. The leaves are unifoliate but may have wings on the petiole. The size of the wings varies with the species.
3. Flower buds are produced following a period of quiescence or stress and usually precede a new foliage flush. The buds are formed without protective scales.
4. The flowers are usually complete and perfect. They are usually white but are tinged pink or purple in some species.
5. The size and quality of the fruit varies with the species. The fruit contains various sugars, acids, and other chemicals to make widely varying flavors. The number of segments varies from eight to 20 (there are usually 10 to 14), and the segments are packed with juice vesicles.
6. The number of seeds formed is highly variable in this group, with some producing no seeds and others producing 50 or more. The seeds have a seed coat, and the cotyledons are white (tinged green in certain mandarins).
7. Cold-hardiness is variable, but in general *Citrus* is damaged when the temperature falls a few degrees below freezing. This greatly influences distribution.

In summary, *Citrus* is a group of evergreen trees of medium size that are sensitive to below-freezing temperatures, which limit outdoor cultivation to subtropical and tropical climates. Growth flushes are produced in response to favorable environmental conditions and may occur several times each year. Flowering is induced under favorable conditions following a period of quiescence or environmental stress. In subtropical climates, most flowering occurs in spring as temperatures increase following a dormant period in winter. In tropical climates, flowering is usually initiated after a release from stress (commonly water deficiency) and may occur at any time of the year and more than once a year.

Citrus trees are long-lived under favorable conditions, and many live over 100 years. The average age in commercial plantings is somewhat less, since trees may be removed because of disease or a decline in productivity before they are dead. Orchard trees are often pruned to keep them at a convenient height and width in a hedgerow configuration. Pruning also tends to maximize the leaf surface area of the planting and to increase the production of fruit per unit of land area. Leaves can persist for up to three years, but their longevity varies according to insect and disease pressures and the shading of older leaves by the subsequent growth of new leaves. Various environmental stresses, diseases, and pests affect tree vitality and fruit production.

of a bad choice by diversifying and planting a spectrum of varieties. Such diversification may also be beneficial by spreading out caretaking, harvesting, and marketing over a longer period.

Care of Young Citrus Trees

Citrus trees to be planted are produced in nurseries where they receive the intensive care required by young, rapidly developing plants. Propagation and nursery care are discussed in detail elsewhere and are not covered here. Nursery-grown plants, either bare-root or from containers, are planted with care to make sure that all air pockets are removed from the soil and that the tree is at the same level at which it grew in the nursery (or slightly higher than that level, to allow for settling). Frequent irrigation, in the absence of adequate rainfall, is essential to the transplant operation. Fertilization is usually withheld until the trees are established and have put on some new growth. Weed control is important, to eliminate competition for moisture, nutrients, and sometimes sunlight. This can be accomplished by hand, by mechanical removal, or by the application of a herbicide especially formulated for use on very young trees. Very little pruning is required beyond the occasional removal of suckers or water sprouts, which may develop from the rootstock or along the trunk of the scion.

Young citrus trees are nurtured with diligent care, beyond that normally given to mature trees. The object of the intensive care is to rapidly grow the tree to containment size and accelerate the yields as the tree approaches maximum productivity. While the extra care will assist in rapid tree growth, these same practices have undesirable effects on fruit quality. The fruit of very young trees is rarely of good quality, often being insipid and dry and having a thick, sometimes green peel. Within three to four years, young trees begin to lose their juvenility, and growth is less rampant, productivity increases, and fruit quality improves. At this point, the trees are gradually shifted into a production regime for mature trees.

Production Program for Mature Citrus Trees

As citrus trees mature, the production program moves from a regime for young trees, which promotes rapid growth, usually at the expense of fruit quality, to a regime of managed growth with emphasis on high yields of good-quality fruit. Once trees reach containment size, additional growth needs to be removed by pruning, or the trees will become overcrowded. Overcrowding inhibits movement within the orchard and reduces productivity due to shading.

The major elements of a production program for mature citrus trees include fertilization, pest management, irrigation, pruning, and tree replacement. Pest management embraces not only insect, mite, and disease control but also control of weeds and some vertebrate pests.

Fertilization

Fertilizers are supplied to trees in order to sustain growth and replace what is removed with the crop harvest and losses due to leaching, volatilization, and application inefficiencies. The amounts and types of nutrients applied vary greatly with the soil type, climate, and product availability. Application technology is likewise quite varied.

The nutrient content of fruit has been carefully analyzed in a Florida study (see Table 6.2).

Pest Management

Pests of citrus fruit and foliage can result in productivity losses and also create grade-lowering blemishes and defects. The degree of control required in a given location depends on pest and disease pressure and the intended market for the fruit. With little or no pest pressure, control is not required. Likewise, if the fruit is destined for juice processing, in which peel quality is not important, insect and disease control are less critical than if the fruit is destined for the fresh market.

Fruit and foliage pests are usually controlled to one degree or another by one of three management philosophies: biological control, utilizing natural pests and predators to assist in the control of harmful organisms; chemical control, using pesticides to control problems as they occur; and integrated pest management, combining (wherever practical) biological control and the judicious use of pesticides, with emphasis on the so-called "soft" pesticides, which target specific pests and have little effect on biological control agents.

Weeds are managed mechanically by mowing or cultivation or by chemical means. A combination of these techniques is often used. Because of the high cost of mechanical control methods, chemical weed control has become popular with citrus growers throughout the world. There is a broad spectrum of contact herbicides and growth-regulating and soil residual chemicals which have proven effective in contemporary weed management programs.

Irrigation

While not essential in all citrus-growing regions, irrigation has been shown to increase yields in many areas. It has become an important management tool in most locations. Microirrigation and drip systems have found the most favor with citrus producers worldwide.

The principal reason for the popularity of irrigation is the pattern of rainfall an area receives, not just the total quantity. In Florida, for example, the average rainfall of over 50 inches per year is more than adequate for good citrus tree growth, but the distribution is not satisfactory, since it is usually very dry during the spring bloom and fruit set. Supplemental irrigation during these times has greatly increased the productivity of most Florida citrus plantings.

Pruning

The removal of excessive growth is necessary to prevent overcrowding and shading of trees which have exceeded satisfactory containment size. A standard citrus planting is typically pruned into a hedgerow and topped at a manageable height of 12–16 feet. In some areas of the world, the trees are kept much smaller. Hedgerows allow the maximum interception of sunlight, improve fruit yield and quality, and facilitate orchard operations and harvesting. While some growers prune by hand where labor is plentiful and inexpensive, most use sophisticated mechanical equipment to accomplish pruning in a timely manner.

Tree Replacement

Inevitably, some trees in a citrus planting will die. Rapid removal of dead and declining trees is essential in order to keep a full stand and maximize productivity. Each vacant space costs money. Most growers have an annual replacement program to ensure high productivity.

W. S. Castle
F. G. Gmitter
University of Florida
Citrus Research and Education Center, Lake Alfred

CHAPTER 3

Rootstock and Scion Selection

A citrus tree normally has two parts: the *scion,* or fruit-bearing portion, usually propagated by budding on a *rootstock* seedling, which provides the tree with a root system. The selection of a scion and rootstock combination is an essentially permanent decision with lifelong ramifications for tree health and management. This decision, made in the nursery, determines how orchard trees will respond to many soilborne pests and pathogens, environmental stresses, and bud-transmissible diseases. Tolerance for pests, diseases, and environmental stresses is inherent in the compound genetic system. In this way, scion and rootstock selection is a form of natural biological control that operates as long as the tree lives. The impact of the selection ranges from minor to major, as the right choice of cultivars may prevent an orchard from becoming unproductive.

The rootstock has a greater influence on tree health than the scion. Scion cultivars are sometimes chosen out of consideration for tree health, assuming reasonable freedom from bud- and insect-transmitted diseases. Factors associated with tree health in the selection of scion cultivars are vegetative incompatibility, in which the scion and rootstock fail to form a healthy union, and the susceptibility of scion cultivars to endemic pathogens in specific ecological niches. Scion cultivars differ somewhat in disease tolerance, but the differences tend to be more quantitative than qualitative.

The selection of a scion and rootstock combination is an essentially permanent decision with lifelong ramifications for tree health and management.

Some management options affecting tree health remain after propagation and orchard planting. Those options (discussed in other chapters) often cannot compensate for poor selection of scions and rootstocks but are nevertheless important. They are much easier and less expensive to implement than changing the scion or rootstock. In the final analysis, however, all options are relevant, because no scion-rootstock combination is suitable under all conditions.

Scion Cultivars

Sweet Oranges

Sweet oranges are the most widely produced citrus fruits in the world. They can be divided into four subgroups: navels, common oranges, blood oranges, and acidless types. All sweet orange cultivars have arisen as the result of somatic mutations, and none have come from controlled pollination except the recently released Ambersweet orange in Florida. Consequently, contemporary citrus geneticists consider all sweet orange cultivars to be clonal selections that could in theory (but not in practice) be traced back to one original, ancestral form. Therefore, among sweet orange cultivars there is minimal variation in traits associated with tree health. Rather, most of the variation is in attributes associated with fruit quality, including the season of maturity, fruit size and shape, internal and external color, juice quality, and seediness. In spite of the great degree of genetic similarity among orange cultivars, some are better adapted to specific growing regions than others, because of the effects of climate on fruit quality. Cultural and ethnic preferences also have some influence over the choice of orange cultivars grown in a particular region.

The four subgroups of sweet oranges are described briefly in Table 3.1. For more complete descriptions of scion cultivars, consult the references listed in Additional Sources of Information.

Navel Oranges

Navels are grown almost exclusively as a fresh fruit product; high levels of limonin result in bitterness in processed juice. As a consequence, navel orange production worldwide is concentrated mostly in regions of arid, Mediterranean climate, where attractive fruit appearance (good external color and minimal rind blemishing) can be achieved. Navels are also grown in some humid subtropical regions, such as Florida, where ex-

Table 3.1. Attributes of selected citrus scion cultivars and cultivar groups

	Production areas	Utilization[a]	Market[b]	Season of maturity[c]	Advantages	Disadvantages	Remarks
Sweet oranges							
Navel oranges: Washington (Bahia), Bahianinha, Navelate, Navelina, Newhall, Leng, Lane Late, and others	Argentina, Australia, Morocco, South Africa, Spain, U.S. (California), Uruguay	F	E, D	E–L	Large fruit; seedless; fairly easy to peel; season of maturity of various clones ranges from early to late	Low yield; specific climatic conditions required for best quality; processed juice is bitter from limonin content	Clones vary in season of maturity, fruit quality (color, peelability, flesh texture, etc.), vigor, yield, and storage capabilities (on-tree and cold storage)
Common oranges							
Ambersweet	U.S. (Florida)	P, F	D, E?	E	Excellent color; early maturity; easy to peel	Poor performance as young trees; demanding cultural requirements	The only sweet orange cultivar developed by controlled hybridization; current trade regulations may limit its acceptance as a "true" sweet orange
Hamlin	U.S. (Florida), Brazil	P	D, E	E	Productive tree; early maturity; few seeds	Poor color; fair flavor	Some are marketed fresh in the United States
Jincheng	China	F, P	D	M	Good yield, color, and quality; stores well	Oval shape creates problems for mechanized grading	Commonly known as the "goose egg" orange; widely planted in interior provinces
Natal	Brazil	P	E	L	The latest-maturing cultivar in Brazil	Freeze damage to fruit in colder regions	The second most widely planted sweet orange in Brazil
Pera	Brazil	P	E	M	Productive tree (in Brazil)	Fair quality; seedy; prone to set multiple crops	The most widely planted sweet orange in Brazil
Pineapple	U.S. (Florida)	P	D, E	M	Productive tree; good color and flavor	Alternate bearing; seedy; cold-sensitive when heavily cropped	A cultivar of decreasing significance in Florida
Shamouti	Israel, Cyprus, Turkey	F	D, E	M	Easy to peel; seedless; distinctive flavor; good storage capability	Thick rind of coarse appearance; low juice content; juice is bitter from limonin	Fruit quality and productivity are optimum in restricted climatic and geographic regions
Valencia	U.S. (California, Florida), Australia, Argentina, Brazil, Morocco, South Africa, Uruguay	P, F	D, E	L	Excellent color and juice quality; few seeds; late maturity	Lower yield potential; freeze damage to fruit in colder regions	The most widely grown sweet orange in the world
Xuegan	China	F, P	D	M	Large fruit with good quality; stores well	Thick peel	Long history of cultivation in China
Blood oranges: Doble Fina, Maltaise Sanguine, Sanguinelli, Moro, Tarocco, and others	Mediterranean region	F, P	D	M–L	Unique flavor and appearance	Specific climatic conditions required for optimum quality	Regional acceptance only
Acidless oranges: Succari, Lima, Anliucheng, and others	North Africa, Middle East, Brazil, China	F	D	E–M	Acceptable fruit for acid-intolerant consumers	Unsuitable for processing	Regional acceptance only

Grapefruit							
White grapefruit							
Duncan	U.S. (Florida)	P	D, E	M–L	Productive, vigorous tree; large fruit; excellent quality	Excessively seedy	The preferred cultivar for processed grapefruit sections because of its flavor and firm texture
Marsh	U.S. (Florida), Cuba, South Africa	F, P	D, E	E–L	Productive, vigorous tree; nearly seedless; good quality	Flavor less intense than that of Duncan	The most widely planted grapefruit cultivar in the world
Oroblanco (Sweetie)	Israel, U.S. (California)	F	E, D	E–M	Seedless; high sugar content and low acidity allow production of good-quality fruit in cool areas and allow earlier harvest	Thick rind and open core, atypical of grapefruit; yields lower than yields of Marsh have been reported	The first commercially significant triploid cultivar developed by breeding
Pigmented grapefruit							
Flame	U.S. (Florida)	F	D, E	E–M	Good flesh color in early and midseason; good flavor and texture; nearly seedless	Color fades in late season; little commercial experience	Nucellar seedling of Henderson, widely planted in Florida since its release in 1987
Redblush (Ruby Red)	U.S. (Florida), Cuba, South Africa	F	D, E	E–M	Healthy, productive tree; nearly seedless; good quality; attractive peel blush	Less colorful flesh than more recently released cultivars	The Florida standard of quality; it is being replaced by more deeply pigmented cultivars
Rio Red	U.S. (Texas)	F	D, E	E–M	Good flesh color; healthy, productive tree; nearly seedless	Little commercial experience	Likely to replace Star Ruby as the most deeply pigmented grapefruit cultivar in Texas
Star Ruby	U.S. (Texas, Florida), South Africa	F	D, E	E–M	Most intensely pigmented flesh and rind; seedless; retains color late in the season; thin rind	Weak, unthrifty trees; greater susceptibility to *Phytophthora*, herbicide injury, storage rots, and cold	The most difficult grapefruit cultivar to manage
Mandarins and their hybrids							
Clementine (Oroval, Fina, Nules, and many other clones)	Spain, Morocco	F	E, D	E–M	Easy to peel; seedless when grown in isolated blocks; good flavor	Small fruit; seedy with cross-pollination	Clones vary in season of maturity, appearance and quality of fruit, and productivity; the most popular mandarin in Europe
Dancy	U.S. (Florida)	F	D	M	Vigorous tree; good color and flavor; easy to peel	Alternate bearing; seedy; difficult to harvest; susceptible to *Alternaria*	A cultivar in decline
Ellendale	Australia, Argentina, Uruguay	F	D, E	M	Large fruit with good color and flavor; easy to peel; nearly seedless in solid blocks	Prone to splitting of fruit and tree; high acidity; seedy with cross-pollination	Natural tangor of unknown parentage
Fairchild	U.S. (California, Arizona)	F	E, D	E	Attractive appearance; adaptable to arid production areas	Small fruit; seedy; rind is oily and difficult to remove	Hybrid of Clementine and Orlando tangelo

(continued on next page)

[a] F = fresh fruit; P = processed fruit. The more common use is listed first.
[b] D = domestic market; E = export market; ? = uncertain.
[c] E = early-season maturity; M = midseason maturity; L = late-season maturity; Y = year-round maturity.

Table 3.1. *(continued)*

	Production areas	Utilization[a]	Market[b]	Season of maturity[c]	Advantages	Disadvantages	Remarks
Mandarins and their hybrids *(continued)*							
Fortune	Spain	F	E, D	L	Productive tree; attractive fruit, with good color and flavor; later maturity than most mandarins	Rind prone to pitting; susceptible to *Alternaria;* seedy with cross-pollination; high acidity	Hybrid of Clementine and Dancy, with specific climatic requirements
Imperial	Australia	F	D	E	Very early maturity; easy to peel	Alternate bearing; fair flavor	Produced only in Australia
Kinnow	Pakistan, India	F, P	D	L	Very high sugar content; juicy; vigorous tree	Seedy; alternate bearing	The main citrus cultivar in Pakistan, meeting consumers' demand for very sweet fruit
Mediterranean (Willowleaf, Baladi, Avana, and others)	Italy, Portugal	F	D	M–L	Easy to peel; unique flavor and aroma; cold-tolerant	Small fruit with seeds; poor storage and shipping capability; alternate bearing	Declining demand for fresh fruit, but rind oil and oil from foliage are in demand as flavoring and for perfumes
Minneola	U.S. (Florida), Israel, South Africa, Argentina	F	D, E	M	Large fruit with distinctive shape, excellent color, and rich flavor; vigorous, productive tree	Susceptible to *Alternaria;* seedy with cross-pollination	Adaptable to a range of climatic conditions; a hybrid of Duncan grapefruit and Dancy tangerine
Murcott (Smith, Honey tangerine)	U.S. (Florida), Brazil, Israel, Australia	F, P	D, E	M–L	Excellent flavor and flesh color; productive, cold-hardy tree	Seedy; alternate bearing; susceptible to scab and *Alternaria;* terminal bearing habit	Tangor of unknown parentage from Florida
Orlando	U.S. (Florida)	F, P	D	M	Vigorous tree; cold-hardy	Difficult to peel; pale flesh and juice; low acidity; seedy	Hybrid of Duncan grapefruit and Dancy tangerine; of declining importance in Florida
Ortanique	Jamaica, Israel, South Africa, Australia, Cyprus	F	D, E	L	Large, attractive fruit with good color and flavor; stores well; juicy; very late maturity; productive tree	Rind is oily and difficult to remove; seedy with cross-pollination	Spontaneous tangor from Jamaica
Ponkan (Nagpur suntara, Batangas, Warnurco)	China, India, Japan, Philippines, Brazil	F	D	M	Large fruit; very easy to peel; crisp flesh texture; productive tree	Rind becomes puffy; difficult to harvest, ship, and store; seedy; alternate bearing	Probably the most widely grown mandarin cultivar in the world
Robinson	U.S. (Florida)	F	D	E	Very early maturity; good color; cold-hardy tree	Seedy; difficult to peel; susceptible to twig and limb dieback	Hybrid of Clementine and Orlando, with limited future potential
Satsuma (Owari, Miyagawa, Okitsu, Clausellina, and many other clones)	China, Japan, Spain, Turkey, Korea, Argentina, Uruguay, U.S. (California), South Africa	F, P	D, E	E–M	The most cold-hardy tree among the edible citrus cultivars; easy to peel; seedless; tolerant of citrus canker	Poor fruit quality in humid subtropics; rind puffiness shortly after maturity	Numerous mutants have been propagated, varying in season of maturity and fruit quality
Sunburst	U.S. (Florida)	F	D	M	Very attractive fruit with excellent color	Susceptible to mite damage; seedy; difficult to peel	The major early- to midseason mandarin cultivar in Florida
Temple	U.S. (Florida)	F, P	D	M–L	Unique flavor; attractive external color and appearance	Susceptible to cold and citrus scab; seedy; fruit acidity	A cultivar in decline

Pummelo							
Numerous cultivars grown in various production areas: Banpeiyu, Chandler, Goliath, Kao Pan, Kao Phuang, Matoyu, Nakon, Pandan Wangi, Shatianyu, Thong Dee, and others	China, Japan, southwest Asia, Indonesia	F	D	E–L	The largest citrus fruit, available in various sizes, shapes, flesh color (white, pink, red, green, yellow, etc.); vigorous, productive tree; excellent storage capability	Cold-tender tree; high heat required for optimum fruit quality; very seedy with cross-pollination; some cultivars are prone to alternate bearing	The citrus fruit of preference in many Asian cultures and ethnic groups, with an ancient history of cultivation and use
Lemons							
Eureka	Argentina, Australia, Israel, South Africa, U.S. (California)	F, P	D, E	M–L	Few seeds; good production from young trees; moderately vigorous tree	Cold-sensitive; terminal bearing habit, leading to wind scar and sunburn; moderate productivity; graft-incompatible with *Poncirus* and most trifoliate hybrids	The most widely grown lemon cultivar, except in the Mediterranean
Femminello (a group of several clones)	Italy	F, P	D, E	Y	Very productive and vigorous tree; everbearing	Most clones are highly susceptible to mal secco	The major Italian lemon cultivar
Fino (Primofiori, Mesero, Blanco)	Spain	F, P	D, E	E	Early maturity in Spain; productive tree; high juice content and acidity	Small fruit; susceptible to mal secco; thorny	Grown in colder areas in Spain than Verna (the major Spanish cultivar) because of early maturity
Interdonato	Turkey, Italy	F, P	D, E	E	The earliest-maturing lemon; nearly thornless; large fruit; tolerant of mal secco	Low juice content and acidity, relative to other lemons	Plant and fruit characteristics suggest that it is a citron-lemon hybrid
Lisbon (a group of several clones)	Argentina, Australia, Uruguay, U.S. (California, Arizona)	F, P	D, E	M–L	Productive tree; cold-tolerant; compatible with *Poncirus* and its hybrids; bears fruit inside the canopy, and thus is less prone to rind damage	Greater vigor requires wider tree spacing; thorny	Fruit characteristics very similar to those of Eureka
Monachello	Italy	F, P	D, E	M–L	Tolerant of mal secco	Low juice content and acidity; thick rind	Probably a citron-lemon hybrid
Verna (Berna)	Spain	F, P	D, E	L	Nearly thornless; can produce a valuable *verdelli* crop, when no other lemons are on the European market	Alternate bearing; low fruit and juice yields	The most important lemon in Spain
Acid limes							
Tahiti (Persian, Bearss)	Australia, Mexico, Brazil, U.S. (Florida)	F, P	E, D	M	Seedless; nearly thornless tree; more cold-tolerant than Mexican lime	Prone to stylar-end breakdown; susceptible to scab	Triploid variety, possibly of citrus and small acid lime parentage
West Indian (Mexican, Key, Lima, Kaghzi nimbu, Galego, Limunbeladi, and other names)	Mexico, Brazil, Egypt	F, P	E, D	Y, E	Distinctive aroma and flavor; abundant, highly acidic juice	Very cold-sensitive; thorny; seedy; fruit drops quickly upon maturity	Everbearing in the tropics but not in the subtropics

[a] F = fresh fruit; P = processed fruit. The more common use is listed first.
[b] D = domestic market; E = export market; ? = uncertain.
[c] E = early-season maturity; M = midseason maturity; L = late-season maturity; Y = year-round maturity.

ternal color is compromised and the rind is too thin to be removed easily, thus not meeting consumer expectations and demands. There are many different clones, or selections, of navel orange, which vary in tree vigor, productivity, season of maturity, color, flesh texture, peelability, and on-tree and cold-storage capability.

Common Oranges

Common oranges comprise the greatest number of cultivars, including the most widely adapted (and thus most widely planted) cultivar, Valencia. Most cultivars in this group are primarily used for processed juice, although several cultivars are also marketed as fresh fruit. Like navels, common oranges do not show substantial variation in important fruit quality traits. In general, they are more productive than navels, but they are virtually identical in traits related to tree health. Unlike the other sweet orange cultivar groups, common oranges are produced to some extent in all citrus-growing regions of the world, ranging from the humid tropics to the cool subtropics. However, fruit grown in warm subtropical areas, such as Florida, produces the highest-quality juice.

Blood Oranges

The blood oranges are a distinctive group of cultivars characterized by red pigmentation (anthocyanin) in the flesh and rind tissue and a unique flavor and aroma. Greater pigmentation develops in the cool winters typical of Mediterranean climates. The same cultivars grown in the humid subtropics or tropical regions bear fruit with less "blood" (or none) in the flesh. They are predominantly used as fresh fruit, although some are processed for juice. They are especially appreciated in Italy, where they are well adapted and can produce fruit of excellent quality.

A red-fleshed navel cultivar, Cara Cara, is not a member of this group; its red color comes from lycopene (the same pigment is present in red-fleshed grapefruit) and is remarkable in regions with warm winters.

Acidless Oranges

Some sweet orange cultivars have evolved to produce mature fruit with virtually no acidity. In all other characteristics, they are indistinguishable from the common oranges. Acidless fruit has found favor in some cultures and ethnic groups, mostly in Asia and North Africa, and some acidless oranges are also grown in Brazil. These oranges often lack some of the flavor and aroma typically associated with sweet oranges and taste insipid because they do not possess the typical balance of sugar and acidity. They are consumed as fresh fruit and rarely processed, because of their weak flavor. Their low acidity tends to minimize the period during which they may be stored.

Grapefruit

The grapefruit originated in the Caribbean region in the seventeenth century, probably by natural hybridization between pummelo and sweet orange. It is the only citrus species to have a New World origin, although it should not be considered a true New World native. Grapefruit cultivars are divided in two groups, described below, based on their flesh pigmentation.

White Grapefruit

The original form of grapefruit had white flesh and a yellow rind and was seedy. The old Florida cultivar Duncan typifies this form. A seedless form was found as a spontaneous mutant in the mid-nineteenth century in Florida and was named Marsh. Other seedless white-fleshed cultivars have been found, but Marsh is the most commonly grown, by far. White grapefruit trees are generally healthy, vigorous, and productive and produce their best-quality fruit in the humid subtropics, although they are productive in other regions as well. When white grapefruit is grown in a cool-winter climate, the fruit is highly acidic and thus much less palatable.

A fairly new triploid hybrid of pummelo and grapefruit, named Oroblanco, may produce less-acidic fruit than Marsh in cool-winter areas; already a significant industry based on this cultivar has developed in Israel, where it is grown for export and marketed under the name Sweetie.

White grapefruit is used as fresh fruit and may also be processed as juice or canned segments.

Pigmented Grapefruit

Mutations for lycopene production in grapefruit flesh and rind have been found in the last 100 years, and cultivars with this trait have become popular because of their visual appeal to consumers. The demand for pigmented grapefruit has increased, and cultivars with deeper flesh color have been developed from spontaneous or induced mutants. Some of these cultivars are more difficult to grow, because of their greater sensitivity to cold, fungal diseases, and herbicide treatments. Unlike white grapefruit, pigmented grapefruit is produced almost exclusively for the fresh market, because of the instability of its juice pigments when it is processed.

Mandarins

Mandarins (or tangerines, a marketing term applied to some cultivars) are the most diverse group of citrus cultivars, both genetically and in characteristics of trees and fruit. Many mandarin cultivars freely hybridize with other citrus species, and several mandarin hybrid cultivars of commercial significance have been developed; the most notable are the tangelos (hybrids with grapefruit or pummelo) and the tangors (hybrids with sweet orange). Mandarin cultivars vary substantially in resistance or susceptibility to pests and diseases, fruit quality, tree morphology, and climatic adaptation, thus reflecting the diversity of their gene pool. Like sweet oranges, at least some mandarin cultivars produce fruit in nearly every citrus-growing region of the world. Current consumer demands for seedless, easy-to-peel fruit are focusing attention on cultivars with those traits.

Some cultivars require cross-pollination for adequate fruit set, and care must be taken to select appropriate, sexually compatible pollenizers that bloom annually in synchrony with the main cultivar. Cross-pollination, though, results in seedy fruit, which will not always earn premium prices. To produce seedless fruit, trees must be grown in isolation, to prevent cross-pollination. To be successful at this, growers must select cultivars that can provide sufficient levels of parthenocarpy in their growing region. The importance of environment in the successful production of seedless fruit by currently available cultivars cannot be overstated. Some cultivars (Nova tangelo, for example) can produce adequate crops of seedless fruit when grown in isolation in Mediterranean regions, but similar solid blocks in Florida would have poor fruit yields. To avoid failure, growers must base decisions about mandarin cultivar selection on a clear understanding of the projected market as well as on the adaptation of the cultivar to the local climate.

Mandarins require more intensive horticultural input than other citrus scions, in part because of their propensity to alternate bearing. In extreme cases, heavy fruit loads can break scaffold limbs or kill the tree, presumably because of a depletion of nutrients and photosynthate. Pruning, fruit thinning, ringing or girdling, and growth regulator treatments are used to manipulate crop load and to control alternate bearing.

Pummelos

Pummelos are produced primarily in Asia, where they are traditionally preferred by many diverse ethnic groups and cultures. Pummelo cultivars were introduced into the Americas by European explorers and colonists in the seventeenth century, but until recently they have remained virtually unknown in the West except as botanic curiosities. Pummelos are the largest citrus fruit, and they are produced by some of the most vigorous citrus trees. They vary in size, shape, and flesh color. They hybridize freely among themselves and with the other *Citrus* species, and numerous cultivars and local selections have been developed as a result. Most cultivars yield fruit of optimum quality in subtropical climates, although certain cultivars are well adapted to tropical or Mediterranean climates.

Lemons

Most lemon cultivars produce large trees, and pruning is an important element of management, because of the inherent vigor of the trees. Lemons are more sensitive to cold than all other commercial citrus scions, except limes, so commercial production has been restricted to regions where freezes are very rare. However, lemons are not particularly suited to tropical regions, because of pest and disease problems and adverse environmental effects on fruit quality. The ideal lemon climate is one with mild winters and cool summers, to enhance the everblooming nature of the tree and thus distribute crop production more evenly throughout the year. Regions of Italy (Sicily and the Amalfi coast, near Naples), coastal areas of southern California, and parts of Spain, Greece, Cyprus, and Turkey meet these requirements. Fruit quality and storage ability are optimized in such locations, although lemons can be grown in other, less desirable areas.

With few exceptions, it is nearly impossible to distinguish lemon cultivars on the basis of fruit morphology, because there can be greater variation between fruit of different bloom seasons than between fruit of different cultivars from the same bloom season.

Acid Limes

There are only two acid lime cultivars of commercial significance, each of which has different names in local regions. Perhaps the more widely known cultivar is the Tahiti, or Persian, lime. This is the larger-fruited lime cultivar, seedless by virtue of its triploidy, which is grown in Brazil, Mexico, and Florida. It is more vigorous, more disease-tolerant, and slightly more cold-tolerant than the smaller-fruited, seedy West Indian (Key or Mexican) lime, which is grown in Mexico, India, Egypt, and Brazil. Limes are best suited to hot tropical and subtropical regions and thrive in humid regions where lemons perform more poorly. However, Tahiti lime has a lower heat requirement for optimum fruit size (along with greater cold tolerance) than West Indian lime, so cultivation of the latter is restricted primarily to regions with high temperatures year-round. Tahiti performs better than West Indian in areas with occasional brief periods of cool weather.

Other Cultivars

Some citrus cultivars, from a broad range of species and related genera, are grown for highly specific purposes or markets. Ethnobotanists and anthropologists have described many culinary and medicinal uses of citrus fruit, flowers, and foliage, from regions where citrus and its relatives are native. In addition to local utilization by indigenous people, commercial uses have been found for other cultivars, including kumquats (*Fortunella* species) for fresh consumption and processing; citrons for a candied rind product; sour orange for marmalade; bergamot and Bouquet des Fleurs for fragrances; and acidless limes for fresh fruit. Such enterprises are insignificant relative to total worldwide citrus production, but they can represent substantial industries in unique ecological and sociological niches.

Rootstock Cultivars

Rootstock Use

Most citrus trees were grown as seedlings prior to the advent of rootstocks (Plate 2). Seedlings have several disadvantages. Many sweet orange and grapefruit cultivars produce reasonably uniform, true-to-type seedlings, because of nucellar embryony, but important chimeric changes, such as in red-fleshed grapefruit, may be lost. Other citrus types, such as pummelo and some mandarins, produce only monoembryonic seeds in which the single embryo is of sexual origin and is unlikely to reproduce all the desirable parental characteristics. Seedlings are thorny and have a relatively long juvenile period which delays the first commercial harvest. When they enter their bearing phase, cropping can be erratic. Therefore, it takes longer to obtain a positive cash flow and post a profit from trees grown from seedlings than from trees grown on rootstocks. Seedlings of certain types, especially sweet oranges, are also highly susceptible to foot and root rot diseases caused by soilborne pathogens, such as *Phytophthora nicotianae, P. citrophthora,* and certain nematodes. For these historical and practical reasons, rootstocks gradually replaced seedling trees. Rootstocks are now widely recognized for their favorable effects on tree health and horticultural traits.

Sour Orange and Related Types

Among citrus rootstocks, sour orange is the most common. Trees on this rootstock are well known for good yields of fruit with excellent internal and external quality. Sour orange is also an excellent choice for a broad range of soil conditions. In areas affected by periodic damage from cold weather and freezes, trees on sour orange are consistently the least cold-injured.

Tree vigor is moderate regardless of soil type; tree size is medium to large. Sour orange is better adapted to the more fertile loamy or clay soils than to deep sandy soils. In the latter, it develops a deep, widespread root system of moderate fibrous root density, which confers relatively good drought tolerance. Growers faced with sites characterized by some combination of clay materials, calcareous or saline conditions, chronic wetness, damaging levels of *Phytophthora* species, or risk of cold injury, often discover that only sour orange provides satisfactory performance.

Cultivars grown on sour orange generally produce medium-sized fruit with a thin peel and juice with high soluble solids and acid concentrations. A high level of soluble solids is beneficial to growers who sell fruit for processing on that basis and

Table 3.2. Comparative rankings of major citrus rootstocks (Plate 2) [a]

Rootstock	Tolerance of: Salinity	High calcium	Clay soil	Flooding	Poor drainage	Drought	Freezing	Tree vigor	Yield per tree[b]	Fruit size[c]	Juice quality[d]	Compatibility[e]	Remarks
Benton citrange	P	P	G	?	?	G	G	H	H	I–Lg	I, I	G	Compatible with Eureka lemon, unlike other citranges; poor nursery performance
C-35 citrange	?	P	G	?	?	?	I?	I–L	I	I	I, I	G	A new rootstock with promising performance in California and Florida
Carrizo citrange	L	P	G	P	I	G	I	H	H	I–Lg	I, I	G	Questionable long-term compatibility with some mandarins; selections are tolerant of burrowing nematodes (the most tolerant, Kuharske, is available in Florida but is largely unevaluated)
Citrus macrophylla	G	H	L	G	G	G	P	H	H	Lg	L, L	G	Compatibility problems with lemons; very tolerant of excess boron; precocious bearing; yield improves with age
Citrus taiwanica	G	I	G	?	?	?	G	I	I	I–Lg	H, H	G	Seedlings highly variable
Cleopatra mandarin	G	I	H	P	P	I–G	G–H	H	L	Sm	H, H	G	
Gou Tou	?	?	G	?	?	?	?	H	I	Lg	I, I	G	A new rootstock; problems with bud survival in the nursery
Palestine sweet lime	P	I	I	G	G	G	P	H	H	Lg	L, L	G	
Rangpur	H	H	I	?	I	G	P	H	H	Lg	L, L	G	Trees very precocious; excellent in calcareous sites
Rough lemon[f]	I	H	I	G	I	H	P	H	H	Lg	L, L	G	Occasional compatibility problems with midseason orange scions and nucellar clones
Rusk citrange	P	P	?	L	?	P	I	I–L	L–I	I	I, I	G	Comparable to C-35 citrange
Smooth Flat Seville	I	H	G	I	G?	I	G	I	I	I–Lg	I, I	G	Somewhat tolerant of citrus tristeza virus, but not as tolerant as Carrizo; about 30–40% of seedlings are zygotic
Sour orange	I	H	H	I	H	I	H	I	I	I	H, H	G	Susceptible to tristeza-induced decline
Sun Chu Sha mandarin	?	H	H	?	I?	?	G?	H	L–I	Sm	H, H	G	A new rootstock; tested in Texas
Sunki mandarin	G	I	H	?	?	?	G	H	L–I	Sm	H, H	G	Susceptible to citrus exocortis viroid; poor seed production in some climates
Sweet orange	I	I	I	P	P	P	I	I	I	I–Lg	I, I	G	Cultivar Ridge Pineapple is tolerant of burrowing nematodes
Swingle citrumelo	L	P	See Remarks	G?	?	I	H	I–L	I–L	I	H, I	G	Stock sometimes overgrows scion to cause compression girdling; soils or soil layers with clay content greater than 30% may limit growth
Trifoliate orange	P	P	H	I–G	G	P	G–H	L	L	Sm–I	H, H	See Remarks	Cold hardiness is climate-dependent; some problems with delayed incompatibility
Troyer citrange	L	P	G	P	I	G	I	H	H	I–Lg	I, I	G	Similar to Carrizo citrange
Volkamer lemon	I	H	I	G	G	H	P	H	H	Lg	L, L	G	Similar to rough lemon

[a] Cultivars are comparatively ranked as follows, in order from highest to lowest ranking: H = high; G = good; I = intermediate; L = low; P = poor; ? = inadequate information or rating unknown. Ratings are based largely on California and Florida experience, with emphasis on the latter.

[b] Absolute yield, not in relation to canopy volume.

[c] Lg = large; I = intermediate; Sm = small.

[d] The first letter represents the soluble solids concentration; the second letter represents acidity.

[e] Compatibility is a complex matter. Problems may appear in the nursery or much later in the grove. Scion or rootstock overgrowth (i.e., bud union not smooth) is not always a reliable indicator of tree longevity.

[f] Milam lemon is tolerant of burrowing nematodes and is otherwise similar to rough lemon.

to processors, whose costs for extracting and handling the juice are lower per unit of sugar. Fruit maturity is commonly determined by the ratio of soluble solids to acid. In some cultivars grown on sour orange, fruit maturity is often delayed, because of the high acid concentration of the fruit, but the high acidity also results in better on-tree storage and postharvest behavior than that of fruit with lower juice acidity. In other cultivars, fruit maturity is delayed because of naturally low levels of soluble solids; when these cultivars are grown on sour orange, their maturity is advanced because of the higher concentration of soluble solids.

Sour orange no longer has a secure place in today's rootstock portfolios because of its susceptibility to citrus tristeza virus (CTV). This disease continues to spread and has severely reduced or eliminated the use of sour orange in many places throughout the world. Nevertheless, the many excellent horticultural advantages of sour orange still outweigh its deficiencies (Table 3.2) in places where CTV is absent or not considered a major risk. Trees on sour orange are very tolerant of citrus blight, a serious disease in Florida, Brazil, and other places where, unfortunately, the presence of CTV prevents or limits the use of this rootstock.

The susceptibility of sour orange to CTV has created interest in other rootstocks as possible replacements for it. Among these are Bittersweet sour orange and *C. taiwanica,* but they are similar to sour orange in horticultural and pathological traits and offer no known advantage. *C. taiwanica,* which has been planted as a minor rootstock in parts of Mexico, produces highly variable seedling populations; also, trees on this rootstock produce lower-quality fruit than those on sour orange.

Smooth Flat Seville, Kinkoji (*C. obovoidea*), Gou Tou, and Zhu Luan are each probably some hybrid combination of pummelo, sweet orange, and sour orange and thus have aroused interest as replacements for sour orange, principally in Florida. Trees on Smooth Flat Seville have grown and fruited well when infected with severe strains of CTV from Florida, but other evidence indicates that this rootstock does not have the same tolerance as Carrizo citrange. Trees on Smooth Flat Seville produce lower yields and poorer-quality fruit than trees on sour orange, and seedling populations of Smooth Flat Seville contain up to 40% off-types. Gou Tou apparently tolerates some of the world's most severe CTV strains and is purported to be salt-tolerant; trees grown on Gou Tou are vigorous, but the nursery performance of the rootstock does not compare favorably with that of sour orange. Less is known about Zhu Luan. The nursery traits of Kinkoji are excellent, and field trees on this rootstock tolerate Florida severe strains of CTV.

Sweet Orange

Sweet orange was first grown as seedling trees for use as scions. When sweet orange trees on sweet orange rootstock were evaluated commercially, they exhibited the advantages of any tree propagated on a rootstock. They are moderately vigorous, producing large trees with good crops and good-quality fruit. The positive traits of sweet orange (Table 3.2), both horticultural characteristics and disease tolerance (e.g., it is blight-tolerant), would ordinarily suggest a superior rootstock. However, sweet orange is limited by two factors: susceptibility to Phytophthora trunk and root diseases and drought intolerance, despite its ability to develop a deep, widespread root system in sandy soils. The search for better sweet orange rootstocks continues. Many cultivars have been tested, but all display the same two weaknesses, which limit their commercial acceptance.

Lemons, Limes, and Related Types

Citrus trees growing on rough lemon, Volkamer lemon, *C. macrophylla,* Palestine sweet lime, or Rangpur are horticulturally similar (Table 3.2), but there are important distinctions. Also, their behavior is strongly conditioned by climate and soils. In tropical and subtropical regions with deep, well-drained, infertile sandy soils, trees on these rootstocks are vigorous and highly productive but bear low-quality fruit. They also grow reasonably well in saline or calcareous situations and have extensive root systems, but they are not very cold-hardy. In cooler climatic regions with loamy to clay soils, tree vigor can be substantially reduced to the point at which yield becomes commercially marginal. Growers selecting any of these rootstocks may find that they have a relatively short orchard life in tropical or subtropical areas, because of problems with *Phytophthora* diseases or citrus blight.

Rough lemon has been one of the primary rootstocks in several citrus industries in the world, but little is planted now. Yield and large fruit size have been the leading reasons for its popularity. Poor juice quality (low soluble solids and acid) was not a limiting factor, because with fresh fruit, once minimum maturity standards are met, fruit size, external quality, and yield become the controlling factors. Yield is critical when processing fruit is sold by volume (as in Brazil) or weight (as in Australia). However, yield is still important when fruit is sold by the concentration of soluble solids in the juice, because highly productive rootstocks can more than compensate for low juice quality.

Volkamer lemon has compared favorably with rough lemon in Florida, especially as a rootstock for oranges grown for processing into juice. Trees on Volkamer lemon have consistently shown slightly better cold tolerance and are a better choice where Phytophthora root rot is a problem. Volkamer is a minor rootstock in Florida (for oranges), Italy and South Africa (for lemons), and Venezuela and other Caribbean countries as a replacement for trees on sour orange lost to CTV.

C. macrophylla (or Alemow, its common name) has been largely a one-scion rootstock. It is a popular choice worldwide for Eureka, Lisbon, and other lemon types, but lemon clones are short-lived on Alemow in California because of a necrosis of the phloem tissue. Alemow is ordinarily not suitable for any scion cultivars except lemons, but it has recently been used as a rootstock for Tahiti lime in Florida, and in the 1990s it was put to more general use as a rootstock for navel oranges and mandarins in Spain. Trees budded on Alemow tend to produce very coarse, thick-skinned fruit in humid, subtropical areas, with the poorest juice quality among the rootstocks in this group. In regions with Mediterranean-like climates, trees on this rootstock are slower-growing but retain their precocity and their heavy, early-bearing habit and produce the large fruit desired for the fresh market. Alemow is susceptible to CTV and blight. However, it is resistant to Phytophthora foot and root rot, and it tolerates high levels of boron, chlorides, and calcium in the soil but accumulates manganese. These traits make Alemow adaptable to an unusual range of soils and soil conditions.

Palestine sweet lime itself produces an insipid, yellow fruit that is consumed fresh in India and elsewhere. Sweet lime is not a true lime and, outside of India and Israel, is a minor rootstock. Shamouti orange was commonly grown on this stock in Israel because of the characteristics typical of this rootstock group: vigor, high productivity, and large fruit. However, its susceptibility to several virus and viroid diseases often led to tree decline in Israel where tree longevity was prolonged by inarching with another rootstock. Overall, it is less desirable than rough lemon or Volkamer lemon as a rootstock.

Rangpur "lime" is not a true lime but probably a mandarin hybrid. A distinctly mandarin trait is its tolerance for saline and calcareous soils. It is not a common rootstock except in Brazil, where it is the dominant choice because of excellent drought and CTV tolerance. Rangpur is susceptible to viroid diseases, but with clean budwood, trees on this rootstock are early-bearing, grow well, and yield good crops of fruit, with generally superior juice quality, compared to fruit from trees on rough lemon. This difference in juice quality may reflect the apparent adaptation of Rangpur to well-drained, highly leached soils with high levels of iron and aluminum, as in Brazil. The quality of fruit from trees on Rangpur in the sandy soils of Florida is very poor, but navel orange trees grown on Rangpur in California's cooler climate and loamy soils have moderate cold tolerance and produce fruit of satisfactory quality.

Mandarins

The mandarin group is diverse and includes mostly scion cultivars, but some cultivars are used as rootstocks. Many mandarins have been tested for their potential as rootstocks, but only a few have reached commercial status. Among these, their general advantages as rootstocks include virtually no susceptibility to graft-transmissible diseases, cold tolerance similar to that of sour orange, and satisfactory growth in saline and high-pH soils. Most mandarins have two serious weaknesses as rootstocks: budded trees achieve a large size but are slow to come into bearing, and their fruit is often small (although it stores well on the tree and produces juice of excellent quality). As a result, mandarin rootstocks are not suitable for many fresh fruit cultivars, with certain exceptions noted below. The decision to use a mandarin rootstock is frequently viewed as a matter of choice between tree longevity versus long-term productivity.

Sunki mandarin is a major rootstock in China, and it has become a minor rootstock in Brazil because of its supposed blight tolerance. It is notable for the relatively large proportion of zygotic plants in seedling populations, a trait which has been useful in hybridizing.

Cleopatra mandarin is the most widely known mandarin rootstock, and it has been the traditional rootstock for mandarin scion cultivars that do not suffer from inherently small fruit size. It is not widely used for grapefruit or sweet oranges except to provide smallish fruit to particular markets or fruit late in the season. Fruit from trees on Cleopatra has high concentrations of soluble solids and acid and thus matures later than fruit from trees on other rootstocks. The fruit is thin-skinned and prone to splitting before harvest.

Nasnaran (*C. amblycarpa*) is an unpleasant rootstock in the nursery because of its thorniness and its tendency to strike many lateral branches. However, in comparisons conducted in Florida orchards, trees on Nasnaran routinely produced 10–20% more fruit than trees on Cleopatra, with no difference in quality.

Shekwasha mandarin (*C. depressa*) has been a useful parent in breeding new rootstocks, because it is one of the most tolerant of iron chlorosis induced at high pH. It is otherwise horticulturally typical of mandarins.

Sun Chu Sha was released by the U.S. Department of Agriculture in 1988 as a blight-, CTV-, and *Phytophthora*-tolerant mandarin that performed well in magnesium-deficient soil. Subsequent screening demonstrated that this rootstock and all other mandarins are very susceptible to Phytophthora root rot. Among mandarin rootstocks, Sun Chu Sha is probably the most tolerant of iron chlorosis.

Trifoliate Orange

Trifoliate orange has great value as a rootstock and as a breeding parent of other rootstocks. This relative of *Citrus* is a deciduous, thorny plant that produces inedible fruit and is capable of withstanding winter temperatures in temperate zones. The cold-hardiness of trifoliate orange has led to a similar expectation for trees grown on it as a rootstock. In Japan, China, Australia, and Argentina, where mild but consistently cool winters promote a dormant state, trifoliate orange is a successful rootstock for satsuma mandarins and sweet oranges. In subtropical areas, warm winters or warm periods during winter limit acclimation, so that trees on trifoliate orange are often more damaged by freezes than trees on less tolerant rootstocks.

The growth, yield, and fruit of trees grown on trifoliate orange rootstock vary considerably with the soil type. The fruit is of superior quality in virtually all climates. The juice contains high concentrations of soluble solids and acids and is highly colored. However, the fruit is generally small, especially where trees are grown in infertile sandy soils. Trees in such soils are also small, but trees in loam and clay soils achieve standard size. Trifoliate orange is particularly well adapted to clay soils with poor drainage. It does not tolerate calcareous or saline conditions.

The relatively slow growth of trees on this rootstock and the small tree size in sandy soils have led to the incorrect description of trifoliate orange as a dwarfing rootstock. One explanation is the variation among selections of trifoliate orange, which differ in their effects as rootstocks. However, the difference between trees grown on various trifoliate orange rootstocks is not as great as the difference between them and trees grown on rough lemon, for example. One selection, Flying Dragon trifoliate orange, appears to be a true dwarfing rootstock.

Viruses and viroids have contributed to the misleading perception about dwarfing. Prior to the general availability of virus-free budwood, many trees grown on trifoliate orange, which is susceptible to exocortis, were infected and stunted. Small trees are also thought of as early-bearing. Trees on trifoliate orange are not precocious in Florida and other subtropical areas, but in China and Japan, where a cool winter climate favors heavy flowering and slow tree growth, trees can begin cropping heavily when they are only two years old.

Citranges

Although originally bred for their potential as scion cultivars, citranges (hybrids of trifoliate orange and sweet orange) have become important rootstocks. Many of these hybrids were produced in an effort to incorporate the cold-hardiness of trifoliate orange into various sweet oranges. The first citranges proved to be highly variable, and many were eventually discarded as rootstocks because of a lack of seediness and other deficiencies. Only Troyer and Carrizo, which are hybrids of navel orange, reached large-scale commercial status.

Most scion cultivars budded on a citrange are consistent in certain traits but variable in others. Virtually all citranges have inherited from trifoliate orange some degree of tolerance for citrus nematodes and *Phytophthora,* but they cannot be used in saline or calcareous soils or when the citrus exocortis viroid is present. They are tolerant of the other common bud-transmitted diseases. Trees on many citranges are prone to zinc and manganese deficiency.

The most consistent horticultural trait is fruit quality. The fruit is usually medium-sized to large, and the juice quality ranks among the best. Morton citrange is an excellent rootstock, as was demonstrated as early as the 1940s in California and subsequently in Texas. However, Morton trees commonly produce seedless fruit. Trees on Rusk are semidwarf and have limited root systems that do not tolerate drought as well as the more vigorous trees on Morton or Carrizo. Also, Rusk fruit produces few seeds, and this trait has limited its commercial acceptance, as well as that of Morton.

Troyer and Carrizo are well established worldwide either commercially or as new rootstocks under evaluation in areas where the continued use of sour orange is being threatened by CTV. Troyer is widely used in California and Spain. Carrizo has been popular in Florida, but the high incidence of blight has increased interest in other rootstocks there.

Troyer and Carrizo have similar characteristics. Both are excellent nursery rootstocks. Their seeds germinate well and are highly polyembryonic. Seedlings are easy to bud and have a good survival rate, and scion growth is good. Grove trees are vigorous and notable for their unusually good growth and cropping as young trees. Most common scions, including mandarins, yield well on Troyer or Carrizo. Mandarin scions are occasionally affected by overgrowth of the rootstock at the bud union, which decreases tree longevity.

Citranges are not compatible with Eureka lemon, except for Benton citrange, a recent release from Australia. Benton is being tested in several places outside Australia and shows promise. It is tolerant of *Phytophthora,* and scions in Florida are slightly more vigorous on Benton than on Carrizo and are very productive of fruit of typical quality. Benton is a difficult nursery tree, producing thin, floppy seedlings with a strong tendency to lateral branching.

A new (unnamed) citrange with considerable promise, based on evaluations in California and Florida and commercial use in the Dominican Republic, is C-35. Trees on this rootstock have been semidwarf, and its tolerance of citrus nematodes and *Phytophthora* is excellent. It appears to be a good choice for oranges and grapefruit. Dancy mandarin, Murcott tangor, and Minneola tangelo on C-35 are being tested in Florida, with encouraging preliminary results.

Citrumelos

Trifoliate orange has also been hybridized with grapefruit, and the resulting citrumelos, like the citranges, constitute a new generation of rootstocks. Their particular combinations of attributes have generated widespread interest in them as substitutes for such historically significant rootstocks as sour orange.

Swingle is currently the only citrumelo of any commercial importance. It has been the dominant rootstock in Florida in the last 10 years, largely because trees on Swingle are tolerant of cold, citrus nematodes, *Phytophthora* species, CTV, cachexia, and exocortis, and they have low rates of blight incidence.

Site selection is a key factor with Swingle, and the soil is particularly critical. In sandy soils, the root system of this rootstock is not as extensive and the fibrous root density is not as great as that of Carrizo citrange or rough lemon. Trees grown on Swingle require careful management to avoid water and nutrient stresses. They typically display mild to severe iron chlorosis in calcareous soils and in climates where the soil is normally cold during winter.

In suitable sites, Swingle is an excellent rootstock for grapefruit cultivars. The trees are vigorous, yield well, and produce large fruit of excellent quality. Orange cultivars share these characteristics, but tree vigor in sandy soils seems to reflect inherent scion vigor. Valencia trees on Swingle are relatively small, while navel and Hamlin trees are large. Also, sweet oranges on Swingle appear to develop peel color later than sweet oranges on other rootstocks.

Cultivar Selection

Factors Influencing Scion Selection

The most important decision that a grower must make is the choice of a scion cultivar (or cultivars) to be grown, because that is the choice of the product to be produced. The impact of this decision persists through the life of the orchard. The most profitable orchards will likely be those with the most suitable cultivars that address market requirements and are well-adapted to the ecological niche in which the orchard is located. Maximum economic returns for the grower should result, because fruit yield and desired quality will be optimized with the minimum cultural input possible. Prospective growers should carefully consider four factors and their potential ramifications for the longevity and profitability of the orchard: climate, marketing, specific local conditions, and selection of budlines.

The most important decision a grower must make is the choice of a scion cultivar (or cultivars) to be grown, because that is the choice of the product to be produced.

Climate

Citrus trees of the various species and cultivars are widely adapted, and they will survive and grow (although sometimes with difficulty) in nearly any climate that does not kill them outright. However, it should be obvious from the brief descriptions provided above that different cultivar groups are better adapted to some climatic zones than others. Prospective growers should realistically consider the climatic extremes that are possible in their region, as well as the typical weather, and their effects on product quality. Regions in which there is a risk of severe freezes are not suitable for lemon and lime production. Tropical zones are generally not the best areas for mandarins and oranges, because color development is poor in a tropical climate. Growers would be well advised to follow the accumulated empirical experience of others in this respect, because in many cases the most suitable areas for particular cultivars or cultivar groups have already been determined. It is possible, however, that growers can develop profitable orchards with cultivars that are less than perfectly adapted, but only if a specific market will pay enough, for fruit of less than optimum quality, to support the additional cultural inputs required to maintain the planting and produce the crop.

The local microclimate can be as significant to profitability as the regional climate. For example, an orchard may be located in a mandarin-growing region with an ideal climate for tree survival and maximum yield of high-quality fruit, but if a particular site is more exposed to prevailing winds than other sites, windy conditions may lead to significant wind scarring of the fruit or greater dispersal of pathogens, and the fruit will be unacceptable to the fresh market because of its poor appearance. Growers producing more than one cultivar in a topographically heterogeneous orchard may be able to take advantage of differences in microclimate by correctly matching the cultivars to the specific conditions.

Growers cannot change climate or weather in their orchards, but with cultural practices they can modify climate or at least ameliorate its effects. However, climatic management can be

> **The local microclimate can be as significant to profitability as the regional climate.**

costly to the point of unprofitability. Therefore, the most successful growers will select the best-adapted cultivars for their region and site, to minimize costly inputs and to maximize productivity and quality.

Marketing

Orchards that are capable of maximum yields of optimal-quality fruit with minimal input will be unprofitable if the product is unmarketable (because of orchard location) or if market prices are too low (because of oversupply or underdemand) to cover all costs associated with production. Therefore, marketing plans need to be considered as carefully as site selection. Several questions must be answered before the first tree has been planted.

First, existing orchards in the region and their production by cultivar must be assessed, in consideration of current market demand and projected future demand. If a given cultivar has been overplanted for current demand, it makes little sense to put in a new orchard of the same cultivar unless demand is projected to increase substantially. The first plantings of newly released cultivars have the potential to be very profitable, because of limited supply to meet demand. The risk with this approach to cultivar selection, though, is that demand may be slow to develop without adequate promotion of the new cultivar.

Second, the grower should decide whether the crop will be marketed as fresh fruit or as a processed product. This decision may be influenced by the effects of climate on various cultivars. For example, lemons grown for the fresh market are generally produced in areas with mild winters and cool summers, to maximize storage capability and everbearing potential; lemons grown in the tropics are generally unsuited for the fresh market because of a more concentrated ripening period and poor storage capability. However, lemons grown solely for juice processing could be a successful enterprise in the tropics, if other conditions were met (such as sufficient demand for lemon juice and access to processing plants and transportation facilities).

Third, the location of the prospective market and the nature of market demand are relevant to marketing and cultivar selection. Some cultivars have limited general appeal worldwide, or even within a country, but there may be some local market for them; they may have poor shipping or processing characteristics, or they may be unacceptable because of a flaw such as seediness or low quality. Yet, if there is sufficient local demand and the prices paid to growers are greater than production costs, such cultivars could support profitable orchards. So, the prospective grower should clearly define the market to be served and the requirements of that market, whether for export or domestic consumption.

Obviously, there is substantial risk in the decision to grow a particular cultivar, because at planting time the grower can only make a "best guess" as to which cultivars will be profitable over the course of the next 30 years.

Specific Local Conditions

In addition to local climate, preexisting local production, and potential local markets, other local conditions affecting tree health should be assessed. Like citrus cultivars, pests and pathogens may also be well adapted to a particular climate or a particular cultivar, to the detriment of the orchard and the grower. The cost of controlling locally endemic pests or pathogens attacking an otherwise well-adapted cultivar may be sufficient to make the orchard unprofitable, especially if there is another site or region, equally suited to the cultivar, in which the pest or pathogen is not present.

Further, regulations have often been imposed to restrict or prevent the dispersal of pests and pathogens, either by local authorities (to maintain open markets) or by distant authorities (to protect their own local growers and markets). Prospective growers should carefully evaluate the pest and pathogen problems they will likely encounter at a new orchard site, not only for their negative impact on potential production but also in consideration of restrictions that may be imposed by those who control the prospective market. An example is the establishment of fruit fly–free zones in the Florida grapefruit industry, to allow continued export of Florida grapefruit to Japan (a very significant market) with little or no risk of introducing the pest there. Florida grapefruit growers in a fly-free zone have a distinct marketing advantage over growers in an infested area, in spite of the minimal effects of fruit flies on tree health and production.

Selection of Budlines

After climate, marketing, and other local conditions have been analyzed and the cultivars to be grown have been selected, there is one final matter influencing profitability and long-term orchard health, over which the grower has some control: the grower should endeavor to have trees propagated from the best available budlines. Growers should select budlines certified as true to type, free of most graft- or insect-transmissible pathogens, and having proven potential for productivity. In many cases, such budlines are of nucellar origin, because of their relative freedom from disease, and because of the tree vigor that comes with the rejuvenation of citrus cultivars after nucellar embryony. However, budlines that are too young may be excessively vigorous and prone to thorniness, and they may produce large, coarse, unmarketable fruit. Again, the experience of other growers and information gathered and disseminated by universities and public or private organizations should provide the best advice on the selection of budlines.

Factors Influencing Rootstock Selection

Few decisions in citriculture have the permanence and long-term impact of the choice of a rootstock. Selecting a rootstock involves considerably more than simply choosing a root system. Over 20 characteristics of citrus trees can be influenced by the rootstock, and individually or in combination they can determine the profitability (survival) of a citrus venture. These characteristics are listed below with some discussion of their relative importance within each category.

Horticultural Factors

Yield per tree (as opposed to yield per unit of land area), tree vigor, tree size, and the external and internal quality of the fruit (size, shape, peel color, peel thickness, rag, juice content, and concentrations of soluble solids and acids in the juice) can vary with different rootstocks. Scion-rootstock compatibility is also a factor in rootstock selection. Profit has a strong relationship with yield and is the primary selection factor when no other factor is otherwise limiting. This is especially true for processing fruit; for fresh fruit, quality is a major consideration, but not always. For example, despite the well-known poor quality of fruit from trees budded on rough lemon, this rootstock has had a successful history in southern Africa, Australia, and Florida; the advantages of yield and fruit size outweigh the poor fruit quality.

Over 20 characteristics of citrus trees can be influenced by the rootstock, and individually or in combination they can determine the profitability (survival) of a citrus venture.

Certain horticultural factors have direct and indirect effects on tree performance. Rootstocks influence juice quality. The ratio of the soluble solids to acid concentrations determines fruit maturity, and the acid concentration is directly related to the on-tree storability of the fruit and its postharvest shelf life. Fruit from trees on rootstocks that produce high-acid fruit, such as sour orange and Cleopatra mandarin, tends to store better.

Experience quickly indicates scion-rootstock combinations that are not compatible. Often the first evidence is obtained in the nursery, when the bud union fails to develop properly, and scion growth and appearance are abnormal. An example is Eureka lemon on trifoliate orange and its hybrids (except Benton citrange). There are many other examples of unsuccessful combinations, which are generally well recognized locally. Such combinations seem to fail reliably. A more subtle type of "incompatibility" occurs when mandarin scions budded on citrange and citrumelo rootstocks in cool, arid climates become essentially strangled by rootstock overgrowth. Some apparent incompatibilities (i.e., bud union crease) may be virus- or viroid-induced.

Environmental Factors

Tolerance of freezes, drought, and flooding is ordinarily only an occasional concern of relatively short duration. When freezing temperatures, drought, or flooding occur routinely, it is likely that serious tree damage or loss would occur regardless of the choice of rootstock.

Pests and Diseases

Rootstocks vary in their tolerance for Phytophthora foot rot and root rot, burrowing and citrus nematodes, several virus and viroid diseases (principally CTV, exocortis, and xyloporosis), and blight. Their reactions to these pests and diseases are reasonably well known. Virus and viroid diseases are of greatest concern internationally, but their impact has been reduced by budwood registration programs. *Phytophthora* species and citrus nematodes are also universal problems. Burrowing nematodes are a problem mainly in Florida. There are other options for controlling these nematodes and for foot rot and root rot, but rootstock selection is presently the only way to avoid heavy losses from blight, which has been a problem in subtropical areas such as Florida, Brazil, and parts of southern Africa and Australia. All commercial rootstocks are affected by blight, but the incidence of this disorder varies in different cultivars. The cause of blight is unknown, and thus the complete characterization of rootstocks approaching commercial release is seriously hindered by the long-term, large-scale field plantings needed to assess blight tolerance.

Soil

Characteristics of the soil—texture, pH, calcium carbonate ($CaCO_3$) content, and salinity—may be a factor in rootstock selection, but usually only under extreme conditions. Trees on most rootstocks will grow and fruit satisfactorily in loamy soils, but differences become most evident in trees planted in sandy or heavy clay soils. Differences between rootstocks generally increase as soil conditions worsen. Changes in soil pH directly affect nutrient availability. Some rootstocks can forage better, within limits, for certain nutrients and are better adapted to changes in soil chemistry brought about by the presence of large amounts of $CaCO_3$. Citrus is classified as salt-sensitive. Salinity problems result from sodium chloride (NaCl) and other salts in the soil and in irrigation water. The accumulation of ions to harmful or toxic levels in the scion is controlled by the rootstock.

Rootstock Selection Rationale

Choosing citrus rootstocks is basically a simple process, but the outcome is often less than satisfactory for two reasons. First, all rootstocks have certain advantages and disadvantages. Sometimes the combination of advantages and disadvantages is especially frustrating. For example, sour orange is an excellent choice in many circumstances, but it is susceptible to CTV, the one disease that is the most difficult to control among those caused by viruses and viroids. Second, descriptions of rootstock behavior are highly localized, often with emphasis on comparative performance. Local research and commercial experience together, over time, provide reasonably accurate generalized descriptions of rootstock performance, but absolute and relative performances change enough from place to place that rootstock selection should be based strictly on demonstrated responses to local conditions.

For these reasons, rootstock selection must be viewed as *risk management,* which necessarily involves establishing priorities and relative importance among rootstock characteristics. One approach is to choose rootstocks depending on the most critical factor or factors limiting production, such as climate (particularly freezes), soil conditions, or endemic diseases or pests. These factors are generally related to tree survival, which is essential, but their relative importance can be overrated if other options are available to compensate for inherent rootstock contributions. Pests and diseases can be managed with budwood registration and with chemical treatments. Trees can be protected from freezes with careful site selection, tree wraps, and water. A truly *critical* factor would be one that has the potential to damage trees or cause tree loss on a grand scale, such as a freeze, as opposed to a factor that usually affects trees only in small numbers at a time. Therefore, if a reasonably intact grove can be ensured, either by the absence of problems or by tree replacement, rootstock selection should be focused on yield and fruit quality.

Rootstock selection decisions are influenced by objectives and circumstances. Planning the rootstocks for a new grove on a virgin site, for a new grove on a site previously planted with citrus, or for replacing dead or removed trees in an existing grove will probably not lead to the same decisions. In the case of replacing trees in an existing grove, trees lost from any cause can provide a valuable insight into grove problems unknown prior to planting. Where such a problem is encountered, it may be advantageous to select a replacement rootstock with a better response to the problem.

The soundest rootstock decisions involve all available information with the objective of minimizing risk, recognizing that the existence of a risk does not necessarily mean that a problem will occur. Information protects against assigning unjustified priority to risks that may not exist or fail to develop. Simple assays of soil samples, for example, indicate whether problems exist or can be anticipated from soilborne pests or pathogens; likewise, irrigation water quality can be evaluated in advance to identify salinity problems. Systematic sampling and data gathering have the advantage of allowing site variation to be accommodated in rootstock selection.

Comparative performance in terms of tree survival, adaptation to environmental conditions, yield, and fruit quality has been the traditional basis for rootstock selection. However, performance in those terms has rarely been given an economic interpretation. Recent evidence from Florida field trials strongly suggests that *financial* results (profit) may be most closely related to yield, despite large differences in tree survival and juice quality of trees on different rootstocks. Production has been and remains a major determinant of rootstock choice.

R. F. Lee
University of Florida
Citrus Research and Education Center, Lake Alfred

Paul S. Lehman
Florida Department of Agriculture and Consumer Services
Division of Plant Industry, Gainesville

Luis Navarro
Instituto Valenciano de Investigaciones Agricolas
Valencia, Spain

CHAPTER 4

Nursery Practices and Certification Programs for Budwood and Rootstocks

A healthy citrus planting may outlive the grower or grove manager who planted it. Because citrus has the potential to be productive for many decades, it is imperative that growers develop effective long-term pest management strategies. Careful selection of healthy, high-quality nursery stock is the single most important decision any grower or producer will make affecting the financial success of new plantings, and it has the greatest cost-benefit ratio of any management strategy (Plates 3–10). The cost difference between the highest-quality nursery trees and ordinary trees is a single investment of usually less than $100 per acre, or less than $5 per year over a minimum grove life of 20 years. This is well below 1% of total annual production costs. In contrast, mistakes made selecting in nursery stock will either result in significant production losses each year for the life of the planting or will require removal and replacement of the planting. This is costly and can result in a significant delay in reaching full production. In addition, clean budwood is usually necessary for utilization of new rootstocks. Starting with a healthy, disease- and pest-free citrus plant is an essential first step in any successful management strategy.

In established citrus orchards, most diseases caused by bacteria and fungi can be managed with appropriate cultural and chemical methods. However, this is not true for graft-transmissible diseases, caused by viruses, viroids, some bacteria, spiroplasmas, and phytoplasmas, which can be a major factor limiting citrus production. Only preventive measures, such as the use of tolerant or resistant germ plasm, exclusion of potential pests from the citrus area, and planting pathogen-free, high-quality nursery trees, are useful for the control of graft-transmissible pathogens. This is the compelling reason for the adoption of clean stock programs and certification programs for citrus, to exclude graft-transmissible pathogens from new plantings. In nurseries, measures to control root diseases caused by nematodes and fungi and bacterial diseases, such as citrus canker, have been implemented in only a few countries. However, existing data demonstrate that the benefits received by growers from these preventive measures far exceed the cost of administration.

A healthy citrus planting may outlive the grower or grove manager who planted it.

In this chapter, we will discuss the concepts, organization, and interrelatedness of successful quarantine, clean stock, and certification programs for graft-transmissible diseases, some non-graft-transmissible diseases, and root pathogens. Examples are described from different citrus areas, having different problems with specific requirements. The economic returns and benefits of such programs will be discussed.

Production of Propagative Material Free of Aboveground Pathogens

Graft-Transmissible Pathogens

For the production of healthy, high-quality nursery trees, three different but closely related programs must be established:

- A *quarantine* program, to ensure the safe introduction of select horticultural germ plasm
- A *clean stock* program, to conduct testing and provide therapy to identify and produce sources of pathogen-free propagating stock
- A *certification* program, to ensure the development and maintenance of virus-free propagating materials for commercial use

Healthy plants recovered in the first two programs provide the starting material for certificaticn programs.

Quarantine Programs: Safe Introduction of Select Horticultural Germ Plasm

The movement of citrus species and varieties between different citrus areas for commercial and scientific purposes is often desirable. This is especially true in citrus areas oriented to the international marketing of fresh fruit (for example, most of the citrus-producing countries in the Mediterranean area). However, uncontrolled importation of budwood has the risk of introducing new pests and pathogens, which in some instances would cause important economic damage. This risk can be minimized by carefully controlled introduction through quarantine stations to allow for the importation of foreign varieties without the introduction of new pests and diseases. These programs are usually operated by plant protection services, often under the jurisdiction of the ministry of agriculture of a country or the commissioner of agriculture in a state or province.

Two different quarantine procedures are followed for the safe importation of citrus plant material. The classical method consists of propagating imported budwood in quarantine greenhouses located far away from citrus-growing areas. Thereafter, the newly propagated plants may be indexed or submitted directly to shoot tip grafting in vitro (Box 4.1) or thermotherapy to free them of graft-transmissible pathogens and then indexed to verify their pathogen-free status. The terms *pathogen-free* and *virus-free* are often loosely used in both popular and scientific literature and imply greater confidence in the pathogen-free status than is warranted. The correct concept is that a plant is free of the pathogens for which it has been specifically and accurately indexed and found negative. It does not preclude the possibility that other pathogens are present. This introduction procedure requires the availability of facilities and trained personnel knowledgeable of citrus pests, diseases, and cultural practices. It is used in countries having long traditions in quarantine, such as the United States and Australia, but it is expensive and difficult to establish for citrus in countries where citrus is a minor crop.

An alternative procedure is the incubation of budwood in tubes. This procedure was developed for the safe introduction of citrus varieties and has proved to be very efficient in excluding citrus pests and diseases. Bud sticks received from another country are thoroughly cleaned and surface-sterilized to remove pests or pathogens and then are immediately placed in test tubes and incubated in a closed growth chamber. Once a bud stick has developed shoots, a tiny slice of the shoot tip is removed, sterilized again, and grafted to a healthy seedling plant. This process minimizes the possibility of the escape of harmful pathogens and allows rapid processing of new entries.

This tissue culture method has several advantages over the traditional quarantine method. Pests and diseases that might be in the original material are eliminated at the early stages of introduction, thus shortening the quarantine period. For example, in the late 1980s, Rio Red grapefruit was in high demand in Spain; one bud stick was introduced from Texas and subjected to the tissue culture procedure in January 1988, and 10,300 healthy buds were released to citrus nurseries in May 1989. In tissue culture, the test tubes serve as a substitute for greenhouses located in isolated areas. Often shoot tip grafting is used for sanitation of local cultivars, and the needed facilities and personnel are already available in a tissue culture quarantine program. This procedure can easily be established in many countries to safely import citrus vegetative material. Because of these advantages, the tissue culture method is recommended for the safe importation of citrus budwood. In Spain, it has been successfully used to import over 100 varieties from different citrus areas. California has also had this option for the importation of budwood for a number of years.

Programs for the safe introduction of citrus germ plasm have many steps in common with the programs described below for testing and therapy to identify and produce sources of pathogen-free propagating stock, but quarantine programs differ in placing special emphasis on testing for particular diseases existing in the country of origin of the plant material. The main problem related to quarantine is the introduction of material from areas with diseases not having established indexing methods, like citrus blight. In these cases, importation from high-risk areas should be restricted as much as possible, and introduced material should undergo additional cycles of therapy.

Box 4.1

Shoot Tip Grafting

Shoot tip grafting is an excellent method of recovering pathogen-free plants from infected sources. In this procedure a small shoot tip from an infected plant is grafted to a young rootstock seedling growing in vitro. The shoot tip, or meristem, of an infected plant is often free of pathogens, and plants regenerated from shoot tips are usually healthy.

The procedure is carried out under sterile conditions and consists of five steps:

- rootstock preparation (Plate 11)
- scion preparation (Plate 12)
- grafting (Plate 13)
- culture of grafted plants (Plate 14)
- transfer to soil (Plate 15)

Young, growing flushes are collected from infected plants and surface-sterilized. A shoot tip, composed of the apical meristem and two or three leaf primordia, is excised and, with the aid of a microscope, is placed in an inverted T-shaped incision at the top of a decapitated rootstock. The micrografted plant is placed in a tube with nutrient medium and maintained in a culture room at 26°C and exposed to illumination (40 $\mu mol \cdot m^{-2} \cdot s^{-1}$) for 16 hours daily. When the graft takes, and two or three leaves have developed from the scion, the micrografted plant is transplanted to soil or grafted to a healthy, vigorous rootstock in the greenhouse to encourage rapid growth.

The size of the grafted shoot tip is an important factor in determining the number of healthy plants recovered. Increasing the size results in a higher rate of successful grafts but a lower proportion of healthy plants recovered. Shoot tips 0.1–0.2 millimeter long, from the cut surface to the tip of the largest leaf primordium, are recommended for routine application of this technique. Shoot tips of this size form successful grafts and give rise to healthy plants with realistic frequency.

Several factors influence the recovery of healthy plants by shoot tip grafting. Most pathogens are easy to eliminate, whereas others (such as those causing psorosis, concave gum, impietratura, cristacortis, and tatter leaf) are more difficult. The frequency of recovery of plants free of these pathogens is increased by growing the shoot tip source plants under relatively warm conditions; more than 90% of plants recovered by grafting shoot tips from such source plants are free of viruses and viruslike pathogens, including those that are difficult to eliminate. The most important advantages of this technique are that healthy plants can be recovered quickly and do not have juvenile characteristics. Shoot tip grafting allows the recovery of plants free of pathogens that cannot be eliminated by thermotherapy.

Clean Stock Programs: Testing and Therapy to Identify and Produce Sources of Pathogen-Free Propagating Stock

In addition to introducing valuable germ plasm, it is also useful to recover healthy plants from local cultivars. Existing local varieties may be those best adapted to the local climate, soils, and market, and the simple step of removing graft-transmissible pathogens can greatly increase their value and usefulness. In most citrus areas these programs are carried out by research institutions with the joint participation of specialists in horticulture, virology, and tissue culture. A clean stock program involves six basic steps:

- Selection of mother trees of local cultivars
- Indexing of mother trees
- Recovery of pathogen-free plants by shoot tip grafting in vitro and thermal therapy
- Indexing of the plants recovered
- Horticultural evaluation of healthy plants
- Maintenance of healthy plants

Selection of mother trees. Individual trees of the cultivars to be included in a clean stock program should be selected according to documented horticultural criteria, such as superior production, higher fruit color, and early or late ripening. Growers and industry personnel should be involved in the selection process to obtain varieties that will be useful. Consideration should be given to future market demands when varieties are selected. The presence of graft-transmissible pathogens should not be taken into consideration for tree selection, since there are methods of recovering healthy plants from infected ones. Clones having superior horticultural value could be lost if trees having disease symptoms were always avoided. Selected trees should be propagated on vigorous rootstocks (e.g., rough lemon) in the greenhouse for use as a source of material for testing for the presence of pathogens and therapy procedures.

Testing mother trees for the presence of pathogens. Most infections by graft-transmissible pathogens cannot be recognized merely by looking at the plant (Box 4.2). It is important to test or index the selected mother trees in order to know the pathogens that must be eliminated. This is normally done by grafting pieces of bark tissue from the selected mother trees to indicator plants, in which pathogen-specific symptoms develop if the pathogen is present. Commonly used indicator plants and growth conditions for testing for the presence of viruses and graft-transmissible pathogens are listed in Box 4.2. Additional information is given in Chapter 9. Biological indexing is usually supplemented by serological tests, polymerase chain reaction assays, or other laboratory tests which provide quick diagnosis of specific pathogens. The indexing of selected mother trees and supplemental laboratory-based indexing provides accurate knowledge of the pathogens present, and this information is useful for quarantine purposes. The indexing of mother trees often provides an opportunity for establishing a virus collection, which is valuable for research and needed as a source of positive controls for indexing.

> **Most infections by graft-transmissible pathogens cannot be recognized merely by looking at the plant.**

Recovery of virus-free trees. Pathogen-free citrus plants may be recovered by nucellar embryony in vivo or in vitro, by thermotherapy, or by shoot tip grafting.

Nucellar embryony is effective in eliminating viruses and graft-transmissible pathogens because most of these pathogens are not seed-transmitted. However, nucellar plants have juvenile characteristics, i.e., excessive vigor, thorniness, and late bearing. They have to be grown for many years, until these characters disappear, to become acceptable for commercial propagation. In addition, nucellar plants are not always true to type, especially those obtained by nucellus culture in vitro. Seed transmission of psorosis-like pathogens, probably the concave gum agent, may occur in trifoliate orange and its hybrids, and indexing for these pathogens is needed.

Thermotherapy produces true-to-type plants without juvenile characters, but it is not effective for the elimination of pathogens which thrive under warm conditions, such as citrus exocortis, cachexia, and other citrus viroids and the citrus stubborn agent.

In most cases shoot tip grafting is effective for the elimination of viruses and viruslike agents, including those not eliminated by thermotherapy, and it produces true-to-type plants without juvenile characteristics, because does not involve the embryonic stage. However, some viruses are harder to eliminate than others, such as the concave gum and citrus tatter leaf viruses. Shoot tip grafting has been internationally recognized as the best technique for freeing plants of viruses and viruslike agents, especially when combined with partial heat therapy. It is being successfully used in most citrus sanitation programs worldwide.

Indexing of micrografted plants. Plants recovered by shoot tip grafting or by other methods should be carefully indexed for the diseases found in the mother trees. It should never be assumed that a plant is healthy because it has been submitted to a sanitation treatment, even if the disease affecting the mother plant is easily eliminated. In Spain, for example, cachexia has been eliminated by shoot tip grafting in 100% of micrografted plants. However, in the citrus variety improvement program, plants recovered by shoot tip grafting are still routinely indexed for this disease despite the relatively long time (two years) required to carry out the standard biological test on Parson's Special mandarin, the indicator for this disease.

Horticultural evaluation of healthy plants. Plants found to be free of graft-transmissible diseases after shoot tip grafting have been true to type. To our knowledge there have been no reports of abnormal plants obtained by this method. This is to be expected, since shoot tip grafting is simply a grafting technique that does not involve the neoformation of buds in vitro or the use of growth regulators. However, natural mutations occur in citrus, and there is always the possibility of grafting in vitro shoot tips carrying one of these mutations. In addition, the whole process requires considerable labeling in the laboratory and the greenhouse, introducing opportunities for errors. Consequently, horticultural evaluation of healthy plants to ascertain that they are true to type and to determine their horticultural performance is vital. The release of an off-type will have a very negative impact on any program.

In some programs the varietal blocks used for horticultural evaluation include some trees on alternative rootstocks to verify virus diagnostic tests. For example, Orlando tangelo is an indexing rootstock for cachexia, trifoliate orange for dwarfing viroids, and Swingle citrumelo for citrus tatter leaf virus. The immediate goal of the varietal blocks is to verify horticultural traits quickly before budwood is released and, with the appropriate rootstocks, to verify freedom from selected viruses by

Box 4.2

Summary of Diagnosis of Graft-Transmissible Pathogens of Citrus

Disease	Pathogen	Indicator plants[a]	Temp.[b]	Symptoms in indicators	Other diagnostic methods[c]
Cachexia	Viroid	Tangelo Parson's Special mandarin	Warm	Gum in bark and scion or gum at bud union	Electrophoresis, PCR
Citrus chlorotic dwarf	Virus?	Rough lemon	Cool	V-shaped notch and chlorotic flecking of young leaves, crinkling and variegation of mature leaves	None
Citrus variegated chlorosis	Bacterium	Sweet orange	Warm	Interveinal chlorosis, necrotic lesions with gum deposits on underside of leaf	Serology, PCR
Concave gum	Virus?	Sweet orange Dweet tangor	Cool	Oak-leaf pattern in young leaves	None
Cristacortis	Virus?	Sweet orange Dweet tangor	Cool	Oak-leaf pattern in young leaves, flecking	None
Exocortis and other viroid diseases	Viroid	Citron	Warm	Leaf epinasty, tip browning	Electrophoresis, PCR
Greening (huanglongbing)	Bacterium	Duncan grapefruit Mandarin	Cool or warm	Leaf blotch, chlorosis	TLC for gentisic acid, PCR, EM, hybridization
Impietratura	Virus?	Sweet orange Dweet tangor	Cool	Oak-leaf pattern in young leaves, flecking	None
Indian citrus mosaic	Virus	Sweet orange	Warm	Mosaic on mature leaves, no symptoms on young leaves	PCR
Infectious variegation	Virus	Citron Cowpea	Cool	Distorted leaves, necrotic spots	Serology
Psorosis	Virus	Sweet orange	Cool	Shock, flecking, patterns in young leaves	Serology, PCR
Ringspot	Virus	Sweet orange *Chenopodium quinoa*	Cool	Shock, ring spots, necrotic local lesions	Serology, PCR
Satsuma dwarf	Virus	Satsuma mandarin *Chenopodium quinoa*	Cool	Stunting, boat-shaped leaves, local lesions	Serology
Stubborn	Spiroplasma	Madam Vinous sweet orange	Warm	Stunting of shoots and plants, small leaves with zinc deficiency	Culturing, PCR
Tatter leaf	Virus	Rusk citrange *Citrus excelsa*	Cool	Chlorotic spots, zigzag growth of stems	Serology, PCR
Tristeza	Virus	Mexican lime Sweet orange on sour orange Sour orange Duncan grapefruit Madam Vinous sweet orange	Cool	Vein clearing, leaf cupping, stem pitting Decline Seedling yellows Stem pitting Stem pitting	Serology, PCR
Woody gall	Virus?	Mexican lime	Cool	Galls on leaf veins	None

[a] For a complete list of transmissible pathogens of citrus, biological indicators, and laboratory techniques, refer to C. N. Roistacher, 1992, Graft-Transmissible Diseases of Citrus: Handbook for Detection and Diagnosis, Food and Agriculture Organization, Rome.
[b] Cool = night temperature of 68°F and day temperature of 78–85°F. Warm = night temperature of 85°F and day temperature above 95°F.
[c] EM = electron microscopy. PCR = polymerase chain reaction. TLC = thin-layer chromatography.

traditional biological indexing. Over a period of time an additional benefit of varietal blocks is the accumulation of production records which enable growers to select and propagate superior budlines. The selection of clones based on this production information by growers in Florida accounts for an average yield increase of 10–15% for the clones now being propagated, compared with those propagated 20 years ago.

Maintenance of healthy plants. Healthy plants should be maintained under protected conditions to prevent reinfection with graft-transmissible pathogens (Plates 16–18). In most citrus areas, graft-transmissible diseases are naturally spread in the field (Box 4.3). Therefore, healthy plants should be protected from vector infestations. One approach is to put healthy plants in isolated areas. Screenhouses or greenhouses are preferred as a means of protecting healthy plants, because they can be conveniently located for collection of budwood and periodic testing.

The collection of healthy plants in protected locations should also be made available for use as foundation sources of budwood in a certification program. Plants should also be placed in a secure germ plasm bank for research and for exchange of cultivars with other countries. These plants should be retested for the presence of pathogens at appropriate intervals.

In many countries where citrus is not a major crop, it is difficult to find the facilities, trained personnel, and other resources to develop a local clean stock program. In this situa-

Box 4.3

Graft-Transmissible Diseases of Citrus Naturally Spread in the Field

Disease	Other common names	Vectors
Citrus blight	Declinio Declinamiento Young tree decline	Unknown
Citrus chlorotic dwarf	. . .	Whitefly (*Parabemisia myricae*)
Citrus variegated chlorosis	. . .	Leafhopper species
Greening (huanglongbing)	Likubin (Taiwan) Leaf mottle (Philippines) Vein-phloem degeneration (Indonesia) Yellow shoot (China)	Psyllids (*Trioza erytreae, Diaphorina citri*)
Indian citrus mosaic	. . .	Citrus mealybug
Leprosis	Zonate chlorosis may be similar	Mites (*Brevipalpus* spp.)
Psorosis	. . .	Aphids reported but not confirmed
Satsuma dwarf	Citrus mosaic virus, natsudaidai dwarf virus, and naval orange infectious mottling virus may be similar	Unidentified soilborne agent
Stubborn	Little leaf (Israel)	Leafhoppers (*Scaphytopius nitridus, Neoaliturus tenellus, N. haemoceps*)
Tristeza	Quick decline Stem pitting	Aphids (*Toxoptera citricida,* the most efficient; *Aphis gossypii, A. spiraecola, T. aurantii*)
Witches'-broom	. . .	Leafhopper (*Hishimonus phycitis*)
Woody gall	Vein enation	Aphids (*T. citricida* and *A. gossypii* are the most efficient)

tion, the importation of clean stock from internationally recognized clean stock programs may be considered. In smaller countries, importation can help to establish a certification program for the distribution of clean budwood for commercial use (described below). If there is a need to develop pathogen-free mother trees from local varieties, it may be more cost-effective to send local selections to a recognized clean stock program, to be cleaned up and returned after being tested to ensure that they are free of graft-transmissible pathogens. If the decision is made to import clean stock from internationally recognized programs, there is still a need to confirm by local indexing that the imported stock is free of viruses, even if only by propagation on sensitive rootstocks. The need for re-indexing and horticultural evaluation on a regular basis cannot be overemphasized.

Certification Programs: Development and Maintenance of Virus-Free Propagating Materials for Commercial Use

The purpose of certification programs is to guarantee the sanitary status and true-to-type character of propagating material in commercial propagation in nurseries. They also control the horticultural quality of nursery plants. Certification programs impose legal regulations governing nursery operations and require periodic indexing and inspection of trees used for nursery propagation. Usually they are operated by a state or provincial agency having legal authority to impose restrictions and to inspect nurseries. All steps of the propagation occur in private nurseries.

There are very few continuously operated citrus certification programs in the world—California, Florida, Spain, South Africa, Taiwan, and Australia are about the only ones, although several countries are establishing new programs. Several countries have established sanitation programs, which are the starting point for certification programs.

Certification programs have to be adapted to the specific situation of each citrus area. The most important considerations in the establishment of a certification program are the organization of the local citrus industry and the pests and diseases present in the area. Certification programs only ensure the absence of the pathogens which are actually included in the program. For example, in the California sanitation and quarantine programs, plants are indexed for all known diseases, including woody gall, but the California certification program does not index for this disease, because it is not present in the state. Only recently has woody gall been added to list of diseases for which plants are indexed in the Florida certification program. Although some details will vary, some recommendations for a general scheme for certification can be established.

Certification programs are necessary even in areas with endemic vector-transmitted diseases. For example, in areas where tristeza is endemic, several measures and modification of certification programs can be adopted. Nursery plants should be propagated on tristeza-tolerant rootstocks. In addition, nursery plants and foundation material can be tested and certified free of severe strains of tristeza virus to avoid their dissemination through nursery material, which happened in Florida during the 1980s. As a further step, plants can be deliberately infected with a local cross-protective mild strain, in countries that use mild-strain cross-protection to control tristeza stem pitting. Where this technique is used, plants infected with specific strains, rather than healthy plants, should be maintained in foundation blocks.

In areas where greening or stubborn is endemic, it is essential that nursery plants be free of these diseases. Specific cultural practices or spray programs may have to be adopted to

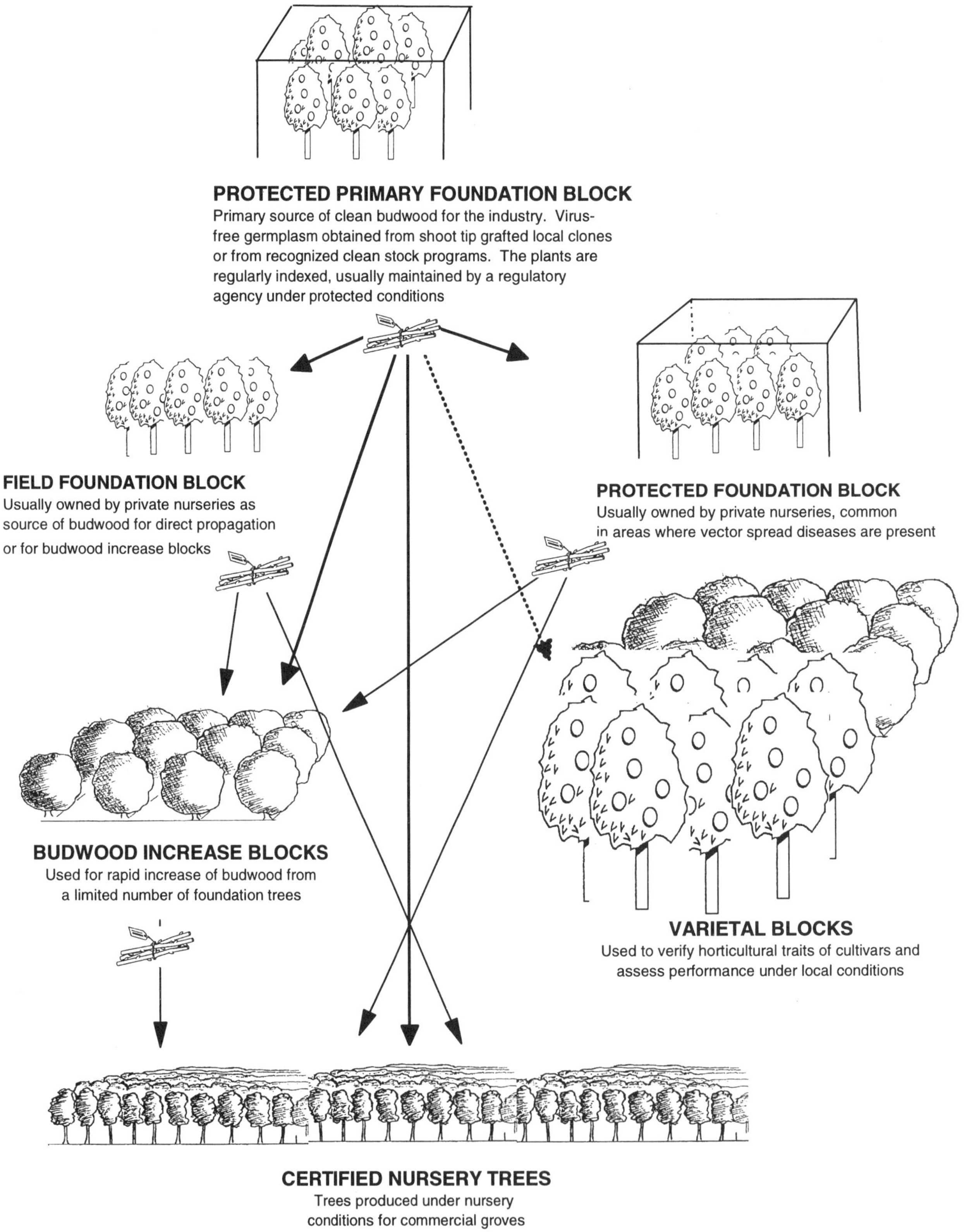

Fig. 4.1. Relationships of various propagation sources in a typical certification program.

control vectors and delay the reinfection of new plantings. This may require producing all budwood under protected conditions, such as in greenhouses or screenhouses.

Propagation components of certification programs. It would be extremely time-consuming and expensive to index each citrus plant propagated in a country for all possible graft-transmissible pathogens. Instead, careful attention is given to indexing and verifying the high horticultural quality of the source trees used to supply budwood for plants to be placed in the field. Thus, the benefit of high horticultural quality and freedom from graft-transmissible pathogens is enjoyed by all the resulting propagations. Certification programs usually are based on four blocks of trees (Fig. 4.1):

- protected primary foundation blocks
- foundation blocks, which may be located in the field or in a protected area, depending on the local situation
- budwood increase blocks
- blocks of certified nursery trees

Protected primary foundation blocks are usually maintained by a public agency. They contain virus-free plants recovered through clean stock and quarantine programs and verified to be of high horticultural quality, which are grown in containers in insect-proof screenhouses to avoid recontamination with vector-transmitted diseases. These trees are the primary source of budwood for the establishment of foundation blocks. The plants should be indexed periodically to examine their health status and inspected to detect any growth or fruit abnormality. In Spain two plants are maintained per accession, propagated on Troyer or Carrizo citranges. They are pruned every year and produce fruit regularly. Some of these plants are 17 years old and are growing normally.

Foundation blocks usually belong to private nurseries, although some are maintained by public agencies. They are usually field plantings propagated with budwood from the protected foundation blocks, but in Florida, Spain, and some other locations foundation blocks are being established in screenhouses to protect trees from vectors of citrus tristeza virus and other vector-borne diseases (Fig. 4.1). These trees must be re-indexed periodically. Generally trees are indexed annually for vector-transmissible pathogens, every three years for mechanically transmitted pathogens, and every six to 10 years for other pathogens. Any infected trees should be removed immediately. The trees are inspected annually by horticultural specialists for identification of possible off-type branches or trees with abnormal growth that should be removed. Budwood should be collected from these trees only after they have produced enough fruit to verify that they are true to type. The amount of budwood collected from each tree should be limited to allow some fruit production. Usually only two to six trees are maintained per accession, as indexing costs can be high. As an example, in Spain only four trees per accession are maintained in the foundation block for an industry that produces three to five million certified trees per year.

Budwood increase blocks generally contain nursery plants propagated directly from foundation trees to increase the number of buds for propagation of certified trees. They allow an exponential increase in the amount of high-quality budwood that can be produced from a limited number of foundation trees. This method enables better inspection and indexing of the limited number of foundation trees. Increase blocks may be established under normal field conditions or in greenhouses, where faster and better growth can be obtained, and where they are protected from vector-borne pathogens. Buds should be collected from these blocks for a maximum of up to three years, to avoid the propagation of undetected mutations. Consequently, new budwood increase blocks have to be periodically established with buds from foundation trees. Increase blocks should be inspected to detect any growth abnormalities and indexed at random for vector-transmitted diseases.

Certified nursery trees are propagated with budwood from increase blocks or foundation blocks. They may be produced under regular field conditions or in greenhouses, according to the specific needs and technology available at each nursery. These trees are inspected to guarantee that they meet the requirements for horticultural quality in the certification regulations, and nurseries must also keep records to show they have complied with the regulations.

Seed production. Even highly nucellar cultivars may have some gametic seedlings, and it is important that the seed parent trees be of nucellar and not gametic origin. These trees must be tested to ensure that they are free of concave gum and similar diseases that are occasionally transmitted through seed. It is desirable that they should be free of other diseases as well.

Foliar Pathogens

Most certification programs focus on graft-transmissible pathogens, but these programs can also effectively reduce the spread of other diseases and pests, including foliar diseases, such as citrus canker, scab, and Alternaria brown spot. The best means of producing trees free of foliar pathogens is to grow them in a greenhouse without overhead irrigation. Keeping the leaves dry virtually eliminates the reproduction of foliar fungal and bacterial pathogens. Eliminating a disease from nurseries will not solve the problem if the disease is endemic in the area, but it can prevent an early buildup of inoculum. If foliar fungal and bacterial pathogens are absent from a planting initially, chemical and cultural controls are greatly simplified.

The production of disease-free trees also avoids the introduction of pathogens into new growing areas. Citrus canker has been introduced into Florida on several occasions, first in the early 1900s, when infected trees were received from Japan. Inspection of citrus nurseries as a component of a citrus certification program is important to prevent the widespread dissemination of new pathogens, such as the canker bacterium. Uruguay controls citrus canker in its certification program by allowing the propagation of nursery trees only in the southern area of the country, where the disease does not occur. This precaution ensures that growers will not be planting canker-infested trees.

The causal agent of mal secco, *Phoma tracheiphila,* may be present in xylem tissue and transmitted in budwood. Importation of budwood from areas free of the disease can help prevent the introduction of this pathogen.

Production of Propagative Material Free of Root Pathogens

Root Pathogens: The Hidden Enemies

Root pathogens, while larger than viruses, are usually not seen, since they are hidden in the soil, and often a microscope is needed to see them. However, once established in a grove, they are difficult and costly to control and almost impossible to eradicate. Chemical treatments for the control of nematodes and other root pathogens are diminishing as a result of environmental concerns. Furthermore, the costs of chemical and cultural control may exceed the potential return. Extensive use of pesticides can lead to the development of resistant strains of the pathogen, as in the case of the fungus *Phytophthora,* which causes foot rot and root rot. Resistance-breaking biotypes have developed for economically important citrus nematodes. Resistance-breaking biotypes can greatly limit the value of rootstocks for perennial crops such as citrus. Quarantine, clean stock, and certification programs should address the exclusion of these pathogens and pests from new areas in which citrus is planted.

The sugarcane root weevil, *Diaprepes abbreviatus,* was introduced into Florida many years ago. In recent years this pest, whose larvae feed on roots, has spread rapidly throughout the state, resulting in the infestation of about 160,000 acres in 20 counties, including 30,000 acres of citrus. The root weevil is having a serious negative impact on citrus, because the feeding of larvae on the roots of Swingle citrumelo rootstock breaks the resistance of this rootstock to *Phytophthora palmivora* in areas where this soilborne fungus is present, and thus the roots are subject to serious damage resulting in tree decline. The rapid spread of this pest was due to the transport of containerized nursery material carrying the weevil in its larval stage. By careful monitoring of nurseries for root weevil infestations, this pest is no longer spreading rapidly. More information on *Diaprepes* is given in Chapter 14.

Citrus production has increased threefold worldwide in the past 35 years, mostly as a result of an increase in the area planted. In almost all citrus areas, clean stock and certification programs, if they exist at all, do not exclude root pathogens. Growers may be unaware that nursery stock is infested with root pathogens, such as nematodes, because often there are no obvious root symptoms in the nursery. Furthermore, the damage that root pathogens cause in the field may not be readily apparent to growers, because it results in minor losses (5–10%) on all trees or because the losses are not immediately apparent. For these reasons, certification programs for root pathogens have not been established in most countries. Consequently, citrus nematodes such as *Tylenchulus semipenetrans* and *Phytophthora* species have been distributed in most of the new citrus areas planted worldwide in the past three decades. A notable exception is Florida, where nematode certification programs were implemented in the 1950s.

The Citrus Nursery Nematode Certification Program in Florida

The Florida nursery nematode certification program is one of the regulatory programs developed to prevent the spread of the citrus burrowing nematode (BN), *Radopholus similis.*[1] In the early 1950s the Florida citrus industry was threatened with a serious decline problem, the cause of which was at first unknown. Once the disease was recognized as being caused by BN, the Florida Department of Agriculture, in cooperation with citrus growers, implemented regulatory programs to control the dissemination of this serious pest. These included 1) *surveys* to define infested areas; 2) an *eradication program,* in which BN-infested trees were removed and burned, and infested areas were fumigated and kept fallow before replanting was allowed; 3) a *barrier program,* in which the perimeter of an infested area was fumigated and kept fallow to prevent the growth of roots and movement of nematodes to healthy trees; and 4) *certification programs,* including approval of nursery sites, approval of soil pits (sites providing soil for fill), and citrus nursery programs implemented to ensure that BN and other citrus nematodes would not be disseminated with nursery material or spread when roads and power lines were constructed in or near groves.

The eradication and barrier programs were mandatory at first, but later they became voluntary. Since the mid-1980s, the push-and-treat and the chemical barrier programs have been discontinued because of environmental concerns. A few barrier programs are being continued, using nematode-resistant rootstocks such as Milam lemon and mechanical root pruning.

These programs have greatly reduced the dissemination of BN and have been cost-effective. In 1954, surveys indicated 5,346 acres infested with BN, and it was estimated that without the regulatory control measures up to 53,000 acres would have been infested by 1967. With time, the cost and benefits of the nursery certification program for nematodes have become apparent, and now the certification program is recognized as the keystone of nematode management in Florida. When the program began in the early 1950s, many citrus nurseries were infested with BN; from 1953 to 1958, 278 citrus nurseries were found to be infested. With the implementation of the certification program, this number declined rapidly. Since 1970, BN has not been found in a single citrus nursery.

The cost and benefits of the nursery certification program for nematodes have become apparent, and now the certification program is recognized as the keystone of nematode management.

An additional benefit of the nematode certification program has been a reduction of the percentage of the citrus area in Florida that has become infested with other nematodes attacking citrus. Records from citrus groves sampled for BN in the mid-1950s and early 1960s show that about half of them were infested with the citrus nematode (CN), *T. semipenetrans.* Because this nematode was not yet widespread, although it was being spread through nursery stock, it was included as a regulated nematode when the nursery certification program for BN was established. Because nematodes have not infested the new citrus areas planted over the past 40 years, only about 20–30% of the citrus acreage in Florida is now infested. In states and countries without certification programs, it is common for more than 90% of citrus acreage to be infested with CN.

Pratylenchus coffeae, commonly called the coffee lesion nematode, was added as a regulated nematode in the citrus nursery certification program in the early 1980s, after it was observed causing a decline called citrus slump in a small number of groves in Florida. Presently this nematode has not spread much beyond limited local areas in Florida. More recently, *P. coffeae* has been associated with declining citrus in Brazil, which lends further support to regulation of this nematode wherever nursery certification programs are established.

Long-Term Benefits of the Nematode Certification Program in Florida

While the nematode certification program was initially established because of the threat of BN, the benefits derived today relate as much to the exclusion of other nematodes from new citrus areas. The benefits are much greater than could have been envisioned when the program was established, for the following reasons.

[1] The race of the burrowing nematode that causes spreading decline of citrus was proposed as a new species, *R. citrophilus,* in 1984. However, recent research indicates that the citrus-specific burrowing nematode does not constitute a unique species (D. T. Kaplan, 1997, Sequence tag site and host range assays demonstrate that *Radopholus similis* and *R. citrophilus* are not reproductively isolated, J. Nematol. 29:421–429; D. T. Kaplan, 1997, Genome similarity implies that citrus-parasitic burrowing nematodes do not represent a unique species, J. Nematol. 29:430–440).

Increasing production and shift of production into new areas. The nematode certification program has made available a consistent supply of nematode-free citrus trees, which have been planted in new groves as citrus production has expanded into new areas. When certification began in 1958, there were about 648,000 acres of commercial citrus. This had increased to about 857,000 acres by 1996 (Table 4.1). Urban and suburban developments have replaced citrus groves in the northern citrus areas of Florida, and following three severe freezes in the 1980s, production has shifted to the warmer south, where citrus was not previously grown. The nematode certification program prevented the infestation of this new citrus area with nematodes. Thus, for many years to come, growers will continue to receive this economic benefit.

Changing cultural practices. From the time the nematode certification program began in 1958 until 1994, the total citrus acreage in Florida increased by a factor of 1.3, while the total number of trees increased by a factor of 2.7. Tree densities increased from an average of 60 trees per acre in 1958 to 122 trees per acre in 1994, and many newer plantings have up to 225 trees per acre. High-density plantings favor increases in nematode populations, because there are more roots per unit of volume of soil, providing a greater food base, and nematodes can spread faster, because tree roots overlap at an earlier age. High-density plantings may also be subjected to greater stress due to competition for water and nutrients, thus contributing to the economic losses incurred due to the reduced root growth caused by nematodes.

Tree replacement. Groves that are free of regulated nematodes will become infested if nematode-infected resets are used. Normally up to 5% of the trees in commercial groves are replaced annually, because of losses due to stress caused by insects, nematodes, fungi, viruses, other microorganisms, and environmental factors. Following freezes, up to 24% of the total number of trees in the industry have had to be replaced (Table 4.1). If certified nematode-free trees are planted to establish a grove, and good sanitation practices are followed to keep it nematode-free, the replacement trees have a very high chance of survival, compared to trees replanted in an older grove infested with CN.

Less selection pressure for resistance-breaking biotypes. Citrus rootstocks resistant to BN and CN are commonly used. For example, Swingle citrumelo rootstock is resistant to CN and *Phytophthora*, is tolerant of tristeza virus and most viroids, and has desirable growth and yield. However, as with all perennial crops, if a rootstock is continuously exposed to a pathogen, selection pressure eventually results in the development of resistance-breaking biotypes that threaten the continued usefulness of the rootstock. In 1995 a resistant-breaking biotype of CN was found on Swingle rootstock in Florida. The exclusion of nematodes from new and established groves by the nursery certification program reduces selection pressure on the pathogen and thus lowers the probability of the development and spread of resistance-breaking biotypes.

Reduced need for chemical control. Certification programs which manage nematodes by exclusion eliminate the need for chemical control. This lowers production costs and is favorable to the environment. Eliminating the need for chemical treatment is important at a time when the number of chemicals available for the control of citrus nematodes has been greatly reduced.

Table 4.1. Citrus acreage in Florida, 1980–1996[a]

Census year	Total acres	Change in acreage over the previous biennium: Gross loss (acres)	Change in acreage over the previous biennium: Percentage loss	New acres planted in previous biennium
1980	845,283	25,925	3.1	39,973
1982[b]	847,856	51,942	6.1	54,515
1984[b]	761,365	159,719	18.8	73,228
1986[b]	624,492	185,598	24.4	48,725
1988	697,929	52,240	8.4	125,677
1990[b]	732,767	85,858	12.3	120,696
1992	791,290	74,704	10.2	133,227
1994	853,742	45,214	5.7	107,666
1996	857,687	42,904	5.0	46,849

[a] Data from 1996 Commercial Citrus Inventory, U.S. Department of Agriculture.
[b] January freezes occurred in 1981, 1982, 1985, and 1986. December freezes occurred in 1983, 1985, and 1989.

Cost-Benefit Analysis

Because citrus is a perennial crop, the exclusion of graft-transmissible pathogens and other pests, such as nematodes, has a long-term economic impact (Fig. 4.2). Certification programs for perennial crops require long-term commitments. There must be uninterrupted and adequate funding for these regulatory programs, and growers and regulatory scientists

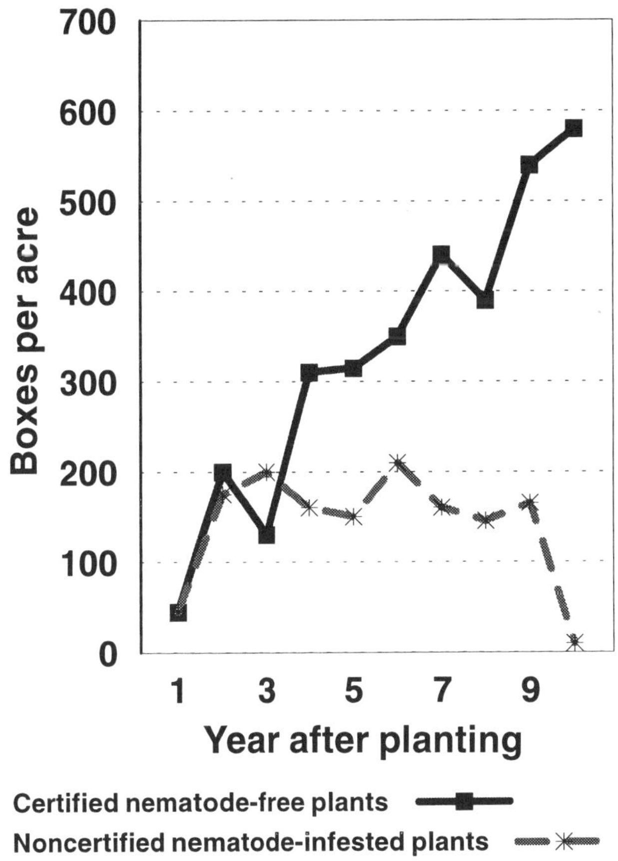

Fig. 4.2. Production of citrus groves planted with seedlings certified to be free of the burrowing nematode and groves planted with seedlings infested with the nematode over 10 seasons in Florida. (Data from C. Poucher, H. W. Ford, R. E. Suit, and E. P. DuCharme, 1967, Burrowing nematode in citrus, Fla. Dep. Plant Ind. Bull. 7)

Box 4.4

Economic Value of Regulatory and Certification Programs

Cost-Benefit Analysis of the Burrowing Nematode Regulatory Program, 1960–1994

Because of the implementation of nematode eradication and barrier programs and nursery certification in the 1950s, the nematode-infested area in Florida citrus groves was only 9,386 acres by 1984. It has been estimated that 44,500 acres would have become nematode-infested within 10 years if this regulatory management program had not been implemented in 1958.

If a grower planted an acre of nematode-free plants in 1960, by 1994 (a 35-year period) the cumulative value of on-tree yield would have been $55,466, based on average published values per acre indexed to 1995 dollar values, provided that the recommended cultural practices for control of the burrowing nematode were followed. However, if this grower had planted nematode-infested plants and after 10 years replanted with infested plants, the value of on-tree yield during the same 35-year period would have been only $24,130, a 57% reduction of return (Fig. 4.2).

This difference in yields due to damage caused by the burrowing nematode, applied to the estimated 44,500 acres which did not become infested because of the regulatory program, amounts to cumulative benefits of almost $1.4 billion for Florida citrus growers during this 35-year period.

The actual cost of the regulatory program from 1954 to 1994 was about $100 million, include the costs of pushing and fumigating 15,200 acres, maintaining buffers around 5,500 infested acres for 30 years, and evaluating 3,000,000 samples to test for the presence of nematodes.

Thus, the regulatory management costs were only about 7% of the benefits to the growers, or *for every $70,000 invested in the regulatory program there was a return of $1 million.* The cost-benefit ratio would be even lower if collateral benefits from the exclusion of other nematodes were included in the calculations.

Cost-Benefit Analysis of the Nursery Certification Program, 1994–1995

The high return on investment in certification programs is illustrated by the following summary of costs and benefits of the program for excluding the citrus nematode and the burrowing nematode in fiscal year 1994–1995.

The benefit gained from control of the citrus nematode alone as a result of the nursery certification program was about $32.5 million in 1994–1995. This estimate is based on a 5% loss of on-tree yield value applied to 75% of the citrus area. The loss estimate takes into account that about 50% of the area planted after the freezes of the 1980s was planted with citrus on nematode-resistant rootstocks.

The benefit gained from preventing the spread of the burrowing nematode as a result of the nursery certification program was about $16.7 million in 1994–1995. This estimate was calculated from the total on-tree value of Florida citrus in the 1994–1995 season ($1 billion), the estimated 44,500 acres which remained free of the burrowing nematode but would have become infested within 10 years if the program had not been implemented, and a 25% loss of on-tree value, which this nematode would likely cause with current management practices.

The total annual cost to the Plant Inspection and Nematology Section for the certification program was $68,698. Growers contributed $15,840 in fees charged for sampling required by the program. Thus the cost of the program was $58,825 after the growers' fees were deducted.

The on-tree benefit from the certification program for the exclusion of these two nematodes is thus estimated to be $49 million for the crop year 1994–1995. *For every $1,403 invested in the administration of the certification program, the return to growers was $1 million.* Additional benefits were derived from exclusion of the coffee lesion nematode and other nematodes but were not included in this calculation.

The total economic impact of the Florida citrus industry on the general economy through jobs and revenue from processing, transportation, and sales is about seven times the on-tree fruit value. Thus, the positive economic impact of this citrus nematode certification program has been far greater for the general public than just the increase in on-tree fruit value for growers. The public also receives indirect benefits in the form of lower food costs, and growers receive additional benefits in the form of lower pesticide costs as a result of the exclusion of pests by the certification program.

must cooperate in conducting these programs with integrity and consistency. A cost-benefit analysis of a 35-year investment in the nematode program in Florida demonstrates this (Box 4.4).

Other Considerations

Funding of Certification Programs

In some countries citrus certification programs are voluntary, whereas in others they are mandatory. The question of voluntary or mandatory participation should be decided according to the situation in each particular country, but mandatory programs are more effective and are recommended, especially where vector-borne pathogens are present. When a certification program is being established, the availability of resources to maintain the program over the long term should be considered. Will the program be financed by a tax on the industry, by governmental budget resources, or by contributions and support from the private sector? Substantial benefits can be realized from a certification program that ensures the production of pathogen-free trees. A certification program should be designed to provide maximum benefit from the available resources and to be self-sustaining. A certification program which runs well for a few years but then cannot continue to function because of a lack of resources does not really contribute to the long-term productivity of the local citrus industry.

Operation of a Certification Program

The most important component of a certification program is well-trained and dedicated personnel with extensive experience. There are other considerations that go along with a certification program, such as cultural practices, pest control, labeling, and planting location. Plants should be grown with the best available cultural practices. Special precautions should be taken to control pests and fungal diseases. All pruning and grafting tools should be adequately disinfested prior to any fruit picking, grafting, or cutting of any tree or nursery plant, to prevent the spread of mechanically transmitted diseases. Minimum requirements for certified plants should be established to meet the needs of different growing areas. They can include plant size, height of budding, stem diameter, and other criteria.

A very important aspect of a certification program is careful record keeping and labeling of plants during the whole process of propagation. Personnel must be well trained in this regard. Special attention should be paid to the collection, treatment, and packing of seeds. Budwood cutting should be witnessed, buds properly labeled, and plat maps used to track propagations. This should prevent mix-ups of rootstocks or cultivars and allow tracing back any abnormality or pathogen found at any stage in the propagation. All nursery operations should be continuously under inspection.

Nursery Site Selection

Planting locations should be approved by the agency in charge of the program. They should be in areas with minimum risk of infection and should be suitable for growing citrus plants of good quality. The citrus nursery nematode certification program in Florida consists of two key components that can serve as a model for other countries. First, prior to the establishment of a field-grown citrus nursery, the site must receive approval from a regulatory agency. Second, after establishment, all nurseries which sell citrus trees are required to comply with sanitary regulations for field-grown plants and container-grown plants on benches. The site selection criteria and sanitary regulations for Florida citrus nurseries are summarized in Box 4.5.

Certification programs should take into consideration special local requirements and conditions. For example, if a disease is limited to one area, it may be desirable to establish nurseries and propagate citrus only in regions where the disease is not present. For example, in Uruguay nurseries are located outside the area where citrus canker is present, and in California nurseries operate in areas free of stubborn and tristeza.

Summary of Principles for Establishing a Successful Citrus Nursery Certification Program

Graft-transmissible pathogens lacking vectors are easily controlled by a certification program. Graft-transmissible pathogens spread by vectors can be effectively controlled by a certification program if it is part of an integrated pest management strategy. Citrus nematodes have been effectively controlled in Florida by a nursery certification program, and a similar approach is being used to reduce the spread of root weevil larvae. Root pathogens often spread with the distribution of containerized nursery material.

The biology and host range of the pathogen or pathogens to be controlled must be understood. The distribution of pathogens in native areas and citrus groves must be determined.

Box 4.5

Site Selection Criteria and Sanitary Regulations for Citrus Nurseries in Florida

Criteria for Approval of Sites Prior to the Establishment of Field-Grown Citrus Nurseries

1. Growers or owners apply to the regulatory agency for approval and provide information on the location and boundaries of the proposed nursery site. It is highly recommended that foundation blocks, increase blocks, and the nursery be located away from established citrus orchards where possible, to avoid contamination with pests.
2. The site is inspected by a regulatory specialist, who reviews the history of the site. The proximity of citrus or crops known to be infested with regulated nematode pests, the proximity of public or private thoroughfares, slopes that influence drainage, and the probability of contamination from adjacent areas are considered.
3. A map of the proposed site is prepared, and the site is sampled for nematodes. Host plants within 100 feet of the proposed site must be free of regulated nematodes. A greater distance may be required if there is a threat because of slope, drainage, standing water, root encroachment, or other factors. If the samples are negative for regulated nematodes, the site is approved with the following setback areas: a) 100 feet from known infestations or where noncertified plants or soil have been introduced and b) 50 feet from all thoroughfares, rights-of-way, and croplands.
4. If an approved site is not planted within 12 months, it must be resampled annually to remain eligible for the establishment of a field citrus nursery.

Sanitation Requirements for Established Field-Grown and Raised-Bench Nurseries

1. In a field-grown nursery, the minimum required setback areas around the site must be maintained free of hosts.
2. All soil, growing media, mulch, and manure introduced into a raised-bench growing area or field site must be certified to be free of regulated nematodes or must be obtained from an approved source. Requirements for soil source site approval and sanitation are the same as those for field-grown citrus nursery sites.
3. Soil or growing media and containers used to produce citrus on raised benches must be stored at least 18 inches above the ground or on asphalt or concrete above ground level and not subject to splashing, flooding, or runoff.
4. Citrus plants must be produced from seed or cuttings taken 18 inches above the ground. All other plants or plant parts introduced into a field nursery or on raised benches must be determined by the regulatory agency to be free of regulated nematodes.
5. In all nurseries, citrus plants are routinely sampled (preferably each crop or annually) by the regulatory agency and must be found to be free of regulated nematodes.

Appropriate and cost-effective sampling techniques must be available.

It is important that sites approved for nurseries are free of regulated pests. Physical barriers, such as screenhouses and greenhouses, may be required to control pathogens transmitted by insect vectors.

Certification programs that are mandatory are more successful than voluntary programs. The agency operating a program should have the legal authority to enter any site where citrus is grown and to have powers of enforcement and certification as established by the rules.

Growers and members of the industry must support the certification program and cooperate with it. The most successful programs are the ones devised by growers and operated by a local regulatory agency.

Certification programs must be established for long-term operation. As illustrated in this chapter, the benefits of a citrus nursery certification program are compounded over time.

PART THREE

Crop Production

T. A. Wheaton
W. S. Castle
J. D. Whitney
D. P. H. Tucker
University of Florida
Citrus Research and Education Center, Lake Alfred

CHAPTER 5

Horticultural Practices for Citrus Health

The design and management of citrus plantings varies around the world. Citrus is grown successfully under a wide range of climatic conditions, in soils ranging from very poor sands to heavy clays, with or without irrigation, and with varying degrees of mechanization. This chapter on grove design and tree size control is based on research from around the world, but it most completely describes the Florida citrus industry. The concepts presented are appropriate for all citrus-producing areas, but some of the examples, practices, and equipment described are unique to Florida citrus. This chapter introduces site selection, land preparation, and the planting and care of young trees, focusing on tree spacing, orchard design, and pruning practices. Irrigation, nutrition, and other horticultural practices are considered in other chapters.

The design of a citrus grove begins with site selection and a plan for the physical arrangement of trees in the planting. Decisions about the spacing between trees and the orientation of rows are based on a number of factors, including soil and climatic characteristics of the site, the vigor of the variety and rootstock combination to be planted, the expected life of the grove, and the allocation of space for the operation of production and harvesting equipment. After these decisions are made, the site is prepared, irrigation is installed when necessary, and trees are planted. When trees mature and grow beyond their allocated space, pruning is started to restrict tree size. Other horticultural considerations include flower induction and fruit set, crop load and alternate bearing, special pollination requirements for some varieties, the use of plant growth regulators, and environmental and biotic factors affecting fruit quality.

Site Selection and Preparation

The sites used for citrus production in Florida include the well-drained soils of the Central Florida Ridge, the poorly drained soils of the coastal areas, and the flatwoods and marsh soils of south Florida. The danger of freeze damage is the major factor in site selection. Citrus production originated in north Florida, moved to central Florida, and is now concentrated in the warmer coastal areas and in south Florida. In central Florida, the preferred sites are the higher portions of the Ridge and locations on the south side of large lakes. In the coastal areas and south Florida, water management and soil type are more important in site selection. Conversion of native lands to citrus production and the development of water resources for irrigation are highly regulated in Florida and require permits from water management districts or other governmental agencies.

Site preparation usually includes clearing and burning existing vegetation. In areas where the water table is near the surface, it is necessary to install extensive drainage systems. Laser leveling is often used in flatwoods to ensure adequate drainage. Movement of soil during leveling and bedding, however, may remove or bury the most productive top layer of soil. Thus the variability of soils is increased. Raised beds are built to provide an adequate root zone above the water table. Beds typically are about 18 inches high and about 50 feet wide. The swales between beds provide drainage and connect to larger ditches for removal of water from the site (Plate 34). Drainage tiles may be used in addition to or instead of elevated beds.

Soils in citrus production in Florida often have low fertility and low water-holding capacity. Many native Florida soils are acidic and require the application of limestone prior to planting to bring the soil pH within the range of 5.5 to 6.5. Calcareous soils and very white sands devoid of organic matter present other problems and may be best left unplanted. However, successful citrus production is possible on calcareous soils with proper selection of rootstocks and attention to special nutrient requirements. Newly cleared soils may also be deficient in phosphorus and other essential plant nutrients. Phosphorus is often applied prior to planting. Old grove land being cleared and replanted may have other problems, including high copper levels from previous applications of copper pesticides and the presence of diseases and nematodes. Soils with excessive copper must be adjusted to pH 6.0 or higher prior to planting, to reduce the availability of this element.

Rainfall is adequate for production in some citrus-growing areas. In Florida, however, optimal citrus tree growth and production can be achieved only with supplemental irrigation. Several types of irrigation systems are used for citrus production. Microsprinklers are commonly used to provide both irrigation and cold protection. An irrigation system may also be used for the delivery of fertilizers and pesticides (see Chapters 6 and 7).

Planting and Young Tree Care

After site preparation is completed, young grafted trees grown in a nursery for one to two years are planted in the grove. Nurseries may provide either bare-root or container trees. Bare-root trees are grown in a field nursery and are supplied for planting with all soil removed from the root system. Roots must be kept wet from the time of digging until planting to prevent desiccation and poor survival of trees. Trees are placed in holes dug in advance, along with several gallons of water, and are "mudded in" by working a slurry of soil and water around the roots. Container trees are generally grown in greenhouses in small plastic pots. The pots are removed immediately prior to planting, and the trees are usually mudded in like bare-root trees.

Trees must be supplied with water frequently during the first few weeks after planting to ensure survival. Young citrus trees can be successfully planted in the grove any time of the year. Optimum first-year growth, however, can be obtained when trees are planted early in the spring. In regions where there is potential freeze damage, planting should be delayed until the danger of frost has passed. In regions with no potential cold damage, planting during the winter, when the trees are less actively growing, is recommended.

The primary objective during the first years in the grove is to fill the space allocated to each tree with canopy as soon as possible. Substantial fruit production is not possible until canopy development occurs. Fruit production and fruit quality become considerations only after trees have reached a size large enough to support a substantial crop. Water and nutrition are the most critical factors in the growth of young trees (these subjects are discussed in more detail in Chapters 6 and 7). Vegetative growth is particularly susceptible to water stress, so it is important to maintain adequate soil moisture at all times to ensure maximum tree growth. The amount and frequency of irrigation depend on a number of factors, including climate, rainfall, temperature, the size of the trees, soil characteristics, and characteristics of the irrigation system. During the cool winter days in Florida, irrigation may be required once every 10 days to 2 weeks. However, it may be required more frequently during May to maintain optimal soil moisture for young tree growth.

Rapid tree growth can be obtained with the application of dry fertilizer four or five times per year or injection of liquid fertilizer into the irrigation system more frequently. Fertigation is a common practice in Florida. Some growers inject small amounts of fertilizer every time they irrigate; others inject fertilizer weekly or monthly.

During the first years in the field, wraps may be placed around the trunks. They may be either insulating water-jacketed Styrofoam wraps or simple corrugated plastic wraps. Wraps serve several functions. Insulating wraps are often used for freeze protection in colder locations. All types of wraps provide protection from herbicides, mechanical damage, and rodent damage. They also inhibit sprout formation, to maintain a trunk without branches up to the heading or branching height developed in the nursery. However, wraps may be detrimental by harboring insects and maintaining a damp atmosphere around the trunk, which increases disease problems in some cases.

Tree Spacing

Tree spacing is increasingly important in the design of groves to maximize economic returns (Plate 19). Optimum tree spacing changes with age, so any decision about spacing is a compromise. Groves of closely planted trees produce an earlier return but cost more to establish and must be managed precisely to ensure that problems do not arise over time. Groves with more widely spaced trees require more time to reach maximum yield but may be as productive and easier to manage at maturity. Grove design must integrate biological and management aspects to maximize average lifetime economic returns. Decisions about tree spacing and management practices to achieve optimum lifetime returns are complex, challenging, and subject to many considerations.

Factors influencing grove design in various citrus-producing regions include

- Land value
- Labor costs
- Available equipment
- Climate—temperature and rainfall
- Soil quality
- Type of operation—corporate or family
- Available resources—water, fertilizer, pesticides
- Market—fresh or processing, local or international
- Available financing
- Suitable varieties and rootstocks

There are tremendous differences in many of these conditions at different locations. For example, differences between the climate of citrus-growing areas in Japan and that of areas in Central America result in very different growth rates in these areas and dictate very different planting systems. A planting density of over 1,000 trees per acre may be appropriate in Japan, whereas fewer than 100 trees per acre may be optimal in some tropical regions.

Some principles of grove design, however, are valid for a range of conditions. Concepts important in grove design include

- The relationships between the interception of sunlight and productivity
- The relationship between canopy bearing volume and productivity
- The compromise between the better productivity of young trees with close spacing and the easier management of mature plantings with wider spacing
- The importance of matching tree vigor to the space allocated for each tree
- Careful financial analysis of the various options

These principles can be used to define optimal grove design for all citrus-growing areas. This discussion will focus on the design of groves for Florida conditions, but the concepts are universal and may be adapted to any citrus region.

Light Interception

Sunlight is the source of energy for tree and fruit growth. Much of grove design can be thought of as arranging canopies to capture sunlight efficiently. Sunlight falling on the ground between trees does not produce fruit. The grove design should be based on optimizing the average interception of light over the life of the planting. This requires minimizing the amount of light that falls on the ground and providing canopy structures where the most leaves receive optimum light intensity. Citrus leaves do not use direct sunlight efficiently; maximum photosynthesis occurs somewhere between full shade and full sunlight.

Canopy Bearing Volume

The yield potential of a citrus grove is closely related to the canopy bearing volume. The bearing volume of a citrus tree is defined as the outer 3 feet of the canopy, since fruit bearing is largely limited to this zone, and because 90% of the solar radiation intercepted by the tree is absorbed in this outer 3 feet. Production per acre is a function of bearing volume per acre. A grove in which a large portion of the space is available for bearing volume produces more fruit than a grove in which a high percentage of space devoted to structural tree support, middles, or other functions. The bearing volume, and thus the production potential, can be increased by

- Increasing tree height
- Reducing tree spread
- Reducing the width of middles
- Increasing the number of trees per acre in a young grove
- Forming continuous hedgerows in a mature grove

There are limits on increasing bearing volume, however. Enough space must be left in the middles to allow for equipment and harvesting operations. Tree height should also be limited to prevent loss of tree skirts due to shading and to increase the efficiency of labor and equipment. Tree spacing must also be in balance with natural tree growth, so that trees can be maintained in a predetermined space without loss of productivity. Since citrus yield potential is a function of the canopy bearing volume, it is important to design a grove that provides for the maximum amount of canopy but still provides adequate access to the trees.

In practice, bearing volume per acre at tree maturity (when the trees have filled their allocated space) is similar for a range of tree densities and spacing arrangements. Thus, potential production at maturity is similar over a range of tree spacings.

The average tree density in new plantings has almost doubled since 1960 (Fig. 5.1). In 1960, new plantings spaced 25 × 25 feet were common, allowing about 75 trees per acre. Planting density increased during the 1960s and again during the 1980s. By 1994, the average density of new plantings was 140 trees per acre. Tree spacing is commonly 12.5 × 25 feet to achieve this density. However, many new groves are planted at even higher densities. For example, spacing of 10 × 20 feet is not uncommon, allowing 218 trees per acre.

Tree Spacing Concepts

The allocation of space in a grove is determined at the time of planting. It consists of space allocated for each tree and the clear space or drive middle between rows of trees. The designed tree height (topping height) is part of the allocated space. The spacing selected is always a compromise, because

- Closer spacing (higher tree density) improves early yield
- High-density plantings cost more to develop
- High-density plantings provide little or no yield advantage at maturity
- Yield may decline at maturity if spacing and vigor are not matched

Maximum production is achieved when about 70% of the land area is covered by the canopy. The height of the tree affects the level of light reaching the lower tree skirt. A conservative estimate of maximum tree height is about twice the width of the row middle, so that tree height and row middle width are in the ratio of about 2:1. For the common 8-foot clear middle, a height of 16 to 18 feet meets this standard. This significance of the ratio of tree height to row middle width for adequate light penetration is illustrated in Figure 5.2.

Grove design requires integration of these concepts along with site-specific information, climatic data, and the relationship between vigor and nematodes or other pests. The expected life of the planting is also important and may depend on freeze damage, disease incidence, or land development for other purposes.

Young Trees

In a young grove, the canopy covers only a small area of the ground and captures only a fraction of the available sunlight. Yield per acre is low, because canopy volume per acre is low. Canopy volume per acre increases more rapidly when more trees per acre are planted and when the trees are maintained at the maximum growth rate. The combination of high tree density and improved management results in a substantial in-

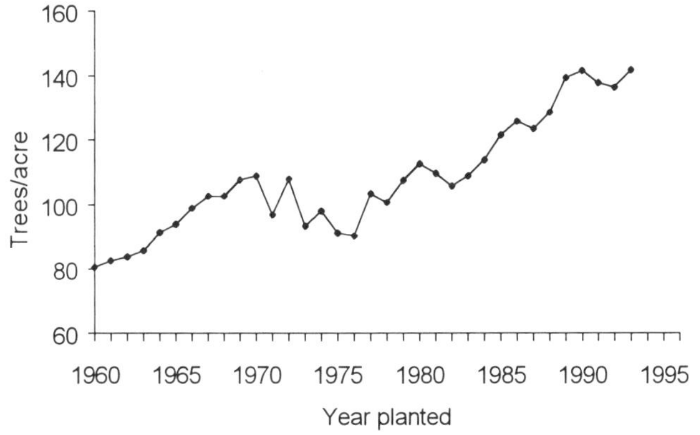

Fig. 5.1. Planting density of citrus trees planted in Florida from 1960 to 1994. (Data from Florida Agricultural Statistical Services, Citrus Tree Inventory)

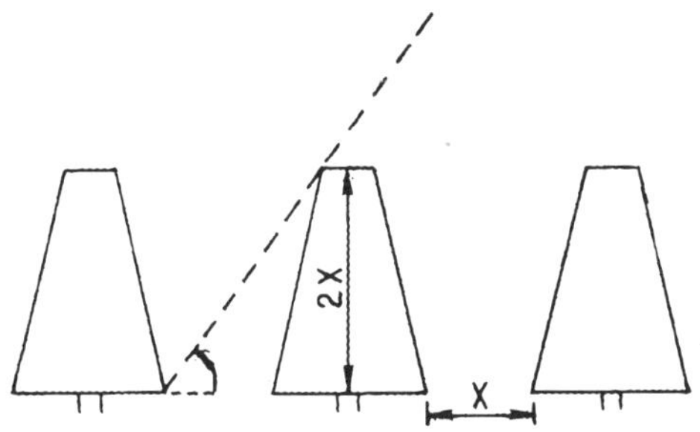

Fig. 5.2. When tree height is approximately twice the width of the row middle, adequate light is allowed to reach the lower skirts of the tree, to maintain leaves and fruit production. The clear spaces between rows (the middles) must be 7 to 8 feet wide for access by production and harvesting equipment. With this width, the 2:1 ratio allows tree heights of 14 to 16 feet. Optimal production is obtained when about 70% of the land area is covered by the tree canopy.

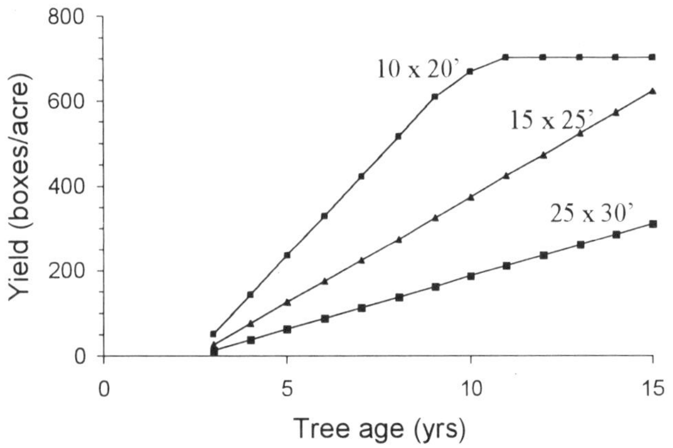

Fig. 5.3. Effect of tree spacing on the yield potential of a citrus grove from planting to 15 years of age. Yield per acre increases with tree density in a young grove. The yields projected here can be achieved by Hamlin orange in Florida, grown at a good site with excellent management. Actual yield varies with the variety, climate, and management. Tree spacing has little effect on yield potential after the trees have filled their allocated space.

crease in early production. There are, however, increases in cost when tree density and management inputs are increased. Careful analysis is required to determine the tree density and management strategy that will provide optimal returns.

It is possible to predict the annual yield of trees planted at different densities if the rate of canopy development and yield of an individual tree are known. Although yield per tree will vary with variety, climate, soil, and other factors, the prediction is useful in enlarging or replanting an existing grove where the growth rate and yield of individual trees are known. This approach is used in Figure 5.3 to compare, during their first 15 years, the annual production of trees planted at three different spacings. The planting with the highest density (tree spacing of 10 × 20 feet) reaches maximum yield in eight to 10 years, the planting with spacing of 15 × 25 feet approaches maximum yield in 15 years, and the planting with spacing of 25 × 30 feet requires many more years to reach maximum yield.

Mature Trees

In a mature grove, the canopy covers most of the land area and collects a high percentage of available sunlight used for growth and fruit production. The trees have filled their allocated spaces and are maintained at that size by hedging and topping. Most leaves are located in the outer 3 feet of the canopy, where there is adequate light for fruit production, and yield per acre is closely related to the volume of this part of the canopy. The canopy bearing volume per acre of mature groves has been calculated for various grove designs in Table 5.1, which demonstrates that grove design and tree spacing have little effect on bearing volume in a mature grove. The authors have observed that maximum production is similar for a range of grove designs and tree densities.

The location of fruit on the tree is affected by tree spacing. In widely spaced trees, a high percentage of the fruit is located close to the ground. With decreased row spacing, however, a greater portion of the fruit is produced at the top of the tree (Fig. 5.4). This change in the distribution of fruit affects harvesting efficiency. Fruit that can be reached from the ground is more easily harvested than fruit higher in the tree. On the other hand, high-density plantings of small trees produce a wall of easily accessible fruit and provides attractive harvesting. Platform harvest aids are also a possibility for plantings with close row spacing.

Table 5.1. Canopy bearing volume per acre of mature trees in various grove designs with different tree spacings

Spacing (feet)	Bearing volume per acre (thousands of cubic feet)
Individual trees	
20 × 25	210
25 × 25	213
25 × 30	213
30 × 30	213
Hedgerows	
20	248
25	224
30	205

Hedgerows or Individual Trees

Citrus trees planted close together in a row rapidly grow together to form a continuous hedgerow. The close spacing increases tree density, so that canopy bearing volume and yield increase rapidly in the young grove. It is commonly believed, however, that the best production of citrus is obtained when all sides of the tree are open and exposed to sunlight. Thus, it has been traditional to grow citrus as individual trees, and there have been questions about the suitability of hedgerows for citrus production.

We have demonstrated that the hedgerow is an optimal production system under Florida conditions. Attempts to maintain or create spaces in the row by cross-hedging or removal of every second, fourth, or fifth tree generally reduced yield. A dense, solid hedgerow shades out competing weed growth and reduces the need for herbicide applications. In addition, it allows tubing used for microirrigation to run along the row on top of the ground. With very close spacing in the row, movement across the row for harvesting or other operations becomes difficult, and the placement of large containers for harvested

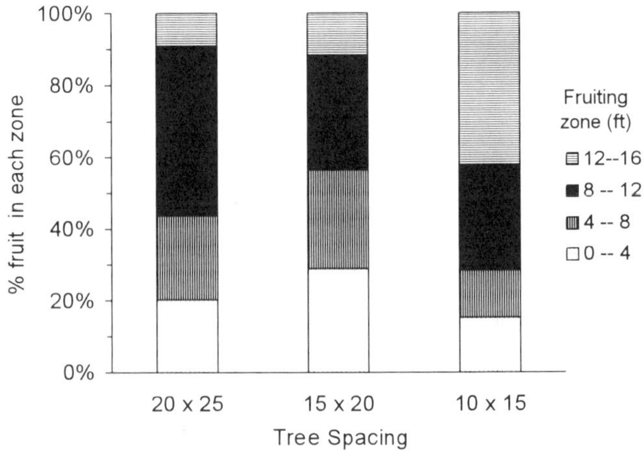

Fig. 5.4. Effect of tree spacing on the vertical distribution of fruit in a mature grove. One of the disadvantages of high-density plantings is the concentration of fruit in the top center of the tree, which makes harvesting more difficult. In this example, only about 10% of the fruit is in the top 12–16 feet with tree spacing of 20 × 25 feet. However, 40% of the fruit is in this highest zone with tree spacing of 10 × 15 feet.

fruit is also more difficult. However, the advantages of hedgerows are greater than the disadvantages, and virtually all new plantings in Florida are designed to rapidly form a hedgerow.

Excessive Vigor

Declining yield after trees reach containment size has been observed in some blocks. These groves grow rapidly and produce very well during the early years, but their yield declines after hedging and topping are begun (Fig. 5.5). The trees appear to be healthy, and there is no obvious explanation for the reduction in yield. Groves with this problem are typically planted at average or above-average tree density, with trees grown on a vigorous rootstock. They have received high levels of nutrients and irrigation since planting, to increase canopy volume and develop their fruiting potential as soon as possible. The problem may be more severe in trees planted in better soil types. The trees appear to be in a vegetative state and produce excessive regrowth following hedging and topping.

The probable cause of this declining yield is an imbalance between the natural growth habit of the tree and the space that has been allocated to it (Fig. 5.6). Trees in this situation may benefit from a reduction in nutrient inputs and irrigation. Hedging or topping two times per year may also improve fruiting. In severe cases, it may be necessary to remove every other tree to provide a better balance between natural tree size and space allocation. The problem can be avoided at the grove design stage by selecting a tree spacing appropriate for the inherent vigor of the variety and rootstock, along with consideration of the soil and climate at the grove site.

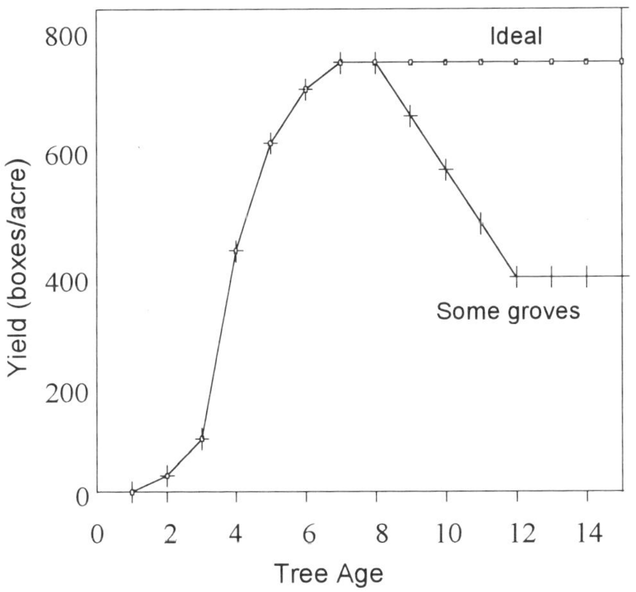

Fig. 5.5. Yield decline observed in some groves after the trees have filled their allocated space and hedging and topping have begun, compared with steady production in an ideal grove. This decline occurs when trees are too vigorous to be maintained satisfactorily in their allocated space.

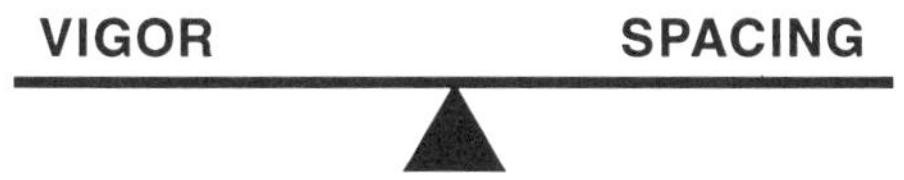

Fig. 5.6. Tree vigor must be in balance with the space allocated for each tree. Vigorous trees produce poorly when confined to a space that is too small.

Irrigation, Nutrition, Freeze Damage, and Fruit Quality

Irrigation and nutritional requirements do not vary with planting density after the first few years. Initially, fertilizer is applied at a particular rate per tree, so that more fertilizer per acre will be used in high-density plantings. As the trees get older, however, fertilizer requirements per acre are similar for a range of tree spacings and should be based on fruit production and leaf analysis. Tree density does not affect the irrigation requirements of a mature grove. When the trees have grown to fill their allocated space, nutrition and water requirements are similar because the canopy volume and production potential are similar.

Several observations in Florida indicate that freeze damage is reduced in high-density plantings in both calm and windy freezes. In calm (radiation) freezes, the continuous canopy traps more heat radiated from the ground. In windy (advective) freezes, hedgerows may act as an effective windbreak and reduce heat loss from the site. Thus, higher densities may be better for colder locations. Grove design in such cases should also ensure adequate air drainage on cold nights.

The effects of tree spacing on fruit quality are minor as long as tree vigor is matched to the space allocated to the tree. Under Florida conditions, fruit color and other quality factors are not adversely affected in high-density plantings. No consistent effects of tree spacing on internal or external fruit quality have been observed. Therefore, fruit from high-density plantings is suitable for the fresh market as well as for processing. In California, however, reduced in-row spacing has resulted in poorer fruit quality.

Spacing Recommendations

Spacing of 20–25 feet between rows is generally recommended for citrus groves. Vigorous varieties such as grapefruit and some tangelos require more space than oranges and mandarins and should be planted with spacing at the upper end of this range. Row spacing of 20–25 feet leaves room for a clear middle of 7–8 feet, which is adequate for production and harvesting equipment in use today. Row spacings greater than 25 feet were once common in the Florida citrus industry, but yield per acre in a young grove is reduced at row spacings greater than 25 feet. Some commercial and experimental plantings with 15-foot row spacing have been successful, and the advent of rootstocks that provide moderate control over tree size may make this closer spacing more attractive. However, it is difficult to conduct conventional harvesting operations with less than 20 feet between rows. Production and harvesting also require more miles of travel per acre in groves with close row spacing, since they contain more rows per acre.

Spacing of 10–15 feet between trees in a row is desirable. Within this range, the influence of climate, soil, variety, and rootstock on tree vigor must be considered. With modern production practices, trees spaced at distances in this range grow together in the row and form a hedgerow in a few years in most citrus-producing areas. There is little advantage in spacings of less than 10 feet, because trees planted less than 10 feet apart begin competing for space before producing any significant yield. Planting trees at regular intervals in a single row is preferable to planting closely spaced trees in groups that are more widely spaced in the row. For example, regular 10-foot spacing is more desirable than grouping two trees 5 feet apart and leaving 15 feet between groups, even though the number of trees per acre is the same. Trees planted at regular intervals form a hedgerow more rapidly.

These recommended spacings allow tree densities ranging from 116 trees per acre (with 15 × 25-foot spacing) to 218 trees per acre (with 10 × 20-foot spacing). Table 5.2 shows the tree spacing and number of trees per acre in several common grove designs.

Tree densities outside this range may be justified in some situations. Some variety and rootstock combinations can be successfully grown at densities of up to 350 trees per acre, but such high densities appear to have little or no economic advantage under Florida conditions. Spacings greater than 15 × 25 feet may be appropriate where tree growth is very vigorous or where equipment for hedging and topping trees is not available.

Row direction or orientation is relatively unimportant in many citrus areas, but it becomes more important as latitude increases. The distribution of light is more uniform on both sides of the trees in rows running north to south. Rows generally should run in the direction most suited to a particular site, which may depend on row length, air or water drainage, or the need for erosion control.

> **Recommendations for tree spacing**
>
> **Minimum 10 × 20 feet**
>
> **Maximum 15 × 25 feet**
>
> **Match spacing and tree vigor**

In older established groves in which trees are spaced 20 to 30 feet apart in the row, it is common to interset trees, or place additional trees in the row. This can improve production if the trees have not already grown together in the row. However, unless there is 10 feet or more of open space, competition from adjacent trees may be so severe that young interset trees never develop adequately. Removal of mature trees in a hedgerow due to disease also leaves space for intersets. Intersets should be planted only when there is adequate space for development, and they should be evenly spaced. The practice of resetting with two trees less than 8 to 10 feet from each other is not warranted.

Citrus is adaptable to a range of planting densities. A final decision on tree spacing requires careful economic analysis. Optimal spacing requires maximizing returns averaged over the life of the grove. Complete financial analysis includes appropriate discounting of money over time and must include additional factors such as loss and replacement of trees and the expected life of the grove (see Chapter 15).

Table 5.2. Tree spacing and trees per acre in common grove designs

Spacing between rows (feet)	Spacing within rows (feet)	Trees per acre
20	10	218
20	15	145
22	11	180
22	22	90
25	10	174
25	12.5	139
25	15	116
25	20	87
25	25	70
30	25	58
30	30	48

Tree Size Control

The ideal citrus tree is one that grows rapidly to fill its allocated space and then stops or greatly reduces its rate of growth. Thus, tree size control begins with selecting a tree of appropriate vigor for the space allocated. Methods of tree size control include

- Pruning
- Using size-controlling rootstocks
- Inducing heavy cropping
- Reducing nutrition and irrigation levels
- Inoculating with citrus viroids that reduce growth

Hedging and topping are the most common forms of tree size control in Florida, but other methods of size control should also be considered. A wide range of dwarfing rootstocks is not available for citrus. However, available rootstocks differ in vigor, and new rootstocks are being developed. Swingle rootstock has been used extensively for the past decade and results in smaller trees. Selected strains of viroids can be introduced into citrus and reduce tree size in a predictable manner. These viroids are used commercially in Australia but are not used in the Americas at this time. A heavy crop load reduces the amount of vegetative growth and thus effectively reduces tree size. Management strategies that reduce fertilization and irrigation may assist in tree size management.

Pruning Young Trees

Removal of sprouts from the trunk is required during the first several years in the grove. The number of sprouts is reduced if wraps are present or by application of a sprout inhibitor. Vigorous branches extending beyond the rest of the canopy are frequently cut off, but there is no evidence that this is advantageous. Trees may produce a little fruit during the first two years in the grove. This fruit is removed in some cases, so it will not compete with and limit tree growth. This growth inhibition occurs only with very heavy crops, however, and the removal of fruit from young trees prior to commercial production is seldom justified.

Pruning Mature Trees

Trees require little pruning until they exceed containment size. The practice of removing dead wood from trees every few years to reduce melanose damage is now rare. However, it is important to begin hedging and topping as soon as trees exceed their allocated space and before severe cuts are necessary. The larger the wood that is cut, the more ineffective the pruning operation. When large wood is cut, the regrowth is very vigorous and rapidly extends to or beyond the original canopy. In addition, this "juvenile" growth produces only a small amount of poor-quality fruit. Two or three years may be required after severe pruning to restore good production of quality fruit.

Hedging

Hedging is the practice of cutting back the sides of trees to maintain a clear alley for equipment access and to provide good distribution of light to the lower parts of the tree (Plate 20). This is usually done with several saws on a fixed or rotating boom mounted on a large mover designed especially for that purpose. Hedging operations cut off all twigs or branches extending beyond the allocated space. New sprouts form at the cut surface and form a dense layer of canopy, which becomes the primary fruiting zone of the tree.

Hedging should be started as soon as trees extend beyond their allocated space, so that only small branches are cut and crop reduction is avoided or minimized. Hedging may be required in some groves as early as five years after planting if the trees are on a vigorous rootstock and tree growth is rapid. The operation is also less expensive when started early, because brush removal is not required. Small branches and twigs may be left on the ground, but removal is required if large wood is cut.

A hedging cut is usually made at an angle, so that the tree is narrower at the top than at the bottom, to allow more even distribution of light to the lower parts of the tree. An angle of approximately 10° is recommended, and angles between 5 and 15° are common. Tree height and row spacing may influence the hedging angle to some extent. With closer row spacing and taller trees, hedging at a greater angle cuts more severely into the top of the tree, and can result in a "Christmas tree" look. This may have a negative impact on yield, since the top center part of the tree is the most productive.

Hedging may be conducted annually for vigorous trees and less frequently for trees of reduced vigor. Annual hedging of every middle is popular, because it avoids large pruning cuts, there is no brush disposal problem, it does not stimulate excessive vegetative growth, and good fruit quality is maintained.

Groves should be hedged about the same time each year. Hedging shortly after the harvest of early and midseason fruit is common. Selecting a time for hedging Valencia orange is more difficult, since there is overlap between the mature crop and the new crop for next year. Thus, fruit at some stage of development may be removed no matter when hedging is conducted. In practice, there is little yield loss associated with hedging at almost any time of year if it is practiced annually at about the same time of year. In colder locations, however, trees may be more susceptible to cold damage if hedging or topping carried out during the fall or winter results in a new growth flush.

Topping

Tree topping should also be started when trees grow taller then desired. Topping maintains tree height for ease of harvest and for pest control, and it allows adequate light to reach the lower parts of the tree (Plate 21). It usually causes a yield reduction, however. Vigorous trees may need annual topping to avoid excessive tree height. However, if trees of lower vigor are planted in a grove design that provides adequate space and height, topping may be required less frequently or not at all. Trees that grow rapidly to the desired height and then stop growth are ideal, because canopy development is rapid but tree size is easily managed at maturity.

Trees may either be flat-topped or topped at an angle of 15–30° from the horizontal, to form a peak about 2 feet or more above the shoulders. Rows that are closely spaced and hedged with a sufficient angle are relatively narrow at the top and can be flat-topped with a single pass of the topping boom. The cost of topping may be reduced if a machine with a boom on each side can top two rows in one pass. As with hedging, removal of small brush during maintenance topping does not require brush disposal. Severe topping results in excessive vegetative growth and substantial yield reduction.

The rule of thumb for topping height is to maintain a 2:1 ratio of tree height to middle width. This height-width ratio, with a hedging angle of 10°, allows good distribution of light to all parts of the tree. Topping heights of 14 to 18 feet are common. Trees at these heights are easily managed and can be harvested with standard ladders. Topping trees at 12 feet or less is questionable. Trees in young plantings are sometimes topped at heights in this range, but this practice can only delay the production of optimal yield.

Topping height affects harvesting efficiency. Most of the fruit from short trees can be harvested from the ground (Plates 22 and 23). Because harvesting from the ground is more rapid, there may be a small reduction in the picking cost associated with shorter trees. However, topping to maintain trees of standard vigor at heights of 12 feet or less causes a reduction in yield. The loss in yield is generally more important than the savings in picking costs.

Because of the yield reduction generally associated with topping, it is difficult to make a general recommendation about the timing and frequency of the operation. The timing, like that of hedging, is not critical if the operation is conducted at the same time every year. Again like hedging, topping is commonly done shortly after the harvest of early and midseason varieties but before major bloom development. Valencias may be topped either before or after harvest. Vigorous trees benefit from annual topping or in some cases more than one topping per year. With less vigorous trees, however, it may be possible to top only every two, three, or four years and avoid some of the yield loss due to topping every year.

Skirting

Skirting is the practice of pruning lower branches to remove foliage and fruit in direct contact with the ground. A skirting machine is used to remove the lower 18 to 30 inches of the canopy. Skirting is becoming an accepted practice with citrus grown for the fresh fruit market, because it improves fruit quality. Lifting the skirts results in less mechanical and chemical damage to fruit from fertilizer and herbicide application and less disease from direct contact of fruit with the ground. Skirting is also required for mechanical harvesting of fruit, to allow visibility for trunk shaking and to provide room for the catch frame. Additional benefits include better distribution of fertilizer under the tree and easier maintenance of under-tree irrigation systems.

Crop Load and Fruit Size and Quality

Hedging and topping may also be used as a means of reducing crop load in special cases. Murcott and some other mandarins tend to set excessive crops some years, and the heavy crop load results in poor fruit size and quality and may even cause the death of the tree. A very heavy crop one year can also cause alternate bearing, in which heavy and light crops alternate for a number of years. Both hedging and topping performed early in a heavy crop year are effective in reducing crop load by cutting off a significant portion of the newly forming crop. In contrast, pruning during a light crop year may increase the problem of alternate bearing by reducing a small crop even more. Hedging or topping (or both) to reduce crop load can be done before fruit set if an excessive crop is expected, or it can be delayed until the amount of fruit set can be more accurately determined. However, pruning should be done by May or early June (in the Northern Hemisphere), since later removal of fruit could result in a crop reduction without the benefit of an increase in fruit size. Hedging or topping later in the season may help prevent the collapse of Murcott trees with excessive crop loads.

Pruning that substantially reduces crop load generally increases fruit size and lowers the soluble solids content. The increase in fruit size is usually desirable for fresh fruit, but the loss of solids reduces the value of fruit grown for processing. Severe pruning can create a juvenile growth habit, in which trees produce less fruit, and the fruit is of poor quality, juice content is reduced, internal quality is poor, and external color and peel characteristics are poor; the fruit quality is like that generally found in young trees.

Pruning After Freeze Damage

Decisions about pruning groves with severe freeze damage are best delayed until the full extent of the damage can be assessed. New growth will appear, but dieback may continue for some time. Several months may be required to determine the full extent of the damage. If freeze damage extends into the primary branches down to the trunk, the trees should be removed and the grove replanted. Selective pruning to remove dead wood is desirable but costly. Hedging and topping are used as an alternative to hand pruning. Cuts should made in live wood, either by hand or by mechanical pruning.

Fruiting

Careful selection of a site, along with good orchard design and effective management, should result in a production of an economic quantity of quality fruit. The final crop load, however, depends on the amount of flowering and fruit set during the year and the amount of fruit drop prior to harvest. A number of management decisions affect both the quantity and quality of the fruit.

Flowering and Fruit Set

Citrus is grown in the tropics and subtropics in a range of climates, in soil of different types, and at various elevations, both north and south of the equator. Citrus is relatively insensitive to day length, and induction of flowering usually requires some quiescent period, when conditions are not favorable for growth. This quiescent period is generally induced by a period of water stress or low temperatures. The cool winters in Florida and California provide the needed induction and result in a single major bloom each year. Under more tropical conditions, with no true quiescent period, several minor induction periods may result in flowering more than once a year. In this case several crops may exist on the tree at one time. Sites suitable for citrus production must provide some induction of flowering as a prerequisite for satisfactory production.

Flowers generally form on growth from wood that is one year old, either on a vegetative sprout (leafy bloom) or directly on the older wood (leafless bloom). A higher percentage of set occurs on leafy bloom than on leafless bloom. Young trees and trees with excessive vigor frequently do not flower well. Excessive leaf drop due to diseases or freeze damage will reduce flowering.

Citrus produces many more flowers than needed for a good crop of fruit. Generally less than 5% of the flowers set and produce fruit, and in some cases as little as 0.2% of flowers produce the final crop. The reduction in potential crop load takes place in several steps. The largest drop occurs during and shortly after flowering, when flowers and very young fruit drop as a result of the separation of the pedicel from the twig. This is a natural drop, but it may be increased by water stress or nutrient deficiency.

Natural adjustment of crop load

Early drop
- **Flowers and very young fruit**

May–June drop
- **Pea-sized to marble-sized fruit**

Preharvest drop
- **Fruit near maturity**

Another drop occurs in May (in the Northern Hemisphere), when the fruit ranges from pea-sized to marble-sized. This is often called June drop, because it parallels the well-known June drop of deciduous fruits. For citrus it is the final major drop and largely determines the eventual crop load. This drop involves the separation of the pedicel from the young fruit. Water stress and high temperatures at this time will enhance drop and reduce the final crop. In Florida, this is a critical time for irrigation, since May is usually hot and mostly dry.

Additional fruit drop, called preharvest drop, occurs prior to harvest. The amount of fruit affected varies from year to year and among varieties. It usually ranges from 5 to 15%, but in Murcott it can reach as high as 50%. Fruit at this late stage of development drops for various reasons. Fruit splitting, which begins in late summer or early fall, causes a significant loss of fruit. Fruit drop may also be due to disease or mechanical damage caused by equipment traveling through the grove. In some varieties, a physiological drop occurs, with no apparent cause. Navel oranges go through several periods of drop during the summer and early fall. The cultivar Pineapple orange frequently begins a substantial natural drop when the fruit reaches maturity in January or February. Fruit drop of navel and Pineapple oranges can be reduced the use of an auxin plant growth regulator spray.

The critical time for obtaining adequate fruit set and optimal crop load is during early flowering and June drop. Good soil moisture should be maintained during these periods to avoid water stress. It is recommended that available soil water be maintained above one-third depletion during these periods (see Chapter 7).

Pollination

Most citrus varieties require pollination for fruit production. Self-pollination is satisfactory for most varieties. It occurs naturally by contact between the anthers and the stigma, by pollen that falls on is blown against the stigma, or by the transfer of pollen by bees or other insects. However, certain varie-

Table 5.3. Effectiveness of pollenizer varieties for pollination of selected self-incompatible varieties[a]

Pollinated variety	Pollenizer variety: Temple	Or- lando	Sun- burst	Lee	Nova	Robin- son
Robinson	++	++	+	+	–	–
Nova	++	++	+	++	–	?
Sunburst	++	++	–	?	++	++
Orlando	++	–	+	++	++	++

[a] ++ = Very good; + = good; – = poor; ? = unknown.

ties are self-incompatible or pollen-sterile and require the presence of some other pollen source and bees to transport the pollen, in order to obtain good fruit production. These varieties include Clementine mandarin, Orlando tangelo, Minneola tangelo, and Sunburst tangerine.

In grove design, it is important to include trees as pollen sources if the variety to be planted is self-incompatible. There are a number of different arrangements for including pollenizers in a grove. Generally, a tree requiring another pollen source should be no farther than two trees away from the pollen source. This can be accomplished by placing a pollenizer in every third tree space in every third row. However, this is an inconvenient arrangement for harvesting, since the pollenizers are isolated individual trees, which may mature earlier or later than the variety being pollinated and may be harvested separately from it. Another planting arrangement, which is more convenient at harvest, involves two rows of pollenizer trees alternating with four rows of the variety being pollinated.

Insects are needed to move pollen from a pollenizer to the tree being pollinated, since wind is not effective. Honey bees are the best pollinators available. Generally, one colony of bees per 2 acres, properly spaced on the edge of the grove, is an adequate number. Insecticides hazardous to bees must be avoided during the bloom period, while they are in the grove. The bloom period of the pollenizer trees must overlap with that of the trees to be pollinated.

Table 5.3 lists several varieties requiring pollination with an evaluation of the effectiveness of several different pollenizers. Temple orange is an effective pollenizer for most self-incompatible varieties, but it has limited market potential and is quite susceptible to cold injury and citrus scab. There are a number of other possible pollenizers to select from, and they should provide a satisfactory source for most situations.

Alternate Bearing

Alternate or biennial bearing is the tendency of fruit trees to produce a heavy crop one year (an on-year) followed by a light crop the next year (an off-year). It may occur on only part of a tree, on a whole tree, in a block of trees, or over an entire region. The severity of alternate bearing varies with the variety and the location, but it is observed in all citrus-growing regions of the world. In Florida, alternate bearing is a problem in some mandarin varieties. It is usually a minor problem in oranges, but it can be important in Pineapple and Valencia oranges. It is rare in Hamlin orange.

The causes of alternate bearing are incompletely understood, but a number of climatic and management factors may be important. Environmental stress that greatly reduces a crop one year is an important trigger of alternate bearing. A significant freeze or an unusually hot and dry period during June drop can result in an abnormally light crop. Pests and diseases, such as postbloom fruit drop, can also alter crop load substantially. Alternatively, unusually favorable growing conditions may result in an abnormally heavy crop one year. Either a heavy crop or a light crop may induce alternate bearing.

Management practices clearly affect alternate bearing. Any practice that reduces flowering or fruit set during the on-year reduces alternate bearing. The converse is also true. Increasing flowering and fruit set during an off-year is generally unsuccessful. Several practices are useful in reducing the crop load during an on-year. Fruit may be thinned by hand or by fruit-thinning sprays. Hedging, topping, and selective pruning are effective in crop load reduction. In fact, the primary purpose of hedging and topping of some varieties grown for the fresh fruit market is to reduce crop load and increase fruit size. It is equally important to avoid hedging or topping that would reduce crop load during a light crop year. Fruit-thinning sprays are most effective in reducing crop load when applied during the June drop. This application appears to accentuate the natural drop occurring at that time. Water stress during this period may also cause more fruit drop and reduce crop load. Nutrition should be reduced in a light crop year and increased in a heavy crop year.

The importance of alternate bearing depends on its severity and duration. In extreme cases, such as Murcott, it is so severe that the tree dies. Unusually heavy crop loads result in smaller fruit and delayed maturity, both of which are problems in the fresh fruit market. Alternate bearing on a regional scale reduces not only marketable fruit but also total fruit production.

There are significant economic benefits from regular cropping. A regular cash flow is preferred over one that fluctuates from year to year. Induction of alternate bearing by some climatic event can continue for a number of years before production stabilizes again. Steps should be taken to minimize alternate cropping. Most varieties are self regulating. When regular cropping resumes, special management is not required until some unusual event triggers another light or heavy crop.

Plant Growth Regulators

A few plant growth regulators are available for use on citrus and are used to a limited extent. The use of plant growth regulators and regulations concerning them may vary around the world. A sprout inhibitor applied to newly planted citrus reduces sprout formation along the trunk. This will be more beneficial as the heading height of trees is increased to accommodate mechanical harvesting. Fruit set of a few varieties that set fruit poorly can be increased by the application of gibberellic acid during full bloom. There are also fruit-thinning sprays for varieties that set excessive crops. Gibberellic acid applied prior to harvest also improves peel firmness and color. This is particularly useful on grapefruit harvested late in the season for the fresh fruit market. Fruit drop of navels during the summer and preharvest drop of oranges and grapefruit can be reduced by the application of an auxin spray. The combination of auxin and gibberellic acid is effective in extending the harvest season for grapefruit.

Fruit Quality

Fruit quality is determined by the integrated effects of many factors, including climate, soil, rootstock, nutrition, irrigation, disease, arthropod and nematode damage, tree pruning, crop load, the position of the fruit on the tree, mechanical damage, and tree age. Most of the effects of these factors are discussed in other chapters.

Fruit of very young trees is always of inferior quality. This fruit tends to be large and may have poor external color, a coarse peel, and a low concentration of soluble solids in the juice, in comparison to the fruit of mature trees. Anything that causes trees to revert to a juvenile condition will cause similar effects on fruit quality.

Grove design has minor effects on fruit quality in most cases. However, a grove design in which the allocated space is not matched with tree vigor may be a cause of poor-quality fruit on mature trees. Trees that are too confined by hedging and topping revert to a more vegetative or juvenile state and produce small quantities of poor-quality fruit. Hedging or topping that substantially reduces crop load (or any other factor reducing crop load) increases fruit size and reduces the concentration of soluble solids in the juice.

Planning for the Future

The life of a citrus grove extends over a number of years, so design of a new grove must incorporate the latest information and plant materials and anticipate changes in technology during the life of the grove. Variety selection should be based on predicted market potential. Citrus juice has become an international commodity, so production and projections of future production throughout the world affect the market potential in any local area (see Chapter 2). The local market potential for fresh fruit may also be important. In many cases, however, the popularity of a particular variety results in excessive planting of that variety and a reduced fruit price as recent plantings come into production.

New genetic material should be considered in the design of a new planting. In particular, rootstocks which provide some degree of control over tree size may be considered. With proper matching of the variety, the rootstock, the intrinsic vigor of trees at a site, and tree spacing, it should be possible to minimize requirements for hedging and topping in the future.

Currently, virtually all citrus fruit is hand-harvested (Plates 22–25). Mechanical harvesting of processing fruit is expected to become important in some citrus-producing areas. Trunk shaker and catch frame systems may be used extensively for citrus harvest. This requires a trunk free of branches to a height of 30 inches or more. Thus, it appears prudent to plan for this by growing nursery trees headed at a height of 30 inches or more.

A new grove must be designed to provide maximum economic return averaged over the expected life of the grove. This requires everything from global knowledge of trends in citrus production to detailed knowledge of a particular site and the growth characteristics of the genetic material being used.

Ashok K. Alva
David P. H. Tucker
University of Florida
Citrus Research and Education Center, Lake Alfred

CHAPTER 6

Soils and Citrus Nutrition

Soil conditions, particularly fertility factors and drainage, have a major influence on citrus tree health, production, and fruit quality. The fertility of the soil is highly dependent on the soil type and climate.

The soil provides three major components for plant growth: solid phase, liquid phase, and gaseous phase. The solid phase includes organic and inorganic particles, which contain a reserve of nutrients, in organic and inorganic forms. The mineralization of organic sources supplies nutrients in forms available to plants. The liquid phase of the soil is responsible for nutrient transport in the soil. Nutrients may be applied to the soil as organic or inorganic compounds. The decomposition and mineralization of organic compounds and the dissolution of inorganic compounds convert them to soluble forms, which are maintained in the soil solution. These constitute the pool of nutrients available for plant uptake in the short term. The soil solution, however, is in equilibrium with the reserve pool, so that as nutrients in the soil solution are depleted by plant uptake, they are replenished by the reserve pool. The soil gaseous phase is in equilibrium with the atmosphere. It can influence root growth and the survival of numerous organisms in the soil.

The growth of healthy citrus trees requires soil with an optimal combination of physical, chemical, and biochemical properties. The soil should supply adequate nutrients for citrus trees, either from the reserve pool or from applications of fertilizers and manures. The primary objective of soil management is to provide optimal soil conditions to improve nutrient uptake efficiency while minimizing nutrient losses.

Major Soils in Citrus Production

Citrus is grown in a range of soils, varying with the geographical location of the production region. Citrus trees can adapt to a wide range of soil conditions, but they grow best in sandy to clay loam soils. Planting citrus in soils unsuited for its production, such as poorly drained, highly acidic, or calcareous soils, results in various problems. The properties of the soil at the planting site and, to some extent, cultural practices influence the growth habit of citrus trees, particularly root distribution. For example, some early studies reported citrus roots at depths of 19 to 25 feet in deep sandy soils in Florida, typically Entisols with sand content often close to 96% and no restrictive soil horizons. However, more recent studies have shown that most citrus roots are within the top 6 to 12 inches. With increased adaptation to microirrigation, root distribution is gradually modified so that the roots occupy a rather shallow layer of soil, which is kept well watered and contains most of the nutrients supplied.

Adequate soil drainage is an important factor in the growth of citrus trees. The accumulation of free water in the root zone results in poor aeration and eventual injury to roots. In regions of heavy rainfall and in soils with impervious subsoil horizons, raised beds are necessary to provide adequate drainage. Planting in raised beds is a common practice in Spodosols and Alfisols, which are predominant in the southern and eastern regions of the Florida citrus belt. These soils contain clay or organic pan, and consequently the water table fluctuates and can rise very close to the surface at some times of the year.

Florida

Most Florida citrus is grown in soils with inherently low fertility and low cation exchange capacity, which are thus unable to retain more than minimal amounts of soluble plant nutrients against the leaching action of rainfall and irrigation. Among those soils, 23 soil series are considered to be most susceptible to the leaching of soil-applied agrichemicals and are designated "vulnerable" by the Natural Resources Conservation Service.

Soils classified as Entisols are deep, well-drained sands low in mineral nutrients and organic matter, having no confining soil horizons. The main advantages of such soils are good drainage, good aeration, and a deep root zone. Often drainage is excessive, resulting in soil moisture deficits and a need for frequent irrigation. The uniformity of the soil, the absence of hardpan, and a low water table allow the development of extensive root systems. The extensive rooting zone enables trees to partially overcome the low water- and nutrient-holding capacities of these soils to some extent.

Many of the early citrus groves along the lower coasts of Florida were planted in Alfisols, fine-textured lowland soils with a sandy surface, containing appreciable organic matter and overlying fine-textured, often calcareous marine deposits. Alfisols were initially sought for citrus production because of their high organic matter and clay contents, which provide greater cation exchange capacity, retention of nutrients, and water-holding capacity. However, the calcareous nature of these soils presents problems with micronutrient availability. In recent years planting has been more extensive in Spodosols, which are sandy, acidic, coarse-textured soils containing a spodic horizon (organic hardpan). Spodosols require incorporation of lime, in contrast to the calcareous Alfisols, in which alkaline materials are usually present in the rooting zone in bedded citrus groves.

The principal advantages of these soils are their good natural supply of nutrients, high cation exchange capacity, and high water-holding capacity. The high cation exchange capacity permits better utilization of the naturally present nutrient elements. This is reflected in the vigor usually exhibited by trees in the early stages of growth. Many sites with poor drainage, high alkalinity, and high clay content due to the deposition of subsurface materials over the natural surface horizon during the bedding process can have rootstock-specific problems.

Brazil

Most of the citrus acreage in Brazil is in São Paulo State. Citrus is produced on a range of soils, including sandy (with less than 10% clay) to clay-type soils (50 to 65% clay). The major portion of the citrus acreage is on sandy loam soils characterized by 10 to 30% clay content. They provide a deep, well-drained environment for root growth. As citrus is often grown on rolling hills with slopes of 1 to 10%, minimizing soil erosion is an important consideration during land preparation.

Israel

Citrus production in Israel, in the past, was predominantly on red sandy loam soils of the coastal plain. These soils, of pH 6.5 to 7.0, contain an impervious horizon high in reduced iron. With gradual urbanization of this region, currently only about 30,000 acres of citrus remains in these soils, and these plantings are also vulnerable to increasing development.

At present, most of the citrus production is on heavy-textured soils along the coastal plain and in inland valleys. Yields of crops grown in these soils have traditionally been low but have improved considerably in recent years, with improved irrigation and fertilization management. Commonly used rootstocks include sour orange, Troyer citrange, Volkamer lemon, Cleopatra mandarin, and Rangpur lime. Because of the heavy texture of these soils, drainage is poor, and salinity can be limiting.

Along the Jordan Valley, where soils are usually calcareous, the main problem is small fruit. The general recommendation for plantings in these soils is to prune the tops of trees after the harvest and then begin fertilization.

The main citrus-growing region at present is along the northern Negev region, which is an arid environment, with less than 10 inches of rainfall annually. This area contains two major soil types. The first type is characterized by lack of structure, high pH (above 8.0), and calcium carbonate concretions at depths of 15 to 20 inches; this soil crusts easily and is prone to compaction when wet. The second type comprises sandy soils derived from the Nile Delta, also characterized by high pH (8.0–8.5). In spite of the alkalinity of these soils, high yields can be obtained with proper management.

South Africa

Citrus in South Africa is produced on well-drained soils which provide a well-aerated rooting environment of at least 24 inches. The clay content of the soils in citrus production varies from 7 to 55%, but in most cases it is below 15 to 30%.

Australia

Citrus in Australia is produced on a range of soils, including sands, loamy sands, and clay loams. The major citrus-growing regions are the Sunraysia and Riverland areas along the Murray River, where light-textured soils are common, and the Murrumbidgee or Riverina area in New South Wales, where heavier-textured soils predominate. Land preparation includes the installation of tile drains above impermeable clay layers and raised beds for planting. It is customary to grow green manure crops or apply organic matter. The most common cover crops include lupines, field peas, and oats planted in autumn. Residues are worked into the soil in the spring. Permanent sod culture in citrus groves is also gaining wide acceptance. The sod is planted in the autumn and is mowed periodically, and the cut material is left on the surface to break down and increase the organic matter content.

Belize

Citrus is an important commercial crop in Belize, with close to 57,000 acres of Valencia oranges and grapefruit produced for export. Most of the citrus production is on alluvial Ultisols. The pH of the natural soil is 4.0 to 5.0, and hence liming with about 1.0 to 1.5 tons of lime per acre is necessary to raise the pH to at least 5.5 to 6.0. In recent years, citrus acreage has expanded to marginal acidic, infertile, and poorly drained soils. Such limitations have resulted in low production and low net returns.

Physical Properties of Soils

Soil Water

The amount of soil water directly affects the growth of citrus trees and microorganisms. Water is the substrate for photosynthesis and is also important in the transport of nutrients in the soil and within the tree. The availability of water to citrus trees depends on the total water content of the soil and the extent to which the water is bound to soil particles, referred to as soil water potential. Water in a free state has a high water potential, and it flows from areas of high potential to areas of low potential. Water potential is difficult to measure; therefore, water content is often used as an indicator of the water status of soils. Water content is expressed as the ratio of the mass of water present in a sample to the mass of the solids dried to a constant mass at 220°F. This ratio is generally referred to as the gravimetric soil water content. Soil water content can also be expressed as the ratio of the volume of soil water to the volume of soil in a sample. The volumetric soil water content can be calculated by multiplying the gravimetric water content by the bulk density. For most sandy soils the bulk density is 1.5

grams per cubic centimeter. Therefore, the volumetric water content is generally 1.5 times the gravimetric water content.

The field capacity of a soil is the amount of water held in the soil against gravity after a drainage period of 24 hours following rainfall or irrigation that saturates the soil. The soil moisture content at field capacity is generally considered to be the water potentially available to plants. The water content at field capacity varies considerably, depending on the soil type. Fine-textured soils can hold a greater amount of water than light-textured soils. Most soils used for citrus production, particularly in Florida, are light-textured sandy soils, with very low water-holding capacity. Efficient irrigation management is the key to providing adequate amounts of water available to citrus trees and minimizing the transport of water to depths below the rooting zone. This is especially important in light of increased emphasis on the development of best management practices (BMP) for irrigation to minimize leaching of agrichemicals which may contaminate the groundwater. Guidelines for efficient irrigation management are presented in Box 6.1.

The availability of adequate water to avoid stress is an important consideration in soil management. It is equally important to avoid prolonged periods of excess soil water. Excess soil water causes a lack of oxygen in the root environment, which can cause root injury or death, resulting in stunted trees with sparse foliage, small leaves, and low, small fruit (Plate 26). Deficiencies of nitrogen and other nutrients are also expressed in cases of chronic water damage. Water damage can occur in patches, depending on the distribution of low spots and areas where drainage is imperfect because of the proximity of hardpans to the soil surface.

Soil Structure

Soil structure is the arrangement of solid particles into aggregates. Soil structure affects the soil water content and gas exchange. The formation of soil aggregates is dependent on the colloidal content of the soil. Sandy soils are low in colloidal content, have poor structure, and have low reserves of nutrients. The relative abundance of calcium ions in soils in relation to the pool of other cations is a key factor affecting soil structure, as calcium binds organic and inorganic particles in the soil. A soil with good structure should have approximately 50% pore volume. Citrus tree growth is optimal if the coarse pores are filled with air and the fine pores are filled with water.

Soil Air

Oxygen is an essential component of the soil atmosphere. The respiration of plant roots depends on the oxygen supply in the soil. Respiration provides the energy for various metabolic processes, including active ion uptake by plant roots. Oxygen deficiency can directly affect carbohydrate metabolism and the synthesis of phytohormones in citrus roots. The quantity of soil air decreases with an increase in the water content in the soil, since the air which normally fills the larger soil pores is replaced by water. Accordingly, after a heavy rainfall or prolonged irrigation, the aeration of the soil decreases, especially in heavy-textured soils, favoring anaerobic conditions in the soil.

Soil Temperature

Citrus trees require high ambient and soil temperatures for optimal tree and root growth. The soil temperature is dependent on the ambient temperature, soil water content, and other physical and chemical characteristics of the soil. It is relatively high if the soil water content is high. Sandy soils drain very rapidly, so that the soil moisture fluctuates, and accordingly the temperature may also fluctuate drastically within a short period of time. Soil biological processes may also affect the soil temperature.

Chemical Properties of Soils

Soil testing is generally defined as any chemical or physical measurement of a soil. In common usage, it refers to rapid chemical analyses to assess the status of nutrients available to the plant. The ultimate aim of soil testing is to evaluate the test results as basis for the application of fertilizers and other soil amendments to improve conditions for optimal tree growth. The success of a soil-testing program depends on the availability and quality of research data for the calibration and interpretation of the test results (Box 6.2). Among the 16 elements essential for most crop growth, the three major elements (macronutrients)—nitrogen (N), phosphorus (P), and potassium (K)—are generally deficient in most soils. In extremely sandy soils with low organic matter content, a soil test for nitrogen is of limited value, because of the low concentration of residual nitrogen. However, soil tests are most useful for determining the P and K contents and soil pH.

Soil pH

Soil pH is one of the most important measurements in standard soil analysis, as it influences many chemical and biological reactions. The availability of most nutrients is affected by soil pH (Fig. 6.1). The ideal pH for optimal availability of most nutrients is in the range of pH 6.0 to 7.0. The major effect of soil pH is the substantial decrease in the availability of certain micronutrients, including iron (Fe), manganese (Mn), zinc (Zn), and copper (Cu), at pH greater than 7.0. Accordingly, micronutrient deficiencies are most acute at pH greater than 7.0, as shown in a case of severe Mn deficiency in Hamlin orange trees grown in high-pH soil (Plate 27). In contrast, the availability of molybdenum is lower at pH less than 6.0.

Soil pH is measured in an aqueous matrix, in which soil is mixed with either water or a dilute salt solution. Measurements made with water are closer to the pH of the soil solution in the field, giving the true pH of soils with low electrical conductivity, such as unfertilized soils or extremely sandy soils which are highly leachable. The electrical conductivity of a weak saline solution (0.01 M $CaCl_2$) is similar to that of most agricultural soils. However, the addition of salt to the matrix solution lowers the measured pH by about 0.5 unit, compared to the pH measured in water. In the procedure followed by the University of Florida Soil Testing Laboratory, 20 grams of soil is mixed with 40 milliliters of deionized water (a 1:2 ratio of soil to water). The suspension is stirred and allowed to stand for 30 minutes, and the pH is measured while the suspension is being stirred.

Citrus trees can grow adequately with pH in the range of 5.5 to 7.5, depending on the rootstock and other conditions. In the major citrus production regions of Florida and California, the soil pH is maintained at 6.0. However, it may be raised to 6.5 if the soil contains toxic amounts of Cu (Box 6.3).

The soil pH in citrus production areas in São Paulo, Brazil, ranges from 4.8 to 5.5. Raising the pH of these soils is very difficult, because of their high aluminum (Al) content. High

Box 6.1

Guidelines for Efficient Irrigation Management

In most agricultural soils, field capacity is considered to be the water content at 0.33 bar. However, in extremely sandy soils, such as those in most citrus-producing regions in Florida, in which the sand content is greater than 95%, field capacity is considered to be the water content at 0.1 or 0.08 bar.

A soil moisture characteristic curve represents the relationship between soil moisture suction and gravimetric water content, as shown in the figure below for a typical Entisol (Tavares fine sand) and a typical Alfisol (Riviera sand). At a given level of soil moisture suction, the gravimetric water content is greater in a Riviera sand than in a Tavares fine sand. A clear understanding of the soil moisture characteristic curve is important in irrigation scheduling. The gravimetric water content decreases quite rapidly as soil moisture suction decreases below 0.1 bar. As the curves shown in the figure demonstrate, these soils require lighter, more frequent irrigation than fine-textured soils. Because of the low water-holding capacity of these soils, prolonged irrigation simply moves the water front deeper in the soil profile. Irrigation which pushes the water front below the rooting depth is a wasteful application of water and contributes to the leaching of soil-applied chemicals, carrying them below the rooting depth. This decreases the nutrient uptake efficiency of the plant.

Tensiometers are used in scheduling citrus irrigation. In most sandy soils, the recommended tensiometer set point for scheduling irrigation is 10 to 15 centibars in the top 6 to 12 inches of soil. As shown in the figure below, the gravimetric water content of a Tavares fine sand at field capacity (8 centibars) is 6.2%; at the tensiometer set point of 15 centibars, it is 5.0%. As soil moisture is depleted from 6.2 to 5.0%, irrigation becomes necessary to replenish the water deficit. The gravimetric water content at the driest point in this soil, i.e., the wilting point (15 bars), is 1.1%. Therefore the available soil moisture content (field capacity minus wilting point) is 5.1%. Accordingly, the decrease in soil moisture content at 15 centibars of soil moisture suction corresponds to a 25% depletion of the available soil moisture.

Once we know the depletion level, it is important to determine the length of irrigation needed to replenish the water deficit. Consider mature citrus trees with an effective rooting depth of 36 inches in a typical sandy soil, irrigated with 360°-coverage emitters delivering 22 gallons per hour and wetting an area 16 feet in diameter. At 25% depletion, under these conditions, 3.5 hours of irrigation is necessary to replenish the water deficit, raising the soil moisture content back to field capacity to a depth of 36 inches within the wetting area of the emitter (if the moisture depletion is uniform throughout the entire 36-inch depth).

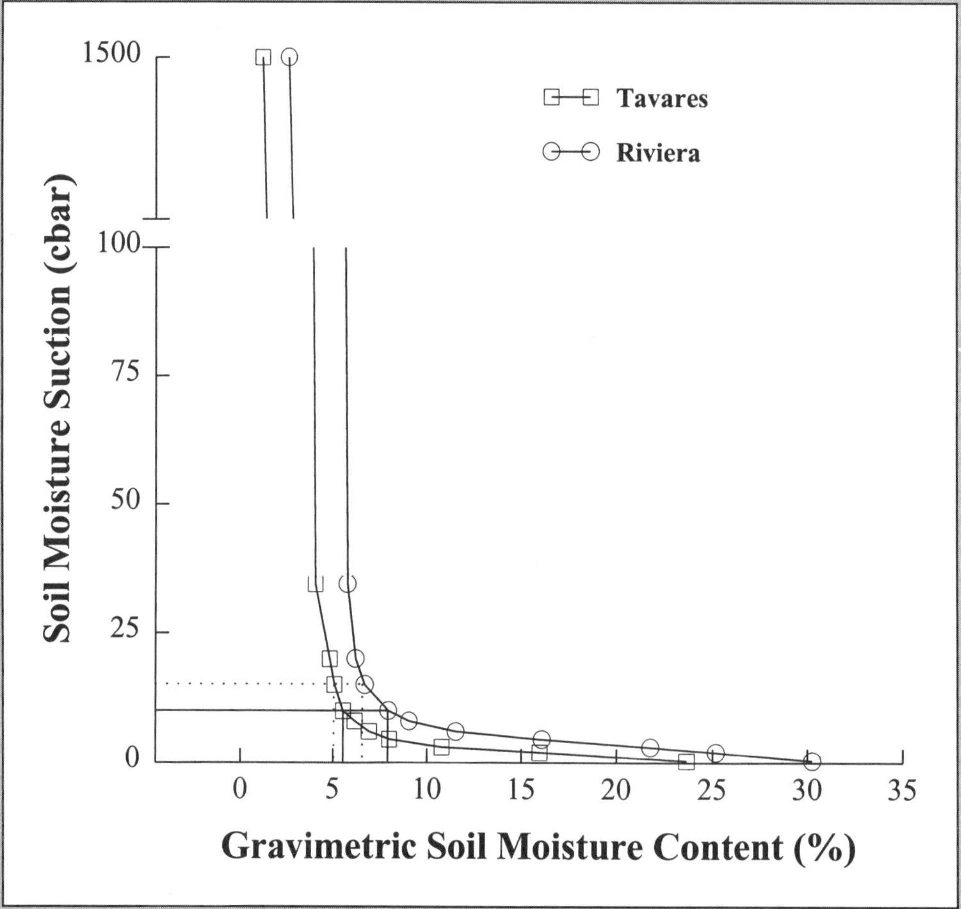

Soil moisture characteristic curves for a typical Entisol (Tavares fine sand) and a typical Alfisol (Riviera sand).

levels of Al restrict root growth, confining many of the fibrous roots to the top 12 to 18 inches of soil. Tree growth and production are satisfactory in soils of pH 5.5 or above, and accordingly no liming is recommended. In new plantings, the application of about 5,000–6,000 pounds of limestone, mixed with the top 24 to 36 inches of soil, is generally recommended.

Citrus in South Africa is grown in soils with a wide range of pH, from 4.0 to 8.5. However, maintaining the soil pH at 6 to 7 is generally recommended. Soil pH is corrected by the application of either dolomitic or calcitic limestone, depending on the calcium and magnesium status of the soil.

Soil Copper

High levels of Cu and other metals in agricultural soils can originate from various causes, including industrial wastes. In parts of Chile, injury to citrus trees has been reported in soils contaminated with Cu from mining operations. In Florida, the Cu content of soils under continuous citrus production is considerably greater than that of similar soil series in their natural state, as a result of the historical application of Cu fungicides for disease control. Copper in the soil is adsorbed by organic matter and iron oxides, and therefore it is not mobile in soil.

The accumulation of Cu in soils results in phytotoxicity. The most common symptoms of trees grown in high-Cu soils is Fe chlorosis, particularly when groves are replanted with trees on Cu-sensitive rootstocks. Raising the soil pH often overcomes Cu toxicity (Box 6.3).

Since Cu in soils is associated with organic matter, most of it is in the surface layers. Therefore, analysis of the top 6 inches of soil is adequate to determine the Cu concentration. Double acid (Mehlich 1) is the common extractant used for analysis of soil Cu in Florida.

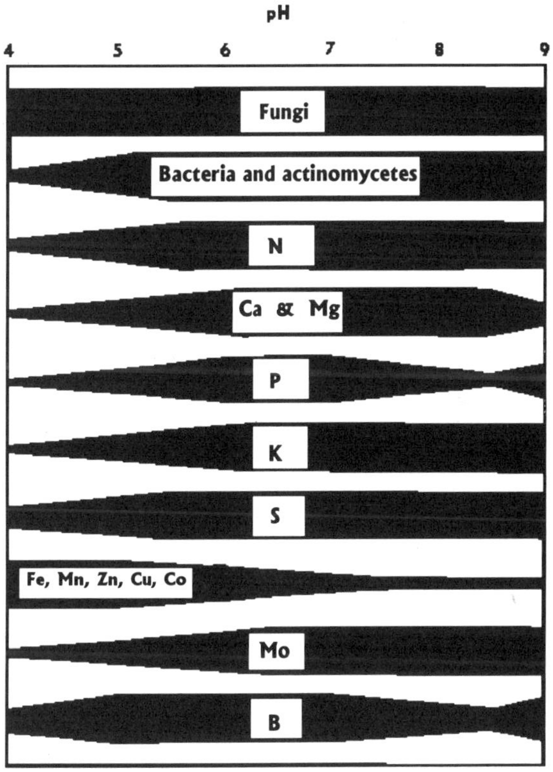

Fig. 6.1. Effects of soil pH on the growth of soil microbes and the availability of nutrients. (From N. C. Brady and R. R. Weil, 1999, The Nature and Properties of Soils, 12th ed., © 1999, reprinted by permission of Prentice-Hall, Inc., Upper Saddle River, N.J.)

Box 6.2

Guidelines for Soil Sampling

The sampling procedure is an important step in ascertaining the value of soil analysis and making valid interpretations of the analytical results. Soil variability is the major factor determining the size of the sampling unit. The general rule of thumb is to divide the grove into sampling units of no more than 20 acres. Each sampling unit should contain similar soil series and trees of similar scion-rootstock combination. In large groves, sampling each 20-acre unit may not be feasible, and indicator blocks may be used to minimize the number of samples. Indicator blocks represent a much larger area and are sampled repeatedly, year after year, to follow changes in chemical properties of the soil in response to corrective measures adopted on the basis of the results of the previous analysis.

Within a sampling unit, soil samples are taken with a core sampler (1 inch in diameter) to a depth of 6 inches along the drip line of about 15 to 20 representative trees. The soil cores are mixed to form a composite sample and are air-dried before being sent to an analytical laboratory.

In microirrigated groves, samples should be taken within the wetting area of the sprinkler or drip emitter. Deep-aquifer water contains high levels of calcium and bicarbonates. The continuous addition of bicarbonates in irrigation water results in elevated pH of the soil in the wetting zone, compared to soil outside that area. Therefore, in microirrigated groves, soil outside the wetting area should occasionally be sampled, to evaluate differences in pH and other characteristics.

Box 6.3

Recommendations for Soil pH

The following recommendations for target soil pH have been developed on the basis of the amount of copper (Cu) extracted from soil samples by Mehlich 1.

Amount of extractable Cu (pounds per acre)	Recommendation
<50	Cu toxicity is unlikely in citrus, even in relatively acid soils (pH < 5.5)
50–100	Cu toxicity is possible if the soil pH is below 5.5. Liming is recommended to raise the soil pH to 6.0
>100	Cu toxicity is likely unless the soil pH is increased to 6.5

Managing Soil pH

Lime is commonly applied to raise the soil pH and overcome the potential adverse effects of high acidity and the associated excessive levels of aluminum and hydrogen ion toxicity (Box 6.4). The effects of aluminum toxicity on plant growth are pronounced only at pH less than 4.5.

Gypsum as a Source of Calcium

From the nutritional standpoint, gypsum supplies both calcium (Ca) and sulfur (S). Citrus is considered a strong "feeder" of Ca. Since most citrus growers follow a routine liming program or use irrigation water containing lime, Ca is normally supplied in sufficient quantity. If the soil pH of a citrus grove is above 6.0, liming is not recommended. However, since citrus trees require an abundance of Ca, trees on these soils may benefit from supplemental Ca. Under these conditions an alternate source of Ca (other than liming materials) which can supply this nutrient with minimum effect on soil pH is desirable. Gypsum is a suitable source in this situation. Liming as a source of Ca is not justifiable when a cheaper and more soluble Ca amendment such as gypsum is available. Another advantage of gypsum is its ability to supply Ca to subsoil horizons. In their natural state, subsoils are usually acidic and extremely deficient in available Ca. The incorporation of lime in the topsoil has little effect on subsoil Ca, since lime generally moves slowly through the soil profile. Gypsum is more soluble than lime, however, and surface application of gypsum can result in improved Ca availability in subsoils.

Box 6.4

Practical Considerations of Liming Programs

In addition to pH control, liming provides a source of calcium (Ca). Symptoms associated with Ca deficiency in citrus include stunted growth, twig dieback, defoliation, multiple buds, leaf yellowing, rounded or heart-shaped leaves, root rot, decreased yields, and small fruit.

Various liming materials are on the market, with somewhat poorly defined circumstances under which one material may be more appropriate than others. Although the word *lime* strictly means calcium oxide, in practice it is used to refer to all compounds of calcium and magnesium (Mg) employed as soil amendments to raise the soil pH. These compounds include carbonates as well as oxides and hydroxides. The most commonly used liming materials are carbonate forms of calcium or magnesium. Calcium carbonate is referred to as *calcitic lime,* while calcium–magnesium carbonate is *dolomitic lime.* A calcite produces a greater increase in soil pH than a dolomite.

In treatments with either the calcitic or the dolomitic form, a soft-rock source of limestone produces a greater increase in pH than a hard-rock source, and soft-rock calcite produces a greater increase in the soil Ca level than hard-rock material. Calcite is an effective source of Ca, and dolomite is an effective source of Mg for citrus. Therefore, dolomite is more appropriate in soils of pH less than 6.0 that also require Mg.

Box 6.5

Measurement of Cation Exchange Capacity

Weigh out 10 grams of an air-dried, ground, sieved (2-millimeter) soil sample into a 100-milliliter centrifuge tube. Add 40 milliliters of 1 M ammonium acetate. Stopper the tube, and shake it for 5 minutes on a reciprocal shaker at 115 rpm. Centrifuge the suspension for 5 minutes at 3,000 rpm. Filter the supernatant through Whatman No. 42 filter paper, and determine the concentrations of calcium, magnesium, potassium, and sodium by analysis with an inductively coupled plasma emission spectroscope. The concentrations of the individual cations, expressed in centimoles per kilogram, represent the exchangeable cations. The sum of all exchangeable cations is the effective cation exchange capacity.

Calcium plays an important role in alleviating toxicities of metals such as Al and Cu. Since gypsum is an ideal source of Ca but does not raise soil pH, the application of gypsum can be beneficial for high Cu soils.

Cation Exchange Capacity

Cation exchange capacity is a measure of the amount of cations that can be adsorbed in exchangeable forms at negative charge sites in soils. In simple terms, it is the total number of exchange positions or the total amount of exchangeable cations expressed as centimoles per kilogram of soil. The clay content and organic matter content of the soil are major sources of cation exchange capacity. The soils in citrus production in Florida are low in clay (2 to 5%) and organic matter (0.5 to 1.0%) and therefore generally have low cation exchange capacity and low nutrient retention capacity, unlike most agricultural soils in the midwestern United States.

The principal method used to measure exchangeable ions is to saturate soil samples with an ion that forces the exchangeable ions already present on the charged surface into solution (law of mass action). Cation exchange capacity can then be calculated as the sum of the individual cations displaced from the soil (summation method). Alternatively, the ion used to saturate the samples, which is termed the index ion, can be displaced with a concentrated solution of a different salt, and the cation exchange capacity is then calculated as the amount of the index ion displaced (displacement method). The most common method used to measure cation exchange capacity is described in Box 6.5.

Anion Adsorption

Soil particles are predominantly negatively charged, but they may also carry some positive charge. The oxide surfaces of iron and aluminum and edges of clay minerals are generally positively charged and are accessible for the adsorption of anions. Anions can be adsorbed by the hydrous oxides of Fe and Al, clay minerals, Fe and Al organocomplexes, and Ca carbonates. Anion adsorption is important for tree nutrition in affecting the supply of ions such as phosphate and sulfate. Anions adsorbed by the soil, in decreasing order of the strength of adsorption, are phosphate; arsenate; selenite and molybdate; sulfate and fluoride; chloride; and nitrate. Nitrate is very loosely bound by the soil and is readily subject to leaching unless taken up by roots. The deep, sandy soils (Entisols) of central Florida do not have confining layers, and the downward transfer of water and soluble nutrients, including nitrate, is unrestricted.

Soil Organic Matter

The organic matter content of soils varies considerably, depending on the soil and environmental conditions, cropping, and soil management methods.

In warm, humid environments, organic matter breaks down rapidly in the soil, and very little of it accumulates from year to year under normal conditions. This is typical in Florida, and consequently most soils have a rather low organic matter content. In bearing citrus groves, the organic matter content is generally greater under the canopy than in the row middles, because of leaf litter and fibrous root decomposition. The recycling of organic residues is very rapid under Florida conditions, with high temperature and rainfall. This rapid turnover is the main reason for the difficulty in raising the organic matter content of Florida soils, despite the application of organic amendments.

In most citrus production, dry leaves and other vegetative parts are returned to the soil under the trees. These materials and fibrous roots decompose on a regular basis. Nitrogen released by the decomposition of organic debris supplies a significant portion of the nitrogen requirement for tree growth and fruit production. For a given climate, soil texture, and tree age, the net nitrogen available to the trees should be considered in developing a nitrogen budget. This budget can then be used to determine best management practices to improve the efficiency of nitrogen use and minimize the potential loss of nitrate due to leaching below the root zone.

For example, mineralized nitrogen under citrus trees in Florida, measured over a two-year period, amounted to 52–75 and 113–137 pounds per acre each year for four- and 20-year-old

In warm, humid environments, organic matter breaks down rapidly in the soil, and very little of it accumulates from year to year under normal conditions.

trees, respectively, planted in a Tavares fine sand (uncoated, hyperthermic, Typic Quartzipsamments, a typical Ridge soil) and 35–57 and 108–113 pounds per acre each year for seven- and 40-year-old trees, respectively, planted in a Wabasso sand (sandy, siliceous, hyperthermic Alfic Haplaquods, a typical Flatwoods soil). The transformation of potentially mineralizable nitrogen (i.e., total nitrogen in dry crop residue and in fibrous roots under the canopy) into inorganic forms (i.e., the percent mineralization) was greater in the Wabasso sand (41 to 86%) than in the Tavares fine sand (17 to 71%). Significant positive correlations were found between measured quantities of mineralized nitrogen and the potentially mineralizable nitrogen. There were also significant positive correlations between the amount of mineralized nitrogen and rainfall and between this amount and average temperature during each incubation interval of the study. This study demonstrated that climatic conditions in Florida are favorable for the rapid breakdown of organic residues, resulting in a rapid release of inorganic forms of nitrogen.

Cover crops can be grown, particularly in groves of young trees, to increase the soil organic matter content and improve related physical properties and to reduce the costs of weed management. For example, perennial peanut varieties offer considerable potential as cover crops in groves of young trees (Plate 28). These varieties produce profuse growth low to the ground and are fairly tolerant of drought. A cover crop grown in strips 4 feet wide on both sides of a row of trees helps to control weed growth. The cover crop should not be grown too close to the trees, to prevent competition for nutrients and water. As a result of the biological fixation of nitrogen by perennial peanut, this cover crop can also supply some nitrogen to the tree. Further research is needed to understand the benefits of cover crops in commercial citrus groves.

Salinity of Soils

Saline soils contain excessive concentrations of soluble salts, which are harmful to plants. These soils have a high degree of sodium saturation of exchange materials. The soluble

Box. 6.6

Estimation of Soluble Salt Content

Electrical resistance and conductance can be quickly and accurately measured and have long been used for estimating the soluble salt content of soil. Electrical conductivity (EC), expressed as mhos per centimeter or, commonly, as millimhos per centimeter (10^3 millimhos = 1 mho), increases with an increase in the salt content. Electrical conductivity can be measured in a saturated paste of soil or in a more dilute suspension of soil in water. The extraction of soil solution is easier in a dilute suspension of soil in water; however, the conductivity measured may not represent the salt concentration in soil solution under field moist conditions.

Salinity is most commonly described in terms of total dissolved salts (TDS): TDS (in parts per million [ppm]) = EC (in decisiemens per meter, which is equivalent to millimhos per centimeter) × 700. If the TDS of irrigation water is 3,000 ppm, 100 gallons of the water contains about 2.5 pounds of salts. The irrigation water quality requirement of bearing citrus trees exceeds the above quality by a factor of 3. Therefore, irrigation water with this TDS can add a substantial amount of salt over a growing season. In well-drained soil, salt is readily leachable, and accumulation of salt is not a concern. However, in poorly drained soil with a restrictive soil horizon, salts in irrigation water can accumulate in the soil above that horizon. If the salt content of irrigation water exceeds 1,000 ppm, precautions must be taken to avoid its direct contact with the foliage.

Excess salt in irrigation water or in the soil solution can adversely impact tree growth, resulting in tree stunting and defoliation. High sodium levels (above 0.25%) may cause leaf bronzing and reduced growth. Chloride at concentrations higher than 0.70% in leaf tissue is considered excessive. Leaf burn and defoliation can occur when the leaf chloride concentration is greater than 1.00%.

salts that occur in soils consist mostly of cations (sodium, calcium, and magnesium) and anions (chloride and sulfate) in various proportions. Constituents that ordinarily occur in small amounts are potassium, bicarbonate, carbonate, and nitrate. The direct source of all salt constituents are the primary minerals found in soils and in the exposed rocks of the earth's crust. During the process of chemical weathering these constituents are gradually released and made soluble. Bicarbonate ions form as a result of the dissolution of carbon dioxide in water. The carbon dioxide may be of atmospheric or biological origin. Water containing carbon dioxide is a particularly active chemical weathering agent and releases appreciable quantities of cation constituents as bicarbonates.

The soluble salt content of the soil is estimated from measurements of electrical conductivity (Box 6.6).

During some storms, salts carried by the wind can cause considerable damage to foliage, with characteristic symptoms of leaf burn, in severe cases resulting in defoliation (Plate 29). The severity of the defoliation depends on the salt load on the foliage. The initial symptoms resemble those of severe drought: leaf curling, loss of vigor, and subsequent defoliation.

Management of Soils for Citrus Production

The purpose of soil and vegetation management is to maintain and improve the physical, chemical, and biological properties of the soil to facilitate increased citrus production and better fruit quality. Proper soil management involves optimization of the following soil properties.

Irrigation and Drainage

For optimal citrus tree growth and production, the soil water content should be 50 to 60% of available soil water holding capacity. In the sandy soils used for citrus production soil moisture is rapidly depleted, and frequent irrigation is required. However, prolonged irrigation can result in leaching of soil-applied nutrients and chemicals and the loss of water below the root zone.

For optimal citrus tree growth and production, the soil water content should be 50 to 60% of the available soil water holding capacity.

In poorly drained soils, the water table frequently fluctuates, causing intermittent anaerobic conditions during the growing season. This can result in root damage in the wet zone of the soil and restrict the depth of rooting. This damage can occur in citrus trees on Spodosols along the east coast and in southern areas of Florida where soil drainage may not be adequate. Drainage is achieved by planting trees on raised beds, the construction of which causes considerable disturbance of the soil and the deposition of subsoil over the natural surface horizon. In calcareous soils, material from the calcareous subsoil if thus deposited over the natural A horizon, resulting in decreased availability of micronutrients. This can be a serious problem for rootstocks, including Carrizo citrange and Swingle citrumelo, which perform poorly under high-pH conditions.

Fertility Management

Essential Nutrient Elements

A comprehensive strategy for citrus health management must maintain the essential plant nutrients in amounts needed for optimal tree growth and fruit production. When an essential element is in short supply, tree function is restricted. Sixteen nutrient elements are essential for most higher plants, including citrus trees. Carbon (C), hydrogen (H), and oxygen (O) make up about 95% of the dry weight of a tree. These three elements are utilized by the chlorophyll-containing cells of a plant to produce carbohydrates with the help of light energy. The remaining essential nutrients are absorbed by plant roots from the soil solution and, to a lesser extent, by leaves, when nutrient elements are applied as foliar sprays.

A comprehensive strategy for citrus health management must maintain the essential plant nutrients in amounts needed for optimal tree growth and fruit production. When an essential element is in short supply, tree function is restricted.

These essential elements are not always present in adequate supply in the soil, and supplements applied to the soil or the leaves may be needed. A severe shortage of nutrients usually produces characteristic deficiency symptoms in the leaves, which can persist until the deficiency is corrected. Sometimes twigs and fruit also exhibit characteristic symptoms. Visual symptoms and leaf and soil analysis are useful in evaluating the nutritional status of citrus trees.

Nutritional Requirements of Citrus Trees

Appropriate fertility management strategies for citrus production are based on the soil type, tree age, variety and rootstock, and nutritional needs of the crop.

Nonbearing Trees

Citrus trees are considered to be nonbearing until they are three to four years old. The management of nonbearing trees requires maintaining adequate nutrition and water and minimizing adverse effects of weeds, diseases, pests, and freezes.

Nitrogen and irrigation are the most important factors in the growth of young trees. Fertilizer guidelines for young trees

Table 6.1. Recommended nitrogen rates and minimum number of applications for nonbearing citrus trees[a]

Year in grove	Nitrogen rate (pounds per tree per year)	Minimum number of applications	
		Dry	Fertigation
1	0.15–0.30	6	10
2	0.30–0.60	5	10
3	0.45–0.90	4	10

[a]Source: D. P. H. Tucker, A. K. Alva, L. K. Jackson, and T. A. Wheaton, 1995, Nutrition of Florida citrus trees, Univ. Fla. Coop. Ext. Serv. Publ. SP 169.

include a range of application rates for trees of different ages (Table 6.1). The following factors must be considered in the selection of a rate within the recommended range:

- Lower levels of nitrogen may be used during the first year if the trees come from a nursery with an intensive nutrition program.
- Trees planted in heavier soils require less fertilizer.
- Nutrient uptake efficiency may be increased by the use of liquid forms of fertilizer applied through an irrigation system.
- Controlled-release formulations may allow for a reduction in fertilizer rate.
- Rates may be lowered when fertilizer is applied within the root zone rather than broadcast over a larger area.
- Application of fertilizers outside the root zone should be avoided, particularly in areas vulnerable to nitrogen leaching.

Application Frequency and Timing

Frequent application of small amounts of fertilizer is desirable in order to minimize leaching, since smaller amounts per application are more likely to be absorbed by the tree roots. If a fertigation program is in operation, the frequency of application can be increased with little extra cost. Application may be curtailed during months of heavy rainfall in humid tropical and subtropical climates, particularly in groves planted in extremely sandy soils, which are vulnerable to excessive leaching of nutrients. Less fertilizer may be required if applications are made more frequently and if the timing is adjusted to avoid periods of high rainfall. A minimum of four to six applications of fertilizer in dry granular form per year and 10 to 15 applications by fertigation per year are recommended. The frequency of application can be decreased to two per year when a controlled-release fertilizer is used.

Fertilizer Mixtures for Nonbearing Trees

Nonbearing trees planted in previously uncropped soils should receive fertilizers containing N (1 unit), P_2O_5 (1 unit), K_2O (1 unit), Mg (0.20 unit), Mn (0.05 unit), Cu (0.025 unit), and B (0.003 unit), until soil and leaf analysis and the appearance of the tree indicate that the proportion of one or more may be changed.

For trees planted in previously cropped soils, the proportion of P_2O_5 in the blend may be reduced or omitted if soil test results indicate adequate P levels, i.e., 30 to 60 pounds per acre in extraction with double acid (Mehlich 1) or 7 to 15 pounds per acre in extraction with 1 M ammonium acetate (pH 4.8).

Zn, Mn, and Fe should be applied if visible symptoms of deficiencies appear on leaves.

Bearing Trees

Nutritional management of bearing trees requires many of the same considerations important for nonbearing trees. Tree growth continues to be an objective for several years after fruit production begins, and fruit quality also becomes very important for both the processed and the fresh fruit markets. When trees reach containment size, further growth is less desirable, and nutrition levels may be stabilized and possibly reduced. Nitrogen continues to be the most important element, but others, particularly K, also have substantial effects on fruit production and fruit quality.

Nutrients removed by harvesting the crop (Table 6.2) must be replaced. The nutrient content of grapefruit, per unit of weight, is lower than that of oranges, and therefore fertilizer rates are lower for grapefruit. Nutrient uptake efficacy from applied fertilizer is not 100%, so more nutrients must be applied than the minimum required by the tree. Several other factors also influence crop requirements. Fertilizer guidelines for bearing trees include a range of rates (Table 6.3). Low to medium rates are recommended for groves producing only average fruit yields.

Under most conditions 200 pounds of nitrogen per acre per year appears to be adequate for optimal fruit production, with good management of fertilizer application and irrigation to minimize excessive leaching below the rooting depth. In bedded groves, the timing of applications of dry granular fertilizer is

Table 6.2. Analysis of the nutrient content of Hamlin orange fruit[a]

Nutrient element	Pounds per 100 boxes of fruit[b]
Potassium	17.6
Nitrogen	12.5
Calcium	4.5
Magnesium	1.9
Phosphorus	1.4
Sulfur	1.1
Iron	0.024
Boron	0.020
Zinc	0.020
Manganese	0.011
Copper	0.006

[a]Source: D. P. H. Tucker, A. K. Alva, L. K. Jackson, and T. A. Wheaton, 1995, Nutrition of Florida citrus trees, Univ. Fla. Coop. Ext. Serv. Publ. SP 169.
[b]Each box contains 90 pounds of fruit.

Table 6.3. Recommended nitrogen rates and minimum number of applications for bearing citrus trees (four years old or older)[a]

Nitrogen rate (pounds per acre per year)			Minimum number of applications	
Oranges	Grapefruit	Other varieties[b]	Dry	Fertigation
120–200	120–160	120–200	3	10

[a]Source: D. P. H. Tucker, A. K. Alva, L. K. Jackson, and T. A. Wheaton, 1995, Nutrition of Florida citrus trees, Univ. Fla. Coop. Ext. Serv. Publ. SP 169.
[b]For Orlando tangelo and Honey tangerine (Murcott) the rate can be increased to 250 and 300 pounds per acre per year, respectively.

critical, since heavy rainfall soon after application may result in surface runoff. This could result in surface water quality concerns and a decrease in nutrient use efficiency. Higher rates may be considered for very productive groves, e.g., 240 pounds of nitrogen per acre per year for orange groves producing over 700 boxes per acre and 180 pounds per acre per year for grapefruit groves producing over 800 boxes per acre per year. Nutrient requirements can be calculated from fruit production data for previous years, if nutrient removal (Table 6.2) and fertilizer use efficiency are taken into account. The amounts of nutrients required increase with tree age and production, but a reduction in the rate of nitrogen application may be advantageous when trees reach containment size, to reduce vegetative growth, especially after pruning.

In Brazil, trees are considered to be nonbearing in the first four years after planting. Nutrient recommendations are based on the N rate, which is increased from year to year, from 0.2 to 1.0 pound of N per tree per year during this period, to account for the tree growth. The rates of P and K are based on

Under most conditions 200 pounds of nitrogen per acre per year appears to be adequate for optimal fruit production, with good management of fertilizer application and irrigation to minimize excessive leaching below the rooting depth.

soil analysis. Bearing trees are fertilized according to productivity; the nutrient requirement is calculated on the basis of 0.33 pound of N per box of fruit, and the most common nutrient ratio is either 1.0:0.5:1.0 or 1.0:0.3:1.0 (N:P:K). The annual fertilizer allocation is applied in three or four equal amounts, usually in the rainy season. Since most groves are not irrigated, fertilizer is applied in the rainy season to take advantage of soil moisture from rain to facilitate fertilizer dissolution and nutrient uptake. Most of soils in citrus production in Brazil are heavier in texture than those in Florida, and therefore leaching of soil-applied nutrients is of less concern in Brazil. Fertilizer is generally applied in a band, either on one side or on both sides of the trees under the canopy, the band extending about 1 to 2 feet outside the drip line. Common sources of N are ammonium sulfate, urea, nitrocalcium, and ammonium nitrate. Single applications of superphosphate and potassium chloride are made to provide P and K.

In the Negev region, the main citrus production area in Israel, drip irrigation is the common practice, and therefore fertigation is the preferred method of fertilization. Nonbearing trees may receive supplemental fertilization by foliar application. For nonbearing trees (in the first four years after planting) nutrients are applied with every irrigation. The N application rate varies from 0.6 to 0.9 pound per tree per year. Phosphorus and K_2O rates are based on leaf analysis. If leaf analysis is not available, the recommendation is to apply 90 pounds of P_2O_5 and 260 pounds of K_2O per acre every two to three years. Potassium is applied as KCl, and P_2O_5 is applied as ammonium phosphate or H_3PO_4. Bearing trees receive 180 to 270 pounds N per acre per year. Fertilization stops after August, since it is believed that fertilization after August may affect color break. Lemons receive higher rates of N than those stated above and can be fertilized year-round. Ammonium nitrate and ammonium sulfate are the most common sources of N.

Fertilizer recommendations for citrus in South Africa are made on a case-by-case basis after a thorough evaluation of the results of soil and leaf analyses, the history of fruit production and quality, the citrus variety, and the intended market. As a generalization, nonbearing trees are fertilized with a target rate of about 0.22 pound of N per tree in each of the first three years. The annual dose is divided into five equal applications, made between August to February. From the age of four years onward, most groves are managed strictly by interpretation of soil and leaf analyses. In most cases, fertilizer is applied by hand at the base of the tree. Nitrogen is applied as ammonium sulfate, urea, or ammonium nitrate; phosphorus as superphosphate or phosphoric acid; and potassium usually as potassium chloride or potassium sulfate.

In Australia, nonbearing trees receive 0.15 pound of N per tree per year until they are three to four years of age. Subsequently the trees are fertilized on per acre rate basis. Available information suggests that fertilization recommendations for nonbearing trees do not include an adjustment of the rate in succeeding years to account for the growth of the trees. The N recommendation for bearing trees is 160 pounds per acre per year for oranges, grapefruit, and lemons. Phosphorus (P_2O_5) and K_2O are recommended at 184 and 96 pounds per acre of oranges and grapefruit per year, respectively, and 245 and 130 pounds per acre of lemons per year, respectively. Commonly used sources of N include urea, ammonium nitrate, and ammonium sulfate. Phosphorus sources include single and double superphosphate. Potassium sources include potassium chloride and potassium sulfate. To supplement ground-applied fertilizer, in some cases, foliar applications of small quantities of urea are made, in low-biuret formulations (less than 0.4%) to avoid potential biuret toxicity.

Fertigation is gaining popularity in Australia with the introduction of modern microirrigation techniques and is considered the most efficient method of fertilizer application, since the nutrients are applied within the irrigation zone, which is the area of maximum root activity. Unlike broadcasting of dry granular fertilizer, fertigation requires less traffic in the grove, which reduces soil compaction and fuel costs.

Most citrus groves in Belize are unirrigated, and therefore fertilizer is applied through the rainy season, i.e., February to May. The annual fertilizer dose is divided into two or three equal applications per year during May–June, November–December, and January–February. Nonbearing trees may receive three or four applications per year. For bearing trees, the target rate is 175 to 200 pounds of nitrogen per acre per year.

Fertilizer Management for Citrus Health and Economic Returns

Tree Growth and Fruit Production

The major goal of fertilization is tree growth and fruit production. When trees are young, it is important to encourage their rapid growth. At eight to 10 years a tree has usually attained sufficient size and filled its allotted space, at which time the fertilization program should place emphasis on fruit production.

Fruit Quality

The goal of developing a tree as rapidly as possible with high nutrition and irrigation inputs is incompatible with the production of high-quality fruit, which is affected by climate, soil, and the scion and rootstock varieties, as well as irrigation and nutrition. Mature trees on a moderate program of nutrition and irrigation produce fruit of acceptable quality and show little response to minor changes in management programs.

In general, fruit quality factors of importance to the processed and fresh fruit industries include juice content, the concentration of solids and acids in the juice, maturity date, fruit size, shape, color, and rind thickness. External appearance, fruit size, and maturity date are most important for the fresh fruit market, while a high concentration of soluble solids and high juice content are more desirable in processed fruit. The most important factors influencing fruit quality are the nutrients N, P, and K and irrigation. Their effects are described in Table 6.4.

Table 6.4. Effects of mineral nutrition and irrigation on fruit quality[a,b]

	N	P	K	Mg	Irrigation
Juice quality					
Juice content	+	0	–	0	+
Soluble solids (SS)	+	0	–	+	–
Acid (A)	+	–	+	0	–
SS/A	–	+	–	+	+
Juice color	+	0	–	?	0
Solids/box	+	0	–	+	–
Solids/acre	+	+	+	+	–
External fruit quality					
Size	–	0	+	+	+
Weight	–	0	+	+	+
Green fruit	+	+	+	0	+
Peel thickness	–[c]	–	+	–	–

[a] Source: D. P. H. Tucker, A. K. Alva, L. K. Jackson, and T. A. Wheaton, 1995, Nutrition of Florida citrus trees, Univ. Fla. Coop. Ext. Serv. Publ. SP 169.

[b] Increase (+), decrease (–), or no change (0) in quality; ? = no information.

[c] Fruit from young trees may have a thicker peel.

Nutrition Management

Fertility management is important to ensure optimal tree growth, fruit production, and fruit quality. Soil and plant analysis provide guidelines for developing a balanced nutrient management program. Most mineral nutrient deficiencies cause characteristic symptoms which can be visually identified (Table 6.5 and Plates 30–32). Mineral analysis of plant tissue is widely used as a diagnostic tool for determining the nutrient status of the tree (Box 6.7). Sampling guidelines should be followed precisely to ensure that the analytical results are meaningful. The ultimate aim of successful nutrient management is to attain optimum levels of mineral nutrients in leaf tissue (Table 6.6).

Table 6.5. Nutrient elements, their importance in citrus health, and deficiency symptoms

Element	Symbol	Importance	Deficiency symptoms
Nitrogen	N	Direct influence on yield and quality	In nonbearing trees, growth is limited, and foliage turns yellowish pale green In bearing trees, premature shedding of older leaves leads to a thinness of the canopy; persistent nitrogen shortage leads to defoliation, fruit drop, and shoot death
Phosphorus	P	Involvement in metabolism and transport of photosynthates, with direct influence on yield and quality	Delayed maturity; leaves are slightly variegated, mottled with light or dark green; fruit may drop before it is mature
Potassium	K	Role in determining yield, fruit size, and quality	Excessive growth of weak, immature wood, which is liable to be injured by frost
Calcium	Ca	Influence on fruit quality, color, and aroma	Rare
Magnesium	Mg	Key element in the chlorophyll molecule; essential role in the formation of some amino acids and vitamins	Characteristic chlorotic pattern and in some cases premature defoliation
Sulfur	S	Required element in chlorophyll formation, although not part of the chlorophyll molecule	Rare
Iron	Fe	Required element in chlorophyll formation	Chlorotic patterns, appearing first on young shoots (Plate 30); very common in plants growing in soils that are high in pH, waterlogged, very low in organic matter, or high in Cu
Copper	Cu	Participation in nitrogen metabolism	Corking of fruit and dieback
Zinc	Zn	Involvement in the development and function of growth regulators, internode elongation, chloroplast development	Severe interveinal chlorosis (Plate 31); new leaves are progressively smaller as the deficiency becomes more severe, and shoot internodes become shorter, resulting in a rosette effect; severe deficiency restricts growth and reduces yield
Manganese	Mn	Required element in chlorophyll formation; participation in oxidation-reduction reactions in cells	Dark green veins and light green interveinal areas on leaves, which may be masked by deficiency symptoms of other elements
Boron	B	Involvement in various enzyme systems and in carbohydrate movement or metabolism	Excessive fruit drop, gum formation on the outside of the fruit, and brown areas on the albedo (Plate 32)
Molybdenum	Mo	Component of nitrate reductase and nitrogenase	Characteristic yellow spots on leaves

Guidelines for interpreting the results of citrus leaf tissue analysis are presented in Table 6.6. The values obtained from analysis of sample tissue can determine adjustments in a fertilizer program to maintain leaf tissue nutrient levels in the optimum range.

Micronutrient Programs

In Brazilian citrus production, boron and zinc are applied in foliar sprays once or twice a year, in the form of borax or boric acid and zinc sulfate. In some cases, manganese is applied as manganese sulfate. A typical micronutrient spray mix consists of 3 pounds of boric acid (17% B), 12 pounds of zinc sulfate (22% Zn), and 8 pounds of manganese sulfate (28% Mn) in 500 gallons of water.

In Israel, the most common micronutrient spray includes magnesium nitrate, zinc oxide or zinc sulfate, and manganese sulfate. The rate of application depends on leaf analysis. In addition, young trees also receive potassium nitrate or low-biuret urea. Iron, if needed, is applied as a chelate at the rate of 10 to 15 pounds per acre of mature trees. For young trees, iron is applied at the rate of 0.07 ounce per tree per year. Iron sulfate can be used as a foliar spray on lemons, but only on nonbearing trees. If sprayed on bearing trees, iron may cause fruit damage.

In South Africa, bearing trees are on a program of routine leaf analysis, and micronutrient application is based on the results of analysis. It is a routine practice to apply a foliar spray monthly, from August to February, consisting of 0.33 pound of zinc oxide (60% Zn), 0.33 pound of manganese sulfate (28% Mn), 0.33 pound of copper oxychloride (30% Cu), 0.22 pound of soluble boron (20% B), and 2.2 pounds of low-biuret urea in 26 gallons of water.

Box 6.7

Guidelines for Leaf Tissue Sampling and Analysis

Sampling units must be grouped according to variety, rootstock, tree age, and soil type. Samples of eight to 10 leaves should be collected randomly from each of 15 to 20 trees in the sampling units. Five- to seven-month-old spring flush leaves from nonfruiting twigs are sampled, taken from about the same height on all trees. Leaves from fruiting terminals generally have lower concentrations of N, P, K, Cu, Zn, Mn, Fe, and B and higher concentrations of Ca and Mg than those from nonfruiting terminals. A thorough washing of the leaf samples is recommended prior to analysis. The leaves are subjected to decontamination, oven-dried, and ground to powder. Subsamples are taken for nitrogen analysis by the Kjeldahl procedure. For analysis of other elements the leaf tissue is either wet-digested or dry-ashed to bring all of the nutrient elements into solution. The concentrations of these elements are measured by an inductively coupled plasma argon emission spectroscope.

Nitrogen Management and Groundwater Nitrate

Water quality issues, particularly the contamination of groundwater by agricultural chemicals, including nitrates, have risen to the top of the agricultural and environmental policy agenda. Over 50% of all fresh water used in Florida comes from groundwater, and over 90% of the public relies on groundwater for drinking water. During the summer months in Florida rainfall exceeds evaporation and recharges the groundwater. The quality of this water is an important consideration, since it comes in contact with a number of chemicals and heavy metals prior to returning to groundwater. In some citrus production regions in Florida and California the soil, climate, and hydrological features are favorable for the downward movement of soil-applied chemicals.

The loss of nitrogen by leaching can be minimized by improving the efficiency of nitrogen use by the trees, by means of fertilizer and irrigation management (Box 6.8).

Table 6.6. Guidelines for interpretation of leaf analysis of four- to six-month-old spring flush leaves from nonfruiting terminals[a]

Element	Deficient	Low	Optimum	High	Excessive
Nitrogen (%)	< 2.2	2.2–2.4	2.5–2.7	2.8–3.0	> 3.0
Phosphorus (%)	< 0.09	0.09–0.11	0.12–0.16	0.17–0.30	> 0.30
Potassium (%)	< 0.7	0.7–1.1	1.2–1.7	1.8–2.4	> 2.4
Calcium (%)	< 1.5	1.5–2.9	3.0–4.9	5.0–7.0	> 7.0
Magnesium (%)	< 0.20	0.20–0.29	0.30–0.49	0.50–0.70	> 0.70
Chlorine (%)	...	...	< 0.20	0.20–0.70	> 0.70[c]
Sodium (%)	...	...	...	0.15–0.25	> 0.25
Manganese (ppm)[b]	< 17	18–24	25–100	101–300	> 300
Zinc (ppm)	< 17	18–24	25–100	101–300	> 300
Copper (ppm)	< 3	3–4	5–16	17–20	> 20
Iron (ppm)	< 35	35–59	60–120	121–200	> 200
Boron (ppm)	< 20	20–35	36–100	101–200	> 200
Molybdenum (ppm)	< 0.05	0.06–0.09	0.10–1.0	2.0–5.0	> 5.0

[a] Source: Source: D. P. H. Tucker, A. K. Alva, L. K. Jackson, and T. A. Wheaton, 1995, Nutrition of Florida citrus trees, Univ. Fla. Coop. Ext. Serv. Publ. SP 169.

[b] Parts per million (ppm) = percent concentration × 10,000.

[c] Leaf burn and defoliation can occur at chlorine concentrations greater than 1%.

Box 6.8

Improving the Efficiency of Nitrogen Use

Leaf Analysis

Leaf analysis is an important tool in monitoring the nutritional status of trees and provides guidelines for determining fertilizer needs. The optimum concentration of nitrogen in six-month-old spring flush of citrus trees is 2.5–2.7%. Higher levels are considered an indication of luxury consumption and will not increase fruit yield. Nitrogen is a very mobile element, and a deficiency of this nutrient can be corrected relatively rapidly. Annual leaf analysis is a well-recognized method of monitoring the nitrogen nutrition of citrus trees. If the nitrogen concentration in spring flush leaf tissue is above 2.7%, the nitrogen rate should be decreased in subsequent applications.

Application Timing

Most conventional dry-soluble sources of nitrogen are extremely soluble, and thus heavy rainfall or irrigation following application results in rapid leaching, carrying nitrates below the root zone. Nearly 60% of annual rainfall in Florida is received between June and September, and application of fertilizer at high rates should be avoided during these months. There is evidence that the leaching of nitrogen is reduced by increasing the frequency of application. In groves with a fertigation program, the frequency of application can be increased with minimum additional cost and effort.

Irrigation Scheduling

Proper irrigation scheduling is the key to efficient utilization of fertilizers. Fertigation should be scheduled toward the end of an irrigation, to minimize leaching of nutrients. A light irrigation may be beneficial soon after a dry-soluble granular form of nitrogen has been spread, by solubilizing the fertilizer and incorporating it into the topsoil. This will enhance the uptake of nutrients by the roots and minimize the loss of nitrogen by volatilization, particularly in high-pH soils. At high soil pH, ammonium forms of nitrogen left on the soil surface may be subject to rapid loss by volatilization.

Sources of Nitrogen

Under soil and climatic conditions which favor nitrate leaching, the application of dry-soluble sources of nitrogen in large doses is not recommended. Fertilizer formulations containing controlled-release materials, which release nutrients slowly over an extended period of time, can be advantageous in decreasing the loss of nitrates by leaching. In some situations, a blend of controlled-release and soluble sources of nitrogen may be used to moderate the cost of the fertilizer program.

Foliar Feeding

Studies in California have shown that up to 50% of the nitrogen required by citrus trees can be supplied by foliar application. With foliar application, the total amount of nitrogen applied to the soil can be reduced, and losses due to leaching can be decreased. Little work has been done on this subject in Florida.

Acknowledgments

We appreciate K. Bevington, New South Wales Agriculture, Australia; D. Woods, Outspan, South Africa; Y. Levy, Ministry of Agriculture, Israel; J. Lima, Brazil; and Marla Holder, Barton Creek Farms, Belize, for sharing information pertinent to their countries' programs on citrus soils and nutrition management.

Brian Boman
University of Florida
Indian River Research and Education Center, Fort Pierce

Yoseph Levy
Agricultural Research Organization
Institute of Horticulture
Gilat Experiment Station, Mobile Post Negev, Israel

Larry Parsons
University of Florida
Citrus Research and Education Center, Lake Alfred

CHAPTER 7

Water Management

Physiological Response to Irrigation and Water Stress

Citrus originated in the wet tropics of Southeast Asia, but it has adapted to a range of conditions ranging from humid subtropical to desert environments. Irrigation is frequently one of the most costly and time-consuming cultural practices in arid and semiarid areas with lengthy dry seasons. It is also an important cultural factor in humid subtropical climates in maintaining yields and fruit quality that would otherwise be lowered during short-term dry periods. Studies have shown that expenditures for improved weed and pest control, pruning, and fertilization do not result in higher yields if irrigation is not satisfactory. It is only when such improvements are combined with good irrigation management that the benefits can be fully realized.

The response of citrus to water deficits depends on several factors. Water stress in citrus trees can result in retarded growth and wilting of leaves, which may abscise and drop. Fruit that reaches maturity usually has a low juice content and may be smaller than normal. When soil moisture deficit is severe, shoot growth will slow and leaf expansion will be limited. Twig and branch dieback may result from prolonged drought stress. Water stress during times of rapid growth should be avoided. Vegetative growth of young trees determines their final size and future yield potential, and mature trees need healthy vegetative growth to maintain and replace fruit-bearing branches.

Citrus trees are very sensitive to water deficits during flowering. Water stress during this period directly reduces fruit set. Continued water stress during fruit set is likely to cause a heavy June drop of small fruit. Water deficit during the early development of fruit (June drop, in the Northern Hemisphere) can increase fruit shedding and reduce the rate of fruit growth. The increase in fruit size from June through December is closely related to water uptake. However, a moderate water deficit in late summer and fall can be desirable to maintain high juice quality. Severe water deficit in the summer followed by rainfall or irrigation often results in the development of a worthless second crop and reduces the yield of the main crop in the following season.

Unlike other evergreen fruit trees, such as avocado and mango, citrus trees continually replace their leaves as they grow. Leaves can remain on the tree for up to two years. The leaves have many xeromorphic characteristics, including an adaxial epidermis covered by a thick cuticle without active stomata. Citrus stomata close when evaporative demand is high, especially when soil water is not readily available. Thus, orange trees in Arizona and Florida have similar transpiration rates and drought stress on summer days, although the evaporative demand is much greater under Arizona desert conditions.

Citrus stomata are also very sensitive to low humidity. They may close when a high vapor pressure deficit occurs, even when evaporative demand is fairly low (as in winter). In Israel, citrus in the desert and on the humid coastal plain consume similar amounts of water, although the evaporative demand is very different. This is the net effect of responses to both evaporation demand and humidity.

The rate of evapotranspiration and effects of drought stress depend not only on humidity but also on leaf age and acclimation to drought stress.

The Root System

Compared to deciduous fruit trees, such as walnuts, citrus trees have a relatively shallow root system. The typical maximum rooting depth of citrus in deep, well-drained soils is 4 to 5 feet, with the main root system reaching a depth of only 2 to 3 feet. Roots can grow much deeper in well-drained sandy soils. In a study of citrus trees planted in the deep sandy soils of central Florida, roots of trees on rough lemon rootstock reached a depth of 11 feet, and roots of sour orange trees were found at a depth of 9 feet. Water depletion studies confirmed that the roots were active at this depth, but most of the water extracted by the trees came from the top 2 feet. Citrus can also survive in very shallow soils, especially when the limiting factor is a high ground water table. Citrus trees grown in the coastal flatwoods of Florida typically have rooting depths of only 18 to 24 inches.

Citrus is sensitive to temperature in the root environment. The conductance of water by roots (hydraulic conductance) continuously decreases as the soil temperature decreases from

95 to 41°F. Drought stress further decreases the hydraulic conductance of roots. Even at low winter temperatures, citrus trees may suffer from drought stress when the soil is cold but climatic conditions bring about high evaporative demand due to low air humidity. Typically, low-humidity winds can cause severe leaf drop or leaf scorch when the soil temperature is low, especially in heavy soils.

Reduced summer irrigation reduces root development in citrus trees. This is especially true when the upper soil layers dry out between irrigations at long intervals. Long periods of drought or long intervals between irrigation force the trees to increase water uptake from deep soil layers.

Canopy Development

The vegetative development of young trees depends closely on the irrigation regime. Under arid conditions, irrigation of young trees at long intervals can cause a delay in canopy development, even though symptoms of drought stress may not be evident. During the first several years after planting, there is generally a good relationship between canopy volume and yield. However, as trees reach full size, excessive growth induced by overirrigation and fertilization can decrease yields, mainly because of shading and the need for severe hedging. Irrigation and fertilization practices that control growth can serve to limit canopy development.

The growth of trunks or main branches can be measured to compare the response of trees to different irrigation regimes. However, various rootstocks develop trunks of different sizes, and there may be poor correlation between trunk size and canopy development among all rootstocks. For example, rough lemon develops a large canopy and a small trunk, compared with Cleopatra mandarin, which has a large trunk but a small canopy during the first years of growth.

Effects of Irrigation on Fruit Quality

Fruit quality is an important factor in determining the value and marketability of citrus, either for fresh consumption or for processing as juice. Fruit quality factors can be divided into two main groups: physical characteristics of the fruit and chemical properties of the juice. The physical factors include fruit size, peel thickness, color, appearance, and juice content, while the chemical factors include acid and sugar contents, the ratio of sugar to acid, and the amount of minor constituents, including bitter or aromatic compounds which may influence fruit palatability.

The amount of sugar in the juice, expressed as total soluble solids (TSS, usually measured with a refractometer) or as a value on the Brix scale (based on the specific gravity of the juice, usually measured with a hydrometer), is an important parameter in determining the price of processed fruit. The acid concentration and the ratio between the TSS and acid contents are important parameters in defining fruit quality and picking time. In general, water shortage causes an increased concentration of TSS in the juice, while excessive rainfall or irrigation results in dilution of the sugars in the juice. Irrigation usually has a greater effect on acidity in grapefruit than in orange.

Fruit size can be increased by irrigation. However, this enlargement is not always linear, since better irrigation can increase the number of fruits per tree substantially enough to cause an overall decrease in fruit size as a result of the heavy crop load. Valencia oranges that develop under luxurious moisture conditions have a more tender peel than fruit from trees that are mildly drought-stressed. Drought increases peel thickness and the peel-to-pulp ratio and decreases the juice percentage.

The effect of irrigation on color break is difficult to assess, since irrigation can affect the nitrogen nutrition of the tree, and increased nitrogen will delay color break and enhance regreening. Color break is usually delayed when irrigation is increased. In Israel, increased irrigation of Ruby Red grapefruit increased the amount of chlorophyll and red pigment (lycopene) in the rind and caused more regreening, but it did not increase the amount of yellow pigment (β-carotene). In humid Japan, color break was accelerated in satsuma mandarin when drought stress was induced by means of plastic sheets covering the soil.

Diseases

High water potential in leaves and fruit greatly increases susceptibility to oleocellosis, which is caused by the rupture of rind oil glands. The usual practice for minimizing oleocellosis is to avoid picking fruit when environmental conditions favor high internal turgor of the fruit, e.g., shortly after irrigation or rainfall, early in the morning, or in weather which reduces water loss (fog, dew, high humidity, or cloudiness). Drought stress has been found to decrease puffiness of satsuma mandarins in rainy areas in Japan, while autumn irrigation decreased russeting and creasing but increased blemishing of Temple oranges in Florida.

Citrus foot rot, a fungal disease caused by *Phytophthora nicotianae* (see Chapter 12), is associated with high soil water content. The best preventive measure is good soil drainage: keeping the soil around the trunk dry and avoiding irrigation regimes that wet the trunk for long periods. Surface irrigation water may contain *Phytophthora,* a common problem where drainage water is collected at the ends of furrows ("tail water") and returned to the water distribution system to be reused. In Texas, microirrigated trees suffered increased foot rot when freeze-protection tree wraps became saturated with irrigation water.

Asphyxiation of roots as a result of excessive water or flooding and inadequate oxygenation after heavy rains can cause severe damage and can even kill citrus trees planted in heavy, poorly drained soils. Some rootstocks, such as sour orange and Carrizo citrange, are very sensitive to this problem, while rough lemon and sweet lime are more tolerant. Citrus is more tolerant of flooding in the winter than in the summer, when air and soil temperatures are higher.

Lime-induced chlorosis is closely related to irrigation and soil aeration and can be aggravated by high soil water content. Frequent irrigation has caused chlorosis in Arizona, and overmoist soil has been regarded as one of the most common causes of chlorosis in California. In Israel, chlorosis increased when the interval between sprinkler irrigations was shortened, and it was reduced in trees watered by drip irrigation rather than by sprinklers. Gumming of trunks and branches of grapefruit trees, similar to Rio Grande gummosis, is common in grapefruit trees suffering from lime-induced chlorosis in the Negev region of Israel. This disorder often disappears when sprinkler irrigation is replaced with drip irrigation or when iron is applied to the trees.

Citrus blight (or young tree decline) is a wilt-like disease of unknown cause, characterized by restricted movement of water through the tree. Studies of this disease have not reported any link between the disease and cultural practices, including irrigation. However, citrus blight causes drought stress symptoms, and one method of diagnosing the disease is to measure the rate of uptake of water injected into the xylem.

Climate Control

Heat spells can damage flowers and fruitlets and increase June drop. Washington navel orange is especially prone to such damage. Reduction in net photosynthesis at temperatures above 86°F may cause low-carbohydrate-induced fruit abscission. Evaporative cooling by overhead irrigation can lower the temperature of the air and the canopy considerably. In Florida, overhead irrigation lowered the midday air temperature from 93 to 82°F and increased the relative humidity from 40 to 60%, compared to an unirrigated control. In the San Joaquin Valley of California, intermittent use of overhead sprinklers reduced leaf temperature from 125 to 85°F when the ambient temperature was 100°F. Overhead sprinkling may increase the risk of sodium and chloride damage to leaves, and only water of low salinity should be used.

Drought Stress as a Means of Inducing Bloom

A dormant period is generally considered essential for flowering in citrus trees. The duration of dormancy is directly related to the number of flowers produced in the subsequent bloom. In the subtropics, dormancy can be induced by low temperatures in the wintertime (less than 55°F). In the tropics, it can be induced by a water deficit. Flower bud initiation occurs during this dormant period, when vegetative growth is at a minimum. Once dormancy has ended, the trees need an adequate water supply. A severe water deficit following the break of dormancy will delay flowering, and there may be excessive flowering when bloom does occur, resulting in an overabundance of small fruit.

In tropical climates, citrus trees can bloom continuously year-round, and rainfall or irrigation after a period of drought can trigger a flush of flowering. In subtropical climates, some citrus species (notably lemon, lime, citron, pummelo, and Minneola tangelo and other mandarin hybrids) are capable of flowering all year. The ability of orange trees to reflower after a severe drought is important in Brazil; cultivars such as Pera (especially when budded on Rangpur lime rootstock) reflower when the summer rains start after the spring flush fruitlets have been destroyed by spring drought. Irregular irrigation can cause out-of-season flowering in oranges and mandarins. The fruit that may be set is usually undesirable, and heavy out-of-season flowering can reduce the amount fruit produced during the main season.

In tropical climates, citrus trees can bloom continuously year-round, and rainfall or irrigation after a period of drought can trigger a flush of flowering.

A traditional orchard management practice in lemon culture in Sicily is *forzatura* (forcing), withholding irrigation for a month or longer in order to cause wilting and induce summer bloom. The fruit that is set from such flowers is picked the following summer as green fruit (*verdelli*). Lemon cultivars differ in their response to this treatment. The cultivars Eureka, Femminello, Villafranca, Verna, and Fino respond well, compared with Lisbon and Interdonato, which tend to flower only in the spring.

The degree of drought stress necessary to induce flowering is important, since excessive stress can be harmful. Some reports indicate about 60% of flowers abort after excessive drought stress, compared with only 20% after moderate drought. Excessive drought stress also harms the development of fruit already on the trees, and it substantially reduces the yield of regular winter lemons. Lime trees seem to be less responsive than lemon to drought-induced flowering.

Irrigation Methods

Irrigation methods differ in how they distribute water over the field. Gravity irrigation, in which water is conveyed to the point of consumption directly on the surface of the soil, is the traditional method used for citrus. However, in many areas surface irrigation is gradually being replaced by pressurized irrigation, using sprinkler systems (overhead and under-tree) or microirrigation systems (drip irrigation, minisprinklers, bubblers, and microsprayers).

Gravity Irrigation

Water may be distributed in furrows or border strips along rows of trees or in flood basins around each tree (Plates 33 and 34). This method provides less uniform distribution of water than alternative methods and usually requires larger quantities of water in order to provide reasonable coverage. In Arizona, drip irrigation used a third less water (with a 20% leaching fraction) than flood irrigation, while the plots produced similar yields. In a five-year experiment in a Valencia orange orchard near Yuma, Arizona, yields were lower and the amount of water applied was considerably higher in flood-irrigated plots (247 pounds of fruit per tree and 67 inches of water per year) than in plots under sprinkler irrigation (322 pounds and 15.4 inches), basin irrigation (512 pounds and 14.2 inches), or drip irrigation (428 pounds and 16.9 inches).

Sprinkler Irrigation

Sprinkler systems include overhead solid-set sprinklers (Plate 35), traveling gun sprinklers (Plate 36), under-tree low-angle solid-set sprinklers, and portable systems. Sprinkler irrigation was once common in citrus groves, but less costly and more flexible microirrigation systems have taken the place of sprinkler systems in many areas. Overhead irrigation may cause foliar damage if the water contains even a low concentration of salts. For this reason, overhead sprinklers (overhead solid-set, traveling gun, and overhead portable systems) have been abandoned in many citrus-producing areas. There is usually more evaporation in above-canopy systems, which keep the whole canopy area wet for a longer period, than in microirrigation.

Microirrigation

Microirrigation (Plate 37), in which water is supplied by microsprinklers or minisprinklers, sprayers, or drip or trickle irrigation, has become common in citrus orchards. The advantages of microirrigation include high application efficiency, low pressure requirements, ease of operation, freeze protection, the ability to perform fertigation and chemigation, the capability for automation, and good control of soil aeration. In converting an existing mature orchard from flood or sprinkler irrigation to microirrigation, consideration must be given to changes in the wetted surface area.

Microsprinklers

Minisprinklers and sprayers are common in new citrus plantings. Microsprinklers can be tailor-made to the size of the tree, and the discharge and the diameter of the wetted area can be increased as the tree grows. The diameter of the wetted soil surface under many minisprinklers ranges from 10 to 18 feet, a distance similar to or slightly greater than normal tree spacing. Likewise, the number of drippers per tree can be adjusted to cover any desired land area. Microsprinklers have an important advantage over drip irrigation in their ability to provide frost protection, especially in young orchards. Conversion from flood irrigation to microsprinkler irrigation in Florida has resulted in yield increases of over 35%. The yield of a Washington navel orange on trifoliate rootstock in a California orchard remained the same when the orchard was converted from furrow irrigation to drip irrigation, but the drip system used 25% less water. In California, the average direct energy use by drip and furrow systems are comparable, about 370 kilowatt-hours per acre, compared with 920 kilowatt-hours per acre used by sprinkler irrigation.

Studies in Israel showed that the minimum surface area that can be wetted without causing yield reduction depends on the irrigation interval. The yield of Marsh grapefruit trees was the same with drip irrigation (wetting 30% of the soil surface) at intervals of three days as with sprinkler irrigation (wetting 70% of the surface) at intervals of seven days. The yield with drip irrigation at seven-day intervals was 7% less than the yield obtained with sprinkler irrigation. In another trial with oranges, there were no differences in yields obtained with microsprayers (wetting 30% of the soil surface), minisprinklers (wetting 67% of the surface), and under-canopy sprinklers (wetting 90% of the surface). In Florida, yields of Valencia orange on rough lemon rootstock receiving supplemental irrigation from two or four drippers per tree (wetting 5 and 10% of the soil surface, respectively) or from one or two microsprayers per tree (wetting 28 and 50% of the surface, respectively) averaged 43 and 65% more, respectively, than yields of unirrigated plots.

Drip Irrigation

Drip irrigation is common in new plantings in Spain and Israel. In Israel, no substantial differences could be detected in yields of Marsh grapefruit irrigated by sprinkler or irrigated by one or two drip lines. Trees under drip irrigation should be fertigated frequently. Drip irrigation necessitates careful attention to water filtration. Chlorination of the water is often needed to prevent blockage of the emitters by bacteria. Advances in filtration techniques make it possible to use surface water and even reclaimed water. With a properly designed drip system, the grower can decide to start with a few drippers near a young tree and add more emitters as the canopy expands. Currently, the usual practice is one drip line per row, with drippers spaced at distances of 2 to 3 feet.

Underground Drip Systems

Early underground drip systems in citrus orchards failed, mainly because of penetration of the emitters by roots. New and better emitters, coupled with chemical prevention of penetration by roots, make underground drip irrigation feasible. This method is beneficial where pests, such as rodents or woodpeckers, damage aboveground plastic lines. Buried drip systems have the potential to reduce water requirements, since little water is lost to evaporation and wind drift.

Maintenance of Microirrigation Systems

One of the biggest challenges of microirrigation systems is the clogging of emitters, resulting in lowered system performance and water stress on the plants. Both surface water and groundwater can undergo rapid changes when drawn into an irrigation system. Changes can occur in pressure, temperature, velocity, and composition, which in turn cause changes in chemical reaction balances and bioactivity. These can result in reactions that cause the formation of scales and slimes, which impair the irrigation system.

Clogging is frequently the result of several independent or interdependent factors. For example, the formation of a precipitate is caused by factors such as bacterial activity or changes in pH and pressure. Bacteria in turn require a food source, and suspended particles do not become attached to the surface unless there is chemical attraction or physical trapping.

The main problem with systems using surface water is slime and bacterial contamination. Even a slight mineral load in the water is able to support a significant bacterial population. This is especially true of surface waters in which an abundant source of carbon is dissolved, since the carbon supports the metabolic processes of microorganisms.

Scale is the primary problem in systems served by groundwater. The formation and the rate of development of a scale layer depend on the mineral composition of the water and the relative concentrations of minerals.

Clogging can best be prevented by frequent water treatment, generally with various organic and inorganic compounds which have been found to prevent the formation of scale and dissolve scale-forming elemental matter. Product stability, injection rates, biological control capabilities, and required dosages vary widely. Shock treatments are directed toward existing problems. The two major shock treatments are acid and chlorine injection. In order for a shock treatment to be effective, the system must be completely flushed after the treatment. Preventative treatment usually requires less manual labor, since little or no line flushing is required.

Chlorination is a widely used and highly effective method of killing bacteria. Active chlorine is commercially available as a liquid, solid, or gas. Solid chlorine (calcium hypochlorite) has a high calcium concentration and tends to increase pH. Liquid chlorine (sodium hypochlorite) is the most widely used product. The optimum pH for chlorination with sodium hypochlorite is between 7.0 and 7.2. A half-unit increase in pH can result in a 50% reduction in the effectiveness of sodium hypochlorite. Gaseous chlorine (Cl_2) is generally the most effective for chlorination. However, because of its hazardous nature, Cl_2 gas can be used only by trained, licensed personnel, and facilities where it is used are subject to local, state, and federal reporting requirements for hazardous materials.

Acidification is sometimes used to dissolve existing scale from the water lines. Acid injection can also make chlorination more effective. Successful acidification may require dropping the pH of the water in the system to below 2.0 to dissolve the existing scale. This low pH may injure young trees and leaves that are wetted by the spray. The most commonly used inorganic acids are nitric (HNO_3), sulfuric (H_2SO_4), and hydrochloric (muriatic, HCl). Urea–sulfuric acid and phosphoric acid fertilizers, which are less dangerous to handle, are available in some areas.

Irrigation Water Requirements

Citrus are evergreens and therefore require water for transpiration throughout the year. Water requirements vary with climate, ground cover, cultivation between rows, weed control, tree size and age, and scion and rootstock cultivars. Grapefruit has higher water requirements per tree than orange or mandarin varieties, because grapefruit trees are larger. The water requirements of mature trees generally range from 35 to 50 inches per year.

Citrus evapotranspiration is largely controlled by climatic factors. Under similar climatic conditions, citrus trees have lower transpiration rates than other crop plants. For example, midsummer evapotranspiration in Israel was found to be 0.28 to 0.31 inch/day in many field crops, 0.33 inch/day in apple orchards, but only 0.18 inch/day in citrus orchards. The reduced transpiration rate is due to low conductance in the citrus canopy. Canopy conductance calculated from soil water depletion data and potential evapotranspiration are typically 0.09–0.11 inch/second, compared with 0.8–1.3 inches/second in a number of field crops.

In Arizona, the daily consumption of Marsh grapefruit ranged from 0.06 inch in January to 0.22 inch in July, and that of navel oranges ranged from 0.04 to 0.17 inch. The total annual evapotranspiration was 48 and 39 inches, respectively, in the two crops. In another report, the daily evapotranspiration of Navel oranges was 0.03 inch in December and 0.30 inch in August. The high values in summer reflect the very dry and hot climate of southern Arizona.

Water use by individual trees is highly dependent on tree size (Table 7.1). In Florida, large trees at low planting densities (60–70 trees/acre) may use 20–30 gallons/day during the winter months and 60–70 gallons/day in July and August. The average annual evapotranspiration (ET) was reported to be 48 inches in a developing grove with bahiagrass cover over a 10-year period in Florida. Daily water use by these trees peaked at about 40 gallons/day during June and July.

Water Use by Young Trees

It takes a number of years (four to 10, depending on planting density) to achieve maximum evapotranspiration in a newly planted orchard. The amount of water applied to a young orchard should take into account the canopy size. Newly planted trees require only a fraction of the water needed by mature trees. Average water use by newly planted Midsweet on Carrizo citrange rootstock has been reported to be about 1 gallon/day over a nine-month period. However, irrigation systems that water grassed areas in row middles and between trees can raise the water requirement considerably. In one study, water use by two-year-old citrus trees with grass cover was nearly double that of trees on bare soil.

In Cuba, the mean water requirements of Valencia orange trees in summer were reported as 0.09 inch per day for newly planted trees and 0.12 inch per day for four-year-old trees. In a Florida study, water use by five- to six-year-old Valencia orange trees on Swingle citrumelo rootstock peaked at about 16 gallons/day in May. During November through January, ET rates averaged only about half the May rate. Water use was found to increase at a rate of about 20% per year in three- to five-year-old trees. With modern microirrigation, it is possible to irrigate individual trees and increase the size of the wetted area or the number of emitters as the tree develops. The recommendation for midsummer applications in Israel is 2.5, 4, 7, 12, and 17 gallons per day in the first year through the fifth year, respectively, and 0.16 to 0.18 inch/day from the sixth year on.

Table 7.1. Water use per tree in various planting densities, with equivalent water use per acre

ET[a] (in./day)	Tree spacing (ft × ft)	Tree area (ft^2)	Tree density (trees/acre)	ET (gal/tree per day)
0.10	8 × 22	176	248	11
0.10	10 × 24	240	182	15
0.10	15 × 25	375	116	23
0.15	8 × 22	176	248	16
0.15	10 × 24	240	182	22
0.15	15 × 25	375	116	35
0.20	8 × 22	176	248	22
0.20	10 × 24	240	182	30
0.20	15 × 25	375	116	46

[a] Evapotranspiration.

Application Volume

Water use rates and irrigation requirements are usually described in terms of depth. With microirrigation systems that water only a portion of the grove floor, irrigation depth must be converted to irrigation volume per tree. A detailed conversion would need to take into account tree planting density and tree size and health. As a rule of thumb, however, ET rates expressed in terms of irrigation depth can be converted to volume by the following formula:

$$\text{gallons (gal)/tree/day} = \text{ET} \times \text{TS} \times 0.622\ \text{gal/in.-ft}^2$$

where ET is expressed in inches (in.)/day, and TS is tree spacing, expressed in square feet (ft^2). The conversion should be used only as a starting point, with actual water estimates based on the size and condition of the trees in each block. Water use per mature citrus tree in a high-density planting will be less than that of trees in a low-density planting, because of the difference in tree sizes. Table 7.1 illustrates the effect of tree density on water use by individual trees.

Irrigation Scheduling

The total seasonal amount of irrigation water needed by a fully grown orchard for optimum yield depends on the daily course of evapotranspiration, the rainfall distribution, and the citrus variety. In a Mediterranean climate, rainfall is concentrated in the winter months, with little or none in summer. During the winter there is normally no need for irrigation except under extreme drought conditions. Hence, there is little reference in the technical literature of such climates regarding irrigation needs during the winter.

In Florida, Texas, Brazil, and Argentina, there may be a considerable amount of rainfall during the summer growing season. Often in these areas, there are periods when rainfall is insufficient to satisfy the water requirements of the crop, and supplemental irrigation is beneficial. Most of the rain in Florida falls from June to September, but during the critical period of January to June, rainfall is low. In Florida, irrigation was found beneficial in eight of nine years, and the average yield increase was 22% with the application of 11 inches of irrigation water per year, in addition to 43 inches of annual rainfall.

The timing of the application and the amount of water applied are governed by the prevailing climatic conditions, stage of growth, soil moisture-holding capacity, and root development. Irrigation during flowering and fruit set in the spring has been found to increase yields of Hamlin, Pineapple, and Valencia sweet oranges in Florida. Subsequent studies showed that irrigating when available soil moisture has been depleted by one-third from January through June and depleted by two-thirds in the remainder of the year resulted in the most efficient use of water, while maintaining yields near the amount produced under irrigation at one-third depletion throughout the year. In studies of young Valencia orange and mature grapefruit grown on sandy Florida soils, microirrigation was found to be most effective when it was initiated at tensions of 15–20 centibars.

In a seven-year experiment with Marrs and Valencia oranges and Ruby Red grapefruit in Texas, irrigation supplied 8.5 to 26.6 inches of water per year, supplementing 29 inches of annual rainfall. Irrigation water was applied when available water was depleted by 30 to 40%, respectively. An optimum irrigation interval of 28 days was recommended for grapefruit and 35 days for oranges during the rainless summer season. The daily water use was 0.16 inch in summer and 0.05 inch in winter. There was no yield response to irrigation when annual rainfall exceeded 24 inches if it was evenly distributed throughout the season. In the Rio Grande Valley, an application of 4–6 inches five to seven times during the year is recommended, with an interval of 21 days between applications in the summer and 60 days in the winter.

Height of the Water Table

The water table fluctuates widely in the coastal flatwoods of Florida during the rainy season, because of the variability of soils, nonuniform rainfall, and high-intensity rainstorms. Soil water drainage is especially important in these areas, since root damage may occur when the water table is high for a prolonged period. Effective water management, which includes both irrigation and drainage of these poorly drained soils, is essential for profitable citrus production.

A shallow water table can significantly augment the water available for root uptake. In permitting the upflux of water from the water table into the root zone, a high water table may decrease the need for supplemental irrigation. Capillary rise may be up to 12 inches above the free water surface in a loamy fine sand but less than 6 inches in a fine sand. In sandy soils, when the water table is at a depth of 20 inches below the mid-root zone, there is generally insufficient upflux to meet the ET demand of the tree.

Irrigation Timing

Trees as Biological Indicators

Trees can serve as biological indicators, since they are continuously "solving" their own water balance equations, and in principle an appropriate physiological parameter is all that should be required. Many such indicators have been suggested and used in citriculture.

Monitoring the growth of the trunk is a good means of following the water relations of trees and can even detect short-term oscillations in their water status. Diurnal changes in trunk diameter are related to leaf water potential (LWP). A period in which trees shrink during the day but do not recover at night indicates severe drought stress. Monitoring trunk growth and shrinkage is fairly simple with modern electronic micrometers. However, care should be exercised to measure at exactly the same point each time.

LWP is a measure of the water status of the tree and varies with leaf exposure and location in the canopy. Ideally, estimates of LWP should be determined before sunrise (when the water potential is at a maximum), when the leaves are unaffected by exposure. Enclosing leaves in aluminum-laminated plastic bags can delay the morning change in LWP and enables the measurement of "predawn" LWP in the early morning hours. Pressure chambers can be used in the field for rapid estimation of LWP.

Systems that measure the velocity of sap in the trunk by the heat pulse method have been perfected and are commercially available. Sap flow systems or other noninvasive methods (magneto hydrodynamic and Doppler systems) should be ideal for estimating water use of trees. These systems theoretically can be used to provide feedback to an automated irrigation system.

The difference between the canopy temperature and the ambient air temperature, ΔT ("delta T"), can be used as a measure of drought stress. ΔT indicates the rate of evaporative cooling due to transpiration and stomatal opening, and it is sensitive enough to detect cycling in stomatal opening. Measured with an infrared thermometer, ΔT has been found to be correlated with LWP. Measurement of ΔT is possible only at high levels of drought stress and is not generally performed in commercial citrus orchards.

Soil Instrumentation

Soil water content can be determined directly by taking soil samples and determining the difference in weights before and after the soil has been dried in an oven (the gravimetric method). It is the simplest and most effective method of monitoring soil moisture. However, it is labor-intensive and time-consuming.

Soil moisture sensors, such as tensiometers, must be used carefully when scheduling irrigation, since typically only a small volume of the soil can be monitored at just a few locations in a large orchard. Special care should be exercised when there are salinity problems, since salt accumulation may reduce water uptake.

A good correlation has been found between tensiometer readings and leaf water potential. The quality of these instruments is improving, and they are becoming suitable for use in commercial orchards, especially where trees are irrigated frequently. Electrotensiometers can be coupled to an irrigation system for automatic scheduling of irrigation. Care should be exercised in determining the depth and placement of tensiometers in order to represent the whole orchard.

Soil water content can be measured by neutron scattering, but this method has been used mainly in research and in large commercial orchards because of the cost and licensing associated with the devices. Recently, advances in neutron probes (lower prices and simpler operation) coupled with advanced software has made the technique popular in several areas. Usually, consultants or cooperatives provide a service to several growers by taking field measurements of soil moisture trends and providing users with tabular or graphical results.

Irrigation and Salinity

All natural waters and soil solutions contain soluble salts. The quantity of salts is commonly reported in units of total dissolved solids (TDS) or electrical conductivity (EC). TDS is measured by evaporating a sample of water and weighing the residue. The results are reported in parts per million (ppm) or milligrams per liter (mg/L), depending on whether the calculation is based on weight or based on volume. For most practical purposes, 1 ppm is equal to 1 mg/L.

Salts in solution exist as ions that conduct electrical current, and EC is greater in solutions with higher concentrations of dissolved salts. EC is measured by means of platinum electrodes and expressed in decisiemens per meter (dS/m, the unit of conductance derived from units of the Système International) or in millimhos per centimeter (mmho/cm). These units are equal in magnitude (1 dS/m = 1 mmho/cm) and are generally in the range of 0 to 5 in waters used for irrigation. The relation between EC and TDS depends on the particular salts present in the solution. In Florida soil extracts and irrigation waters, TDS can be estimated as 700 × EC (in dS/m). In other areas, conversion values of 630 × EC to 640 × EC are commonly used.

The concentration of salts in soil depends on the soil water content, and therefore soil salinity is often related to the electrical conductance of a standard saturated extract (EC_e). EC_e represents the amount of salts in the soil when the soil is saturated. Depending on the soil moisture content, the actual salinity level in the vicinity of tree roots may be several times greater than the EC_e. In sandy soils, where salts are easily leached, management decisions based solely on EC_e measurements are not advised. The EC_e of these soils is only an indication of soil salinity at the time of measurement and can change rapidly following irrigation or rainfall.

Large amounts of salts can be deposited on the soil during continued irrigations with high-salinity water. For example, when the TDS of irrigation water is 2,000 ppm, there is about 1.5 pounds of salt in each 100 gallons. Without proper water and nutrient management, citrus irrigated with high-salinity water can suffer reduced growth and production as a result of salt stress.

Salts in solution exert an osmotic effect that reduces the availability of free (unbound) water through both chemical and physical processes. Roots are therefore not able to extract as much water from a solution that is high in salts as from a solution low in salts. In effect, the tree has to suck harder to move water through its system. This can result in an immediate reduction in root growth followed by reductions in shoot growth and yield.

Citrus is a salt-sensitive crop. The sensitivity is mainly related to the specific toxic effect of the accumulation of chloride and sodium in the leaves. In humid areas, symptoms caused by saline irrigation water are not usually permanent, but affected trees may remain stunted, especially trees that were injured when they were young.

There are distinct differences in rates of chloride and sodium uptake by different citrus rootstocks. In field studies in Texas and California, rootstocks were tested for salinity tolerance according to their ability to exclude chloride from the leaves; in general, from most tolerant to most sensitive, they were Rangpur lime, Cleopatra mandarin, sour orange, sweet orange, Swingle citrumelo, rough lemon, and Carrizo citrange. It is important to remember that the growth and yield of trees on any rootstock can be reduced by excessive salts. Salt-tolerant (chloride-tolerant) rootstocks tend to produce trees that grow more slowly than trees on many salt-sensitive rootstocks. The comparatively high tolerance of sour orange to salinity puts growers all over the world in a dilemma, since the threat of tristeza prevents the use of sour orange, and many tristeza-tolerant rootstocks (such as rough lemon, trifoliate orange, Carrizo citrange, and Swingle citrumelo) are sensitive to salinity.

The critical salinity level varies with the buffering capacity of the soil (depending on the soil type and the organic matter content), climatic conditions, and the soil moisture status. Many salinity-induced symptoms (reduced root growth, decreased flowering, smaller leaf size, and impaired shoot growth) are often difficult to assess but occur prior to ion toxicity symptoms in leaves. Chloride toxicity causes burned necrotic or dry-appearing edges on leaves, one of the most common visible symptoms of salt injury. Sodium toxicity symptoms seldom distinctly appear, but rather an overall bronzing of leaves occurs along with a reduction in growth. Both chloride and sodium can cause nutrient imbalances at much lower concentrations than those required for the development of visible symptoms.

Foliar Damage Due to Sprinkler Spray

Sprinkler irrigation wets the foliage, either partially or fully. Severe damage to leaves in the lower portion (skirts) of trees in orchards with under-tree sprinkler irrigation has been described. Chloride and sodium concentrations in the lower leaves of grapefruit and Valencia and Washington navel oranges have been reported to be about four times greater than those in the upper leaves. The lowest concentration of either sodium or chloride generally associated with leaf burn is about 0.25%. Controlled experiments have shown that citrus leaves easily accumulate chloride and sodium from direct contact with water droplets. The accumulation is greater from intermittent wetting than from continuous wetting and greater from daytime irrigation than from nighttime irrigation. Accumulation is a function of the rate of evaporation, which results in an increased salt concentration in the water film on the leaves. The sensitivity of citrus to injury through direct foliar contact bears no relationship to its tolerance for soil salinity.

Nutrition and Salinity

The frequency of injecting nutrients or applying granular fertilizer has a direct effect on the concentration of TDS in the soil solution. A fertilization program of frequent applications with relatively low concentrations of salts will normally cause less salinity stress than programs of only two or three applications per year. Controlled-release fertilizers and frequent fertigation can be used to economically minimize salt stress from high-salinity irrigation water. Growers using surface water in high-salinity areas generally see a very marked improvement in water quality when the summer rains begin. Under these conditions, fertigation during the wet season should pose no salinity problems.

Selecting nutrient sources that have a relatively small osmotic effect in the soil solution can help reduce salt stress. The osmotic effect of a material in the soil solution is defined as its salt index relative to sodium nitrate, taken to be equal to 100 (Table 7.2). Sources of phosphorus generally have a low salt index, and they usually present little problem. However, the salt index per unit of N and potassium (K) should be considered. The salt indices of natural organic fertilizers and slow-release products are low compared to those of commonly used soluble fertilizers. High-analysis fertilizers may have a lower salt index per unit of plant nutrient than low-analysis fertilizers, since they may be made with a material having a lower salt index. Hence at a given fertilization rate, a high-analysis formulation may have less tendency to cause salt injury.

Selecting nutrient sources that do not add a potentially harmful ion to already high levels in irrigation water can also avoid compounding salinity problems. Materials containing potassium chloride (KCl) or sodium nitrate ($NaNo_3$) add chloride and sodium to the soil solution. High rates of salt application can alter soil pH and thus cause soil nutrient imbalances. Some ions can also induce nutrient imbalances in soil and trees. For example, sodium displaces potassium and, to a lesser extent, calcium in the soil solution, so that irrigation with water containing high levels of sodium can lead to potassium deficiency and, in some cases, even calcium deficiency in leaves. Such nutrient imbalances can compound the effects of salinity stress. Problems can be minimized if adequate levels of nutrients, especially potassium and calcium, are maintained.

Managing irrigation and fertilization with high-salinity irrigation water requires routine evaluations of the water with an EC meter. If excess salts accumulate in the soil, it is best to keep the soil moist so as not to further concentrate the salts. Excess irrigations to leach accumulated salt may become necessary and should be made no less frequently than every other week during the peak irrigation season. Heavy soils and areas with compacted soils or poor drainage may need special attention to salinity management.

In order to minimize salt damage, poor-quality water should be kept off of the leaves, especially under conditions of high evaporative demand. When possible, irrigation should be done at night to minimize evaporative concentration of salts. Fertilizer formulations that have the lowest salt index per unit of plant nutrients should be used. Decreasing the application rate and increasing the frequency of fertilization will reduce the salt content of each application and aid in preventing excess salt accumulation in the root zone. Maintaining optimum but not excessive nutrient levels in the soil and leaves will promote healthy growth and minimize the adverse effects of high-salinity irrigation water.

Fertilization rates should be based on the long-term production of the grove. Application rates for trees under salt stress can usually be lower than for trees irrigated with good-quality water, since the high-salinity trees will probably be less productive. Leaf tissue analysis should be performed to detect excessive levels of sodium or chloride or deficient levels of other elements caused by nutrient imbalances due to salt stress.

Table 7.2. Salt index of inorganic and natural organic fertilizer sources

	Salt index	
Material and analysis	**By equal weight of materials[a]**	**Per unit of plant nutrient**
Nitrogen		
Ammonium nitrate, 35% N	105	3.0
Ammonium nitrate, 20.5% N	61	3.0
Ammonium sulfate, 21.2% N	69	3.3
Calcium nitrate, commercial grade, 15.5% N	65	4.2
Sodium nitrate, 16.5% N	100	6.1
Urea, 46.6% N	75	1.6
Nitrate of soda potash, 15% N, 14% K	92	3.2
Natural organic matter, 5% N	4	0.7
Phosphate		
Normal superphosphate, 20% P_2O_5	8	0.4
Concentrated superphosphate,		
45% P_2O_5	10	0.2
48% P_2O_5	10	0.2
Monoammonium phosphate, 12.2% N, 61.7% P_2O_5	30	0.4
Diammonium phosphate, 18% N, 46% P_2O_5	34	0.5
Potash		
Potassium chloride, 60% K_2O	116	1.9
Potassium chloride, 63.2% K_2O	114	1.8
Potassium nitrate, 13.8% N, 46.6% K_2O	74	1.2
Potassium sulfate, 46% K_2O	46	0.9
Monopotassium phosphate, 52.2% P_2O_5, 34.6% K_2O	8	0.1
Sulfate of potash–magnesia, 21.9% K_2O	43	2.0

[a] Sodium nitrate = 100.

Frost and Freeze Protection

Irrigation can be effective in protecting citrus trees from frost and freezing if it is properly managed. If not managed well, water can cause more damage than benefit. Water applied with overhead irrigation systems has been damaging in windy, advective freezes. Trees irrigated with overhead systems during windy freezes have sustained greater damage than unirrigated trees, because of an inadequate application rate, evaporative cooling promoted by strong winds and low humidity, and limb breakage due to ice loading. Research in the 1980s showed that microsprinkler irrigation could be effective in providing good protection for young trees and partial protection of mature trees, and it is now the most commonly used form of freeze protection in Florida.

Protection of Young Trees

Early work showed that microsprinklers are particularly effective in protecting young citrus trees (Plates 38–40). The spray pattern and application rate influence the effectiveness. Emitters that produce a 90° or 180° pattern concentrate the water on the tree and provide better protection than those with a 360° pattern. In one study, lower leaves that received direct water spray stayed above freezing and were as much as 14°F warmer than dry leaves in unirrigated plots. Other work showed that microsprinklers maintained trunk temperatures near 30°F when the air temperature reached 15°F in a freeze with winds exceeding 20 miles per hour. Ice loading can cause one-year-old trees to be bent over, but trunk breakage is minimal, and the trees usually survive.

The placement of the emitter in relation to the tree is important, particularly in a windy freeze. Microsprinklers must be located upwind of the trees. Young trees sprayed with emitters on the downwind side have sustained significantly more damage than trees receiving no irrigation, because the wind periodically blew the spray away from the trees. The irregular wetting and drying of the trunk causes greater damage as a result of evaporative cooling. Similar damage due to evaporative cooling was observed on mature trees after under-tree impact sprinklers froze and stopped applying water.

Greater protection of young trees can be obtained if the microsprinkler spray is angled upward. In a severe freeze with a minimum temperature of 19°F, emitters angled upward increased the survival height by 1 foot over that obtained with a horizontal spray pattern. Evaporative cooling did not lower the temperature of the trunk below the air temperature in any of the continuous microsprinkler treatments, even when the dew point was as low as 1°F and wind speeds were above 20 miles per hour. Evaporative cooling of as much as 3°F was seen when an intermittent spray mist caused discontinuous, irregular wetting of an unirrigated tree. Hence, if microsprinklers are run continuously at a sufficient application rate, there appears to be little risk of evaporative cooling in the wetted zone. Irregular wetting or a power failure causing a cessation of irrigation can lead to additional cooling and tree damage. Brownouts that caused electrically powered pumps to fail have resulted in major damage in some freezes. With insufficient irrigation, evaporative cooling of as much as 7°F has been reported.

In the wetted area, protection of young trees is primarily by the latent heat of fusion from the formation of ice (80 calories per gram). The application rate in the wetted area determines the level of protection. Several models have been developed to predict the application rate needed to protect trees at a given temperature. Application rates of 0.25 to 0.35 inch/hour successfully protected young trees in advective freezes with minimum temperatures as low as 19°F. Small-diameter spray patterns now available can produce application rates of more than 1 inch/hour, which should give reasonable protection to young trees at temperatures as low as 14°F or possibly lower.

Protection of Mature Trees

While microsprinklers are better at protecting young trees, they also provide partial protection to mature trees. The greatest benefit has been observed in calm frosts, but there have been several examples of improved recovery of mature trees when microsprinklers have been used in windy freezes.

In a relatively calm frost, microsprinklers benefited trees over 13 feet tall. Three months after a freeze, trees that had been sprayed with microsprinklers had recovered better and developed greater canopy density than trees that had not been irrigated. Emitters with greater output provide more protection. High-volume emitters have been found to provide more warming than low-volume emitters at several heights in the tree canopy. The greatest increase in air temperature was found close to the ground. At a height of 12 feet, the temperature in microsprinkler treatments was not consistently different from the temperature in an unirrigated control. Leaf temperature was found to depend on the distance from the microsprinkler emitter and the amount of exposure. Leaves on the outer edge of the canopy in the microsprinkler treatments were 1°F warmer, and inner leaves were as much as 2.5°F warmer than corresponding leaves of unirrigated trees. While an elevation in air temperature could not usually be detected in the upper part of the canopy of mature trees, other observations have shown improved overall appearance of the canopy in some cases.

Microsprinkler irrigation is more effective for protection against cold when high volumes of water are used. Approximately 2,000 gallons/acre/hour is needed to provide reasonable protection. This can be accomplished with one emitter per tree supplying 20 gallons/hour in a grove with 100 trees/acre. Irrigation at rates below this level will provide some protection but not as much as at higher rates. Application rates of 3,000 gallons/acre/hour or more can provide even more cold protection.

It is generally thought that microsprinklers provide little or no protection for fruit, although the percent marketable fruit was slightly higher in one study when the air temperature fell to 17°F. At temperatures below this, microsprinkler systems were found to be ineffective for leaf or fruit protection. Depending on climate and cold-acclimating conditions, citrus fruit will generally not be damaged unless the temperature drops below 28°F for 4 hours or more. Since microsprinklers can raise air temperature slightly, they can modify both the minimum temperature and the duration of the coldest period. Hence, in theory, microsprinklers could potentially reduce fruit damage somewhat in the lower part of the canopy under borderline frost conditions.

Several factors may contribute to the effectiveness of microsprinkler irrigation of mature trees. In addition to the latent heat of fusion from the formation of ice, other factors include sensible heat from the water, the formation of fog or mist, increased humidity, and increased soil heat capacity and thermal conductivity. Well water can have a temperature of up to 70°F. As this water cools, it releases sensible heat. Fog does not always form on freeze nights, but microsprinklers can create fog when the dew point is relatively high. Water droplets, particularly those with diameters in the range of 10 micrometers (10 μm), reduce long-wave radiation, but the contribution of microsprinkler-generated fog to freeze protection has been questioned by some. Increased humidity can also reduce radiation loss and may contribute to warming. Any irrigation system that wets the soil, including microsprinklers, increases the thermal conductivity of the soil and the rate of energy transfer from the soil to the atmosphere near the soil surface. High-density plantings may also help slow the loss of heat from microsprinkler-irrigated groves.

Depending on climate and cold-acclimating conditions, citrus fruit will generally not be damaged unless the temperature drops below 28°F for 4 hours or more.

Elevated Emitters

More recent work has emphasized elevating microsprinkler emitters and positioning them inside the canopy. In several Louisiana studies, emitters that delivered 25 to 30 gallons/hour were mounted in the center of the canopy about 2.5 to 3.3 feet above the soil surface. Water sprayed on the major scaffold limbs and ran down the trunk. In a severe 1989 freeze where the minimum temperature reached 12°F, elevated emitters provided over 94% tree survival and 84% scaffold branch survival. Unprotected trees were completely killed. Even though irrigation of the scaffold branches caused some limb breakage, the advantages of tree survival far outweighed the disadvantages of limb breakage. Because of the small-diameter spray pattern, the application rate in these studies was as much as 1 inch/hour.

Additional studies in Florida during a very severe 1989 freeze showed that conventional microsprinklers elevated to a height of 0.9 meter gave protection to a height of more than 1 meter. Trees 6 to 8 feet tall protected with elevated microsprinklers recovered within one year to heights greater than before the freeze. Emitters that delivered as little as 7 to 10 gallons/hour provided protection for the lower canopy of small trees with a minimum temperature of 17°F. The small-spray diameter, with estimated application rates of 0.30 and 0.45 inch/hour in the wetted area, contributed to the effectiveness of these elevated emitters. Trees with emitters mounted at the traditional position near ground level sustained greater damage.

Unirrigated trees were completely killed. Where the elevated emitter saved part of the canopy, tree recovery was rapid, and fruit yield recovered to moderate levels two years after the freeze. Fruit yield in a grove with emitters at a height of 2 feet was approximately 260% greater than that of a similar, adjacent grove where emitters were mounted near ground level. If microsprinklers are mounted too high and application rates are insufficient, evaporative cooling and limb breakage can occur. Studies to determine optimum height and application rates are continuing.

Chemigation

Fertigation (applying fertilizer through an irrigation system) is a common practice with microirrigation. The success of fertigation in microirrigation may depend on the size of the active root system, which in turn depends on the amount of summer rain in the area. In humid areas (such as Florida) the active root system expands to cover much of the orchard floor during rainy periods. In these areas, microsprinkler coverage should be at least of 80% of the tree canopy area to provide adequate fertilizer distribution to the roots. With drip irrigation, which usually covers much less ground area, very frequent fertigation may be needed.

Reports from Japan indicate that overhead impact sprinklers can also be used to apply fertilizers, fungicides, herbicides, and pesticides with excellent results. Some of the chemicals should be applied without the surfactants or emulsifiers commonly used in spray treatment. Good results were obtained when NAA was applied by sprinklers for fruit thinning of satsuma mandarin. Herbicides such as diuron, DCPA, terbacil, and bromacil were used safely for herbigation by means of overhead sprinklers. Water should be supplied after the application of the chemical in order to wash the herbicide off the citrus leaves.

Under-canopy irrigation is particularly suited for applying herbicides to orchards in arid areas where weeds do not develop outside the irrigated area. Compared with conventional application methods, herbigation with bromacil plus diuron gave better control of bermudagrass and other weeds with lower quantities of herbicides. In Spain, good results were reported for drip herbigation in which herbicides were applied continuously in low concentrations. Under-canopy sprinklers have been used successfully for applying copper-based fungicides to control brown rot of fruit. Systemic pesticides can be applied through the irrigation system, but care must be taken to prevent any contact with drinking water, and all precautions must be taken to prevent contamination of water sources or groundwater.

Fertigation Management

Care must be taken to ensure that injected materials do not react with dissolved solids in the irrigation water so as to form precipitates or deposits in the irrigation system. The chemicals must be soluble and remain in solution throughout the operation of the irrigation system. The fertilizers selected to be injected into irrigation water need to be entirely soluble in water and should not react with salts or chemicals in the water. Most nitrogen sources cause few clogging problems. The exceptions are anhydrous ammonia, aqua ammonia, and ammonium phosphate, which increase the pH of the water and form precipitates with calcium and magnesium. Most forms of phosphorus can result in extensive clogging, but phosphoric acid can be safely injected in most waters, since it acidifies the solution to a point at which precipitation is prevented. All of the common potassium fertilizers are readily soluble and present no clogging problems.

Fertilizers can be highly corrosive and can cause injuries to the skin and eyes. Therefore, all system components, including pumps, injection devices, lines, filters, and tanks, should be inspected prior to use. There should be a routine monitoring program of the fertigation process, with particular emphasis on the start-up and shut-down periods. Injection rates and times should be calibrated and rechecked frequently to ensure proper operation of the system. Leaks, runoff, excessive application, and application to areas with open water should be prevented. All system components should be flushed with clean water following each use.

The salinity of irrigation water containing injected fertilizer should be checked. Heavy dosages of fertilizers can cause leaf burn, even if relatively low-salinity water is used. A program of injecting small dosages of fertilizer frequently is generally preferable to a program of fewer applications at a high rate.

It is essential that proper and legal backflow prevention devices be used in the irrigation system to prevent fertilizers from being back-siphoned into the water supply. The injection device itself should have a screen and check valve. It is recommended that injection take place upstream of filters, so that any contaminants or precipitates can be filtered out.

Fertigation rates and times should be calibrated for each area that is fertigated. The flushing time needs to be at least as long as the travel time in the system from the injection point to the farthest emitter. In many microirrigation systems, this travel time is often 20–30 minutes. Fertilizer injections need to last at least this long, and flush times need to exceed this time, so that nutrients will not remain in the lateral tubing and promote algal growth.

David P. H. Tucker
Megh Singh
University of Florida
Citrus Research and Education Center, Lake Alfred

CHAPTER 8

Integrated Vegetation Management in Citrus Production

Citrus grown in warm climates with relatively high rainfall is subject to more intense pressure from weeds than in drier areas. In a warm, rainy environment, natural vegetation can smother young citrus trees if it is left unmanaged. The management of vegetation accounts for as much as 20–25% of total production costs, and lost revenues have been estimated at about $150–200 million annually for large citrus industries. Such losses reflect the cost of 1) operations to control or suppress weeds, including ditch bank and aquatic weed control, 2) production lost as a result of direct competition between weeds and the trees, 3) reduced efficiency of production and harvesting operations, 4) lower orchard temperatures during freeze events, and 5) other less well defined impacts. Much of current citrus acreage is under some form of integrated vegetation management involving chemical, mechanical, cultural, manual, and biological methods.

A weed is a plant growing in an area where it was not intentionally planted, such as the rooting zone of a citrus tree, whereas indigenous or introduced ground cover in row middles can play a positive role in orchard floor management. Weed species are considered economic pests if they harm the growth or health of the trees or if they delay bearing and ultimately production. Eventually, the trees are deprived of whatever portion of resources is utilized by the ground cover. The more competitive the vegetation, the more adversely it alters tree physiology, growth, and fruit yield and quality.

Attaining early production requires controlling the growth of weed competitors. Weeds alter economic status by competing with trees, particularly young trees, for water, nutrients, and even light (in the case of climbing vines and tall weeds such as guineagrass, which can easily cover young trees if left uncontrolled). Weeds also have indirect effects on tree performance, referred to as interference: they reduce the efficiency of low-volume irrigation systems, intercept soil-applied pesticides, reduce orchard temperatures under freezing conditions, and have less clearly defined effects on insect and mite populations, disease incidence, and vertebrate pest activity.

While focusing primarily on the negative effects of weeds on citrus production, we have paid little attention to their less visible beneficial effects, which have attracted less research activity and funding. More comprehensive understanding is necessary in order to integrate the management of weeds with the management of other pests and diseases.

It is the purpose of this chapter to discuss citrus and vegetation management in terms of tree health, vigor, and production; to consider control practices as components of the overall integrated citrus management system; and to place emphasis not only on the direct effects of weeds but also on interactions with the physical and biological environment and other orchard management practices. Integrated vegetation management (IVM) emphasizes the integration of control factors with all other management practices. It involves the selection, integration, and implementation of effective vegetation management strategies in light of economic, ecological, sociological, and environmental consequences.

Vegetation in Orchards and Basic Principles of IVM

Different species of vegetation are not equally competitive with citrus trees. Grasses, especially sod-forming species, are more aggressive competitors than most broadleaf weeds. Mowed grass swards can be more competitive than unmowed grass, because of the demands of vigorous regrowth. The relatively sparse growth of weeds in a poor sandy soil may be more harmful than the growth of weeds in a heavier soil with more moisture and greater nutrient reserves to be shared between trees and weeds. Millions of seeds of numerous species reside in the surface layers of the soil in which tree roots are established. From these seed reservoirs each species has periods of emergence, ensuring weed cover for much of the citrus growing season. With time and appropriate methods of control

From a practical and economic standpoint the complete elimination of weed seed and plant parts is neither feasible nor desirable.

or suppression, weed seed reservoirs become depleted. However, one season of poor management can all but eliminate the benefits accrued over time. From a practical and economic standpoint the complete elimination of weed seed and plant parts is neither feasible nor desirable. However, in many situations it is desirable to eradicate certain aggressive noxious weeds. Practices consistent with an IVM philosophy will substantially reduce the number of weed seeds and vegetatively reproduced parts, making management programs less costly over time.

Reasons for the introduction, establishment, and spread of many species considered to be weeds of economic importance include the following:

- Weeds can move into orchards from ditch banks and fencerows and from adjacent areas where they reside among the native vegetation.
- Some species have an extremely vigorous and aggressive growth habit and efficient means of propagation from seed or vegetative parts.
- Hand labor for the removal of herbicide-tolerant species and vines established in tree canopies has been eliminated or reduced, so that these weeds can form seeds and eventually disseminate them.
- Weeds can become established where winter temperatures are not low enough to kill them.
- Some weed species tolerate or partially tolerate soil-residual and postemergence herbicides applied at low rates, which nonetheless control many competing weed species.
- Adverse economic conditions have created a trend toward reduced orchard care and greater reliance on postemergence herbicides.
- The practice of applying residual herbicides in a narrow band in the citrus row allows vines to germinate and become established in the tree canopy.
- Seed and weed parts capable of vegetative propagation may be moved among and within orchards by mowers and other equipment.
- The low-skirted nature of citrus trees enables vines to easily climb into the canopy.
- Dislodged weed parts left on orchard floors may be sufficient to support a reinfestation.

Moisture

Weeds dry the soil by extracting moisture through their roots and transferring it to the atmosphere. They also intercept rainfall and irrigation water. In the absence of vegetation, moisture evaporates rapidly from the damp soil surface until a dry surface layer is formed, and then further evaporation is slow. Physiological mechanisms by which trees respond to moisture stress have an impact on crop production costs. As leaves wilt and stomata close for longer periods of time, photosynthesis is reduced, and leaves may eventually drop. Photosynthates are needed for active root growth, and root growth is needed for water uptake, and therefore competition for water can ultimately result in reduced shoot growth and even dieback and eventual death of weed-infested young trees.

The soil type and the depth of rooting are also factors in the competition with weeds, as trees with shallow root systems are more adversely affected by competition than those with deeper roots. Trees with shallow root systems therefore require more frequent irrigation to maintain adequate soil moisture in the surface layers of the soil, if weeds are not adequately controlled in the root zone.

Where orchards are flooded, ground cover can play a positive role by removing excess water by transpiration. Under anaerobic conditions, however, the decomposition of vegetation may release toxic by-products, such as hydrogen sulfide, which is toxic to citrus roots.

Nutrients

Trees and weeds compete for nutrients, but because moisture and nutrient competition are closely interrelated, the actual situation is more complex. As trees and weeds dry out the soil, trees can no longer extract nutrients effectively. Weed control also increases nutrient availability, because decomposing vegetation releases nutrients into the soil. Associated changes in soil moisture and temperature accelerate the decomposition. Fertilizing without first controlling weeds can depress tree growth, as weeds are often more responsive to nutrient application and grow more rapidly. Young trees, in particular, should not be fertilized without first controlling weeds.

Certain leguminous plants are being evaluated as beneficial ground covers for citrus orchards, on account of their overall competitiveness, influence on pest populations, growth habits, and nitrogen contribution.

Light

In addition to the roots competing for moisture and nutrients, the aerial parts of vines and tall grasses can interfere with tree growth and productivity. Aggressive vine species can envelop and smother trees, intercepting sunlight and ultimately resulting in canopy dieback and reduced production.

Recently established ground covers around young trees can also provide some protection to newly established trees exposed to wind, blowing sand, or sunburn.

Temperature

Weeds can influence tree growth by their moderating effects on soil and air temperature. Citrus root growth is quite dependent on soil temperature, although this is a less important factor in humid, subtropical growing areas, such as Florida. Living or dead vegetation insulates the soil, keeping it cooler in the summer and warmer in the winter, and reduces daily variations in temperature. The insulating effect of vegetation reduces heat loss by radiation from the soil surface. Nighttime temperatures under calm conditions are lower over tall vegetation than over compact, bare soil, freshly cultivated soil, or soil with minimal vegetative cover. Trees surrounded by bare soil therefore are more likely to escape cold damage under marginal freeze conditions. Trees growing among dense vegetation during relatively dry periods can be damaged or killed by fire, since weed cover killed by frost can be easily ignited. Orchards with no-till chemical mowing programs have been shown to be 2–4°F warmer during freezes than those with sod cover.

Cultivars

Not much is known about the relative abilities of various scion-rootstock combinations to compete with weeds. However, we would expect that trees on more vigorous rootstocks, such as rough lemon or Volkamer lemon, with their more extensive root systems and greater foraging abilities, would be better competitors and may therefore be less impacted by weed competition, particularly in deep sandy soils. More rapid canopy development will also shade out vegetation on the orchard floor in a shorter time during orchard establishment.

Tree Quality and Planting

Poor-quality trees with relatively weak root systems may be more susceptible to competition after planting if the soil surface is not relatively weed-free. The competitive ability of trees is generally reduced if they have been improperly planted, with poor distribution of roots in the planting hole, desiccation of roots, or excessive root pruning prior to planting.

Identification and Monitoring of Vegetation

There are three categories of weeds: broadleaf plants (including vines), grasses, and sedges. A combination of characteristics should be used in the identification of a species, and several plants should be examined to confirm identification. Management is always easier if weeds can be identified at the seedling stage, but identification at this stage requires a higher level of training. Since a species may behave as an annual or as a perennial depending on its location, characteristics such as aboveground rooting stems or stolons and underground storage organs should be sought for the purpose of identification and management.

> **Observations should be recorded on a weed survey form with special attention paid to newly established species and difficult-to-control perennial weeds.**

A satisfactory vegetation management program with ongoing monitoring and timely control measures is an integral part of IVM. Sources of weed infestation (seed or plant parts capable of vegetative propagation), if identified and controlled prior to dissemination, will greatly reduce the cost of vegetation management operations and other indirect losses. New species, if perceived as potential problems, should immediately be controlled. Many currently troublesome invasive species arrive from sites outside the orchard, such as ditch banks, fencerows, and woodlands, and these areas should also be closely monitored. Vehicles and equipment can move seed and vegetative parts of weeds from one orchard to the next and should therefore be cleaned regularly.

It has been estimated that of the more than 100 annual and perennial grasses, broadleaf weeds, and vines that commonly occur in Florida's groves, only about 30 are considered very undesirable. Of these, perhaps 20 are capable of becoming or have become serious pests as evidenced by their undesirable growth behavior, competitive effects, and cost of control. A good example is milkweed vine, or strangler vine (*Morrenia odorata*): one plant is able to smother a tree, if left unchecked. Table 8.1 lists some economically important citrus weeds by their impact on yield in citrus-producing states of the United States.

Some important weed species in other areas of the world include *Setaria* species, *Digitaria* species, *Echinochloa colona, Raphanus raphanistrum, Sinapis arvensis, Stellaria media, Veronica hederifolia, Poa annua,* and *Colium* species.

The spectrum of weeds in citrus orchards varies with location, climate, soil type, and previous history, including pest management programs. For a successful management program, weeds in the orchard must be identified, their location and their abundance must be determined, and changes in weed populations over time must be monitored. Information collected over several years is invaluable in determining changes in the weed flora, so that adjustments can be made in the weed management program.

Surveys should be scheduled at least three times a year, in fall, spring, and summer, possibly by scouts doing other monitoring work, if they are trained in weed identification. Tree rows (under and outside canopies), row middles, and water furrows should be evaluated, and weed species and their abundance noted. As these locations may receive different treatment combinations and be subjected to various cultural practices (e.g., microirrigation under and between canopies), the weed species present and their vigor and abundance may vary.

Observations should be recorded on a weed survey form (Fig. 8.1) with special attention paid to newly established species and difficult-to-control perennial weeds. Vines, shrubs, and other weeds inhabiting ditch banks, fencerows, and other off-site locations that are sources of further infestation should be included. An orchard map should be used to locate infested areas, heavy soil and wet areas favoring luxuriant weed growth, and off-site sources. Resources can then be concentrated on preventing small problems from becoming large and more costly ones. In new orchards, weed management begins before planting. The site should be surveyed, particularly for hard-to-control perennials that can pose a problem for young trees during the critical establishment period. With experience, the species that pose the greatest challenge, because of their vigor, their competitive effects, and the difficulty of control, will be recognized.

Methods of Weed Management

Mowing, cultivation, and application of herbicides are common methods of weed management widely used in citrus orchards. Each method has its advantages and disadvantages. The choice of which method or combination of methods to use depends on the grower's expectations, the orchard location and planting system, the weed species in the orchard, and available resources.

Cultivation

Annual weeds are killed efficiently and economically by severing the stems from the roots while the plants are young. Each crop of weeds must be killed in order to prevent competition with the trees and the production of seeds. When seeds are produced, however, they are incorporated into the soil during the next cultivation, where they may remain viable for many years. Each cultivation brings seeds to the soil surface, where they can germinate. If environmental conditions are right, a new crop of weeds is produced. The sequence of cultivation followed by weed seed germination may continue indefinitely. The number of cultivations required should be de-

Table 8.1. Economically important citrus weeds ranked by their impact on yield in various citrus-producing areas of the United States[a]

Common name	Scientific name	Arizona	California	Florida	Louisiana	Puerto Rico	Texas
Arizona brome	*Bromus arizonicus*	5	...	...	...	...	...
Balsamapple	*Momordica charantia*	...	...	2	...	...	...
Bermudagrass	*Cynodon dactylon*	1	2	2	...	2	1
Briar	*Smilax* spp.	...	...	4	...	...	...
Buffelgrass	*Pennisetum ciliare*	...	...	...	...	...	3
Camphorweed	*Heterotheca subaxillaris*	...	...	4	3	...	3
Climbing milkweed	*Sarcostemma cyanchoides*	...	...	...	...	...	1
Common lambsquarters	*Chenopodium album*	...	5	4	...	...	5
Common ragweed	*Ambrosia artemisiifolia*	...	...	2	4	...	1
Crowfootgrass	*Dactylotenium aegyptium*	...	...	3	...	...	3
Cutleaf eveningprimrose	*Oenothera laciniata*	...	...	3	...	...	5
Dallisgrass	*Paspalum dilatatum*	...	3	...	...	...	...
Dayflower	*Commelina* sp.	...	...	3	...	2	...
Dieffenbachia	*Dieffenbachia* sp.	...	...	...	3	...	...
Dracontium	*Dracontium* sp.	...	...	...	3	...	...
Field bindweed	*Convolvulus arvensis*	...	4	...	...	...	...
Goatweed	*Scoparia dulcis*	...	...	3	...	...	...
Goosegrass	*Eleusine indica*	...	...	3	...	...	3
Goundsel	*Senecio* sp.	...	5	...	...	...	...
Guineagrass	*Panicum maximum* (broadleaf)	...	...	1	...	5	1
Guineagrass	*Panicum maximum* (narrow leaf)	...	...	1	...	...	...
Hairy beggarticks	*Bidens pilosa*	...	...	2	...	...	...
Hare barley	*Hordeum leporinum*	5	5	...	...	...	...
Horseweed	*Conyza canadensis*	3	3	4	...	...	...
Jamaica feverplant	*Tribulus cistoides*	...	...	...	...	...	3
Johnsongrass	*Sorghum halepense*	4	2	...	2	4	3
Largeleaf lantana	*Lantana camara*	...	...	3	...	...	...
Little mallow	*Malva parviflora*	...	2	...	...	...	...
Littleseed canarygrass	*Phalaris minor*	4	...	...	...	...	...
London rocket	*Sisymbrium irio*	3	5	...	...	...	...
Mexicantea	*Chenopodium ambrosioides*	...	...	5	...	...	...
Morningglory	*Ipomoea* sp.	...	...	3	...	3	1
Nettle	*Urtica* sp.	...	...	...	...	...	5
Nettleleaf goosefoot	*Chenopodium murale*	2	4	...	...	...	...
Nightshade	*Solanum* sp.	...	3	4	...	...	5
Paragrass	*Panicum purpurascens*	...	...	4	...	...	...
Peppervine	*Ampelopsis arborea*	...	...	3	...	...	...
Phaseybean	*Macroptilium lathyroides*	...	...	3	...	...	...
Pigweed	*Amaranthus* sp.	1	3	2	1	...	2
Plains bluestem	*Bothriochloa ischaemum*	...	...	...	...	...	3
Possum grape	*Cissus verticillata*	...	...	...	...	...	3
Purple nutsedge	*Cyperus rotundus*	2	4	4	...	...	...
Purslane	*Portulaca* sp.	5	4	4	2	...	1
Rhodesgrass	*Chloris gayana*	...	...	...	...	...	3
Signalgrass	*Brachiaria* sp.	...	...	5	...	...	...
Southern crabgrass	*Digitaria ciliaris*	...	4	4	3	...	...
Southern cupgrass	*Eriochloa gracilis*	3	...	...	...	...	...
Southern sandbur	*Cenchrus echinatus*	3	...	4	...	...	2
Sowthistle	*Sonchus* sp.	4	4	...	...	...	3
Sprangletop	*Leptochloa* sp.	...	3	...	...	...	5
Spurge	*Euphorbia* sp.	4	2	4	...	...	4
Strangler vine	*Morrenia odorata*	...	...	2	...	...	...
Sunflower	*Helianthus* sp.	...	4	...	...	...	3
Texas panicum	*Panicum texanum*	...	...	4	4	1	1
Torpedograss	*Panicum repens*	...	...	3	...	...	...
Turkey mullein	*Eremocarpus setigerus*	...	5	5	...	...	...
Vaseygrass	*Paspalum urvillei*	...	...	4	...	...	...
Vervain	*Verbena* sp.	...	...	...	...	...	4
Virginia creeper	*Parthenocissus quinquefolia*	...	...	3	...	...	...
Wild mustard	*Brassica* sp.	...	5	...	...	...	3
Yellow nutsedge	*Cyperus esculentus*	...	4	4	...	...	...

[a] 1 = Greatest impact; 5 = least impact.

WEED SURVEY FORM

Orchard/block ______________________________

Monitoring date ______________________________

Herbicide program ______________________________

Application date(s) ______________________________

Comments ______________________________

____

____

Infestation level

1 = Very light 2 = Light 3 = Moderate 4 = Heavy 5 = Very heavy

Perennials (grasses, broadleaves, vines, sedges)		Annuals			
		Grasses		Weeds and vines	
Type	Level	Type	Level	Type	Level

Fig. 8.1. Survey form for monitoring weed species.

termined by the number of flushes of weed emergence. The timing of cultivation for annual weed control is not as critical as the timing of application of soil-active herbicides.

Cultivation at proper time intervals can ultimately kill deep-rooted perennial weeds by exhausting their underground food reserves. After each cultivation, these weeds send new shoots to the surface, consuming stored food in the process. The withdrawal of food reserves may continue for a week or more after the new shoots emerge. When sufficient foliage has developed, the food reserves are replenished by products of photosynthesis. Cultivation is most effective at the time of maximum net food depletion.

> **Each crop of weeds must be killed in order to prevent competition with the trees and the production of seeds.**

Cultivation is no longer widely used as a control method in many citrus-growing areas. It rarely achieves control of many perennials in citrus orchards. Infrequent cultivation provides temporary control, but it spreads and invigorates perennials by increasing the number of buried seeds and by widely distributing rhizome and stolon cuttings, tubers, and bulbs. The result of infrequent tillage is the establishment of solid stands of resistant species such as guineagrass, bermudagrass, torpedograss, and nutsedge. Repeated cultivation in citrus orchards can also result in damage to tree trunks (Plate 41) and the destruction of fibrous surface roots, which normally grow in the undisturbed portion of the soil. The cultivated portion of the soil is therefore of little use to the tree (Plate 42).

Mowing

Mowing between rows and away from the trees is practiced together with the application of herbicides in the rows over the major root zone of the trees (Plate 43). It is now common in Florida, particularly in orchards with bedded flatwoods soils, where cultivation is not recommended because of the shallow root systems of the trees. It is appropriate where a cover crop is desired in a bedded orchard or on sloping terrain to prevent soil erosion. Mowing is easy and gives the orchard a pleasant appearance, but it has a high energy demand and high costs for equipment maintenance. Mowers can also spread weed seed and vegetation, reinfesting tree rows where herbicides have been applied.

Weed control by mowing is useful against tall-growing weed species, but it gives low-growing, prostrate species an overwhelming competitive advantage. In mowed orchards in many areas, a solid stand of perennial grasses rapidly develops and becomes the stable vegetation, and the orchard is thus converted to sod culture. Even frequent mowing will not sufficiently lessen the consumption of water by perennial grass, especially under drought conditions. The yield and growth of trees in sod culture are often reduced in comparison with trees in a chemical management program.

Chemical Mowing

Chemical mowing, including low-rate technology, postemergence herbicide spray applications, and herbicide wiping, has been used in combination with mechanical mowing for the suppression of vegetation in row middles. These practices are referred to as middles management. With the frequency and increasing cost of mechanical mowing required to maintain control of vegetation in row middles, chemical mowing and wiping with glyphosate at low rates have become increasingly popular (Plates 44–46). Under middles management, mechanical mowing is done less often, at reduced cost, and eventually results in the elimination of tall-growing species and the establishment of more manageable sod species such as bermudagrass and bahiagrass. Broadleaf weeds may sometimes become established under such programs requiring modification of management practices. Chemical suppression of grasses and broadleaf weeds reduces their competitive effects on citrus more than mechanical mowing. Recent data from Florida indicated that where soil moisture levels were monitored under different middles management programs, soil moisture was higher under chemically mowed sod.

Chemical Control

Herbicides play a pivotal role in IVM strategies. Chemical weed control in tree rows is now widely accepted in the citrus industry (Plates 43 and 47). The success of chemical weed control is far more likely to be ensured if attention is paid to the following:

- The information on the product label regarding the properties, advantages, limitations, and precautions for the use of the herbicide, including its potential for leaching and movement into groundwater
- The use of herbicides appropriate for the particular weed species to be controlled
- The selection of the lowest application rate, according to the label recommendations, that will be effective against the target species in the particular soil to which it will be applied
- The timing and frequency of application, to minimize leaching during the rainy season and ensure maximum effectiveness against the target species
- The calibration of equipment for the application rate required
- A boom design that achieves uniform coverage of the surface while minimizing contact with the canopy and fruit.

Herbicides can be classified as foliar or soil-applied and as having systemic or contact activity. Preemergence herbicides are absorbed through weed root systems; they are most effective when applied during germination and the early growth stages of seedlings. Systemic herbicides are absorbed by either roots or aboveground parts of plants and are translocated throughout the plant. Contact herbicides act as desiccants, damaging or killing all plant parts actually sprayed, with little if any translocation. To control well-established perennial weeds, a postemergence systemic herbicide should be used with a preemergence material.

The herbicides currently recommended for preemergence control in Florida citrus are bromacil, diuron, simazine, oryzalin, and norflurazon. These compounds control a wide range of grass and broadleaf weeds. Herbicides registered for postemergence control include paraquat, glyphosate, fluazifop-butyl, and sethoxydim. Herbicides used in other areas, in addition to those listed above, include trifluralin, oxyfluorfen, dichlobenil, atrazine, napropamide, thiazopyr, and pendimethalin. The marketing policies of agrichemical companies, the actions of regulatory agencies, and local conditions, including soil types and varieties and environmental constraints, determine to some extent which herbicides are registered for each industry.

For herbicide recommendations, refer to the current *Florida Citrus Pest Management Guide* (University of Florida, Institute of Food and Agricultural Sciences), the *University of California Citrus Treatment Guide,* and publications from other areas, including a new, comprehensive publication from Spain, *Sistemas de Manejo del Suelo en Citricultura.*

Herbicides and Tree Phytotoxicity

Soil-residual herbicides can have phytotoxic effects on citrus trees, causing well-defined symptoms on foliage. The occurrence of symptoms is usually related to the type of herbicide used, rate of application, soil type, tree vigor, and citrus variety. Mild to moderate foliar symptoms of soil-residual herbicides, such as bromacil, diuron, simazine, and norflurazon, will not normally reduce the growth of healthy trees, provided that steps are taken to reduce further uptake by lowering the application rate or rotating herbicides in future applications. Persistent severe symptoms can result in injury or death of the tree. Foliar symptoms of trifluralin and oryzalin injury have not been observed, and inhibition of tree roots has not been documented at normal rates of application. As rates of application have been reduced and application frequency increased for economic reasons and to reduce the potential for groundwater contamination, the occurrence of phytotoxicity has also been reduced.

The use of postemergence herbicides has increased, and some symptoms of phytotoxicity due to these herbicides have been observed. These symptoms occur on foliage and fruit as a result of spray contact. Certain postemergence applications can cause significant injury to young trees and growth retardation, but the effect on bearing trees is usually limited to some fruit drop.

Safety

Misapplication of herbicides can damage or kill desirable vegetation and citrus trees, cause undesirable environmental effects, and result in legal proceedings. To reduce the impact of herbicides on the environment, particularly groundwater and surface water, greater efforts should be directed toward

- Minimizing application rates within recommended ranges for weed control
- Avoiding the use of herbicides with high potential for leaching in areas with a history of groundwater contamination
- Improving irrigation management to reduce leaching
- Monitoring drainage outflows to determine herbicide levels in water
- Improving erosion control to minimize the movement of surface-applied chemicals into ponds, lakes, and waterways
- Improving mixing and loading procedures and thorough on-site cleanup to avoid tree damage and contamination of adjacent water sources
- Minimizing drift by avoiding herbicide application under windy conditions and by using shielded booms, low pressures, and, if necessary, drift control additives
- Maintaining crop residue levels within approved ranges by using recommended application rates and avoiding excessive spray contact with trees and fruit
- Avoiding application close to desirable plants and trees susceptible to the herbicide in use

Susceptibility of Weed Species to Some Citrus Herbicides

In order to achieve optimum weed control, target species should be correctly identified, the appropriate herbicide and rate selected, and a timely application made. Weed species vary in their susceptibility to soil-residual herbicides; they are most susceptible during germination or the early growth stages of seedlings.

Weed susceptibility ratings (Table 8.2) have been compiled based on product label information, experimental data, and field observations. The codes used in the table have been selected to represent the varying degrees of susceptibility and tolerance and are believed to be a reasonably accurate representation of what occurs under commercial weed management conditions.

Generally, all weed species listed as susceptible on a herbicide product label will be controlled by that herbicide if it is applied at an appropriate rate and time. Conditions during and following application are also important, including moisture from rainfall or irrigation. Poor control can sometimes be expected from postemergence applications when weeds are under stress, because of poor uptake and translocation of the herbicide. Failure of a herbicide program is usually due to one of these factors or due to the application procedure, including calibration and equipment design and operation, if the appropriate herbicide or herbicide mixture has been selected for the target weed.

Selection of Herbicides and Rates of Application

The choice of herbicides and application rates depends on the registration status of the materials, label directions, soil type (organic matter content and leaching potential), weed species, stage of weed growth, tree age, and scion variety. Preemergence and selective postemergence herbicides are effective against certain vegetative species, less so against some, and ineffective against others. Therefore, in most situations, one or two preemergence products and a contact or systemic herbicide are used in tank mixtures.

Timing of Application

Timing is important in the application of soil-residual herbicides, which are most active against germinating weed seedlings. When they are used in areas of emerged weeds, they are combined with contact or systemic herbicides in tank mixes to achieve satisfactory control. Some soil-active herbicides have some postemergence activity, but their preemergence residual activity is their greatest asset. If a species such as balsamapple is the major problem, applications should be timed to coincide with the peak emergence of vine seedlings. If applied too early, the herbicide concentration in the soil may have decreased below an effective level in the germination zone as a result of breakdown or leaching. If applied later, when the vines are more mature and already in the tree canopies, they are also unsatisfactory. Lantana is another species which proliferates if herbicide application is poorly timed: it is susceptible to some of the commonly used soil-residual herbicides in its early seedling stages, but larger plants are tolerant. Milkweed vines also become more tolerant to preemergence herbicides at recommended rates after they have become established. Postemergence applications must be made to eventually eliminate established plants. Fall applications are more effective against woody perennial vine and shrub species, such as Virginia creeper and lantana.

Color Plates

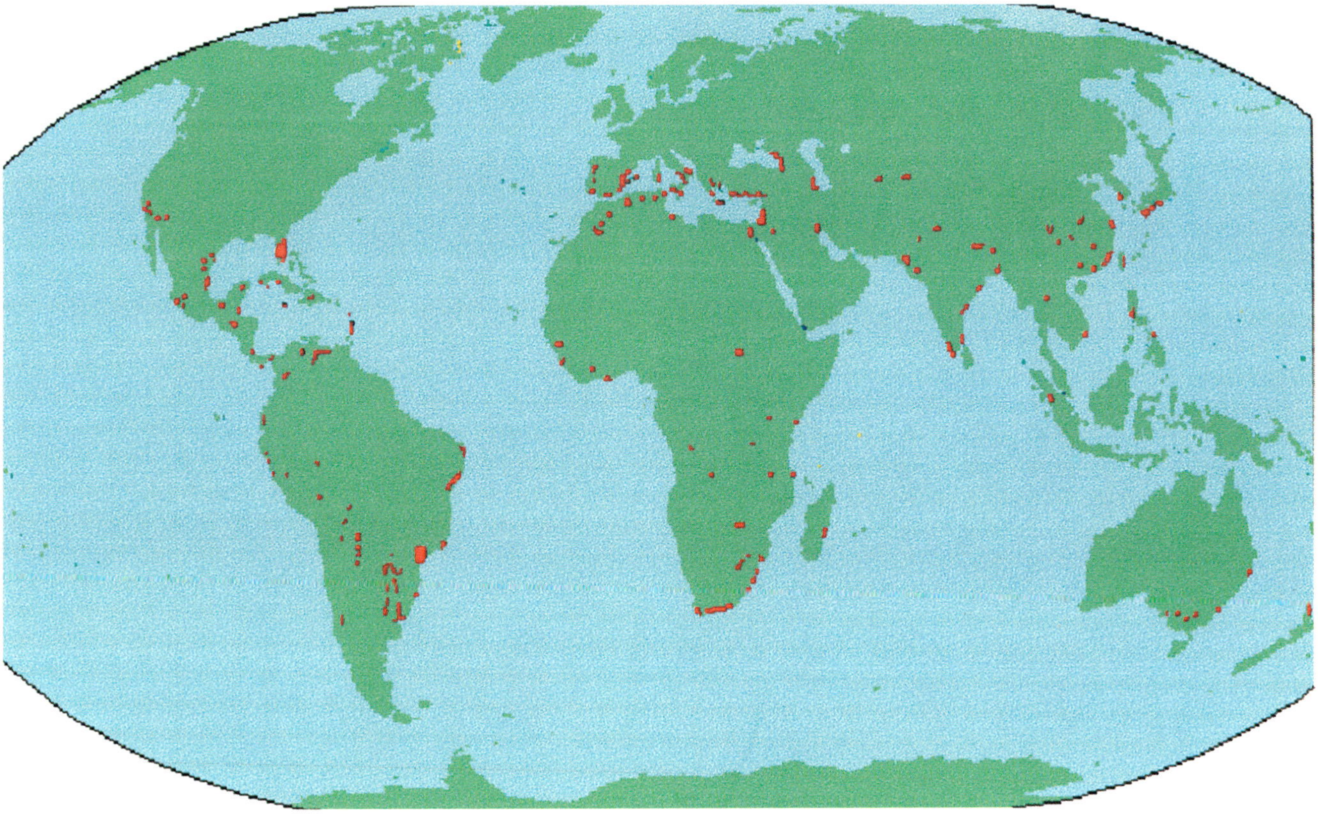

1. Areas of the world where most citrus is produced.

2. Leaves of common rootstock cultivars. **A,** Sour orange. **B,** Volkamer lemon. **C,** Rangpur lime. **D,** Cleopatra mandarin. **E,** Ridge Pineapple sweet orange. **F,** Rough lemon. **G,** (top to bottom) Trifoliate orange, Swingle citrumelo, Rusk citrange, and Carrizo citrange. **H,** *Citrus macrophylla* (Alemow). (Courtesy W. S. Castle)

3. Seed extraction facility. (Courtesy W. S. Castle)

4. Greenhouse production of rootstock seedlings. (Courtesy W. S. Castle)

5. Rootstock seedlings ready for transplant to larger containers. (Courtesy W. S. Castle)

6. Trimming, staking, and tying operation in a field nursery. (Courtesy W. S. Castle)

7. Field production of budded nursery trees. (Courtesy W. S. Castle)

8. Greenhouse production of budded trees. (Courtesy W. S. Castle)

9. Balled and burlapped trees ready for transplant to the grove. (Courtesy W. S. Castle)

10. Bare-root trees ready for transplant to the grove. (Courtesy D. P. H. Tucker)

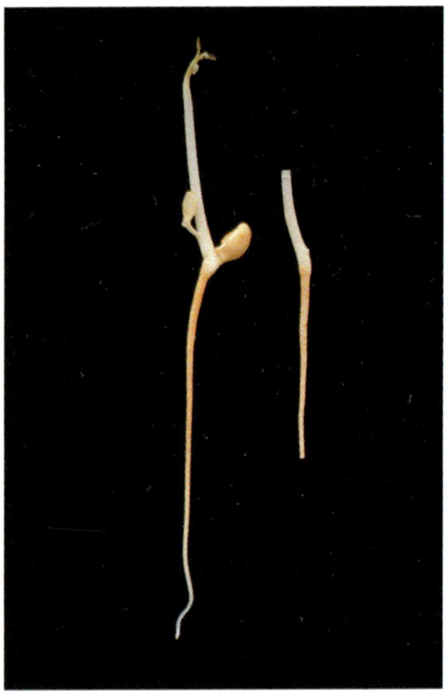

11. Preparation of a rootstock for shoot tip grafting: a two-week-old Troyer citrange seedling, with the epicotyl decapitated, the cotyledons and root tip severed, and an inverted T-shaped incision (not visible) at the top of the decapitated epicotyl. (Courtesy L. Navarro)

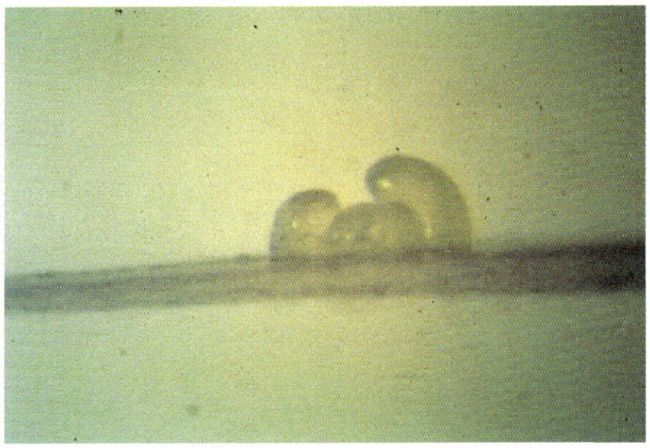

12. Sweet orange shoot tip, composed of the apical meristem and three leaf primordia, 0.14 mm long, on a razor blade sliver attached to a surgical handle. (Courtesy L. Navarro)

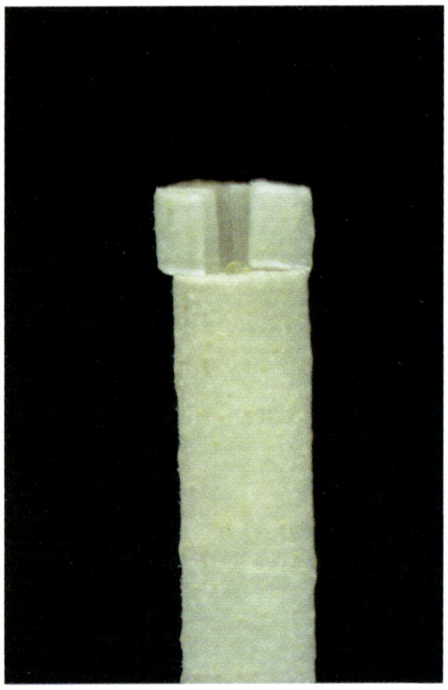

13. Sweet orange shoot tip freshly grafted to a two-week-old Troyer citrange rootstock. (Courtesy L. Navarro)

14. Sweet orange six weeks after grafting to Troyer citrange rootstock. (Courtesy L. Navarro)

15. Two-year-old satsuma mandarin recovered by shoot tip grafting in vitro, on a Troyer citrange rootstock. (Courtesy L. Navarro)

16. A screenhouse for the protection of budwood source trees. The double-entry doorway has a fan to create positive pressure when the house is entered. (Courtesy R. F. Lee)

17. A screenhouse for the protection of parent trees in the Belize Citrus Health Certification Program, featuring a double-entry doorway, a roof with a the large overhang to minimize the splashing of rain, and wind turbines on the roof for cooling. (Courtesy R. F. Lee)

18. A screenhouse in Florida, containing rooms for maintaining certified budwood source trees, a quarantine room for plants being introduced into the facility, and two rooms for indexing. (Courtesy R. F. Lee)

19. Young citrus trees in a tree-spacing experiment. (Courtesy T. A. Wheaton)

20. Mechanical hedging of citrus trees forms a continuous hedgerow as the trees mature. Hedging improves the distribution of light and allows clear access for equipment. (Courtesy T. A. Wheaton)

21. Mechanical topping limits the height of citrus trees to facilitate harvesting and generally results in lower production. (Courtesy T. A. Wheaton)

22. Hand-harvesting fruit. (Courtesy J. D. Whitney)

23. Tubs for fruit collection. (Courtesy J. D. Whitney)

24. Dumping tubs into the "goat" used for the collection of fruit in the grove. (Courtesy J. D. Whitney)

25. Dumping fruit from the goat into trailers to be hauled to the processing plant. (Courtesy J. D. Whitney)

26. Symptoms of water injury in a low-lying grove. (Courtesy A. Alva)

27. Symptoms of manganese deficiency on leaves of a Hamlin orange tree grown in high-pH soil. (Courtesy A. Alva)

28. Perennial peanut grown as a cover crop around young trees. (Courtesy A. Alva)

29. Symptoms of damage due to high salt concentrations following a windstorm. (Courtesy A. Alva)

30. Iron deficiency symptoms on leaves. (Courtesy A. Alva)

31. Zinc deficiency symptoms on leaves. (Courtesy A. Alva)

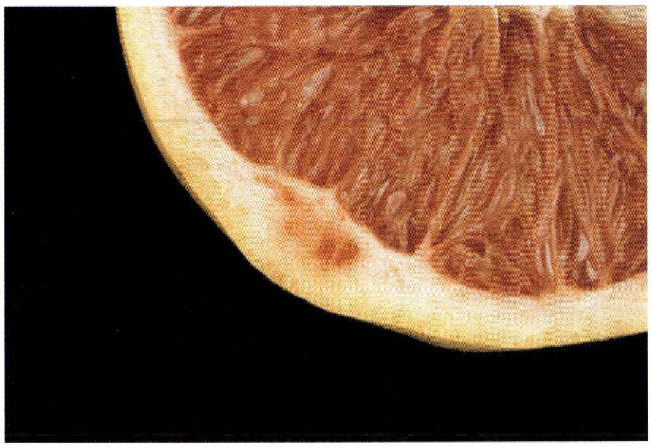

32. Brown discoloration in the fruit albedo, a symptom of boron deficiency. (Courtesy A. Alva)

33. Flood irrigation of citrus in Texas. This is an effective, low-cost method of irrigation, but it requires a high volume of water and appropriate soil types for good distribution to the trees. (Courtesy L. R. Parsons)

A  B

34. A water furrow in bedded citrus in poorly drained soil in the Florida flatwoods. Water furrows provide surface drainage after heavy rainfall. Water can also be diverted into the furrows for irrigation. (Courtesy L. R. Parsons)

35. Overhead sprinkler irrigation in a Florida citrus orchard. This method is effective but has declined in popularity in Florida, because of water restrictions and because it does not provide freeze protection. (Courtesy L. R. Parsons)

36. Volume gun irrigation. This method is effective but declining in popularity because of water restrictions in some areas. (Courtesy L. R. Parsons)

37. Microsprinkler irrigation of a young citrus tree. Microsprinkler irrigation has grown in popularity since 1980 and is now the most common method of irrigation in Florida citrus orchards. (Courtesy L. R. Parsons)

38. Microsprinkler irrigation for frost protection in a Florida citrus orchard. Microsprinklers can provide partial protection against frost in both young and mature trees. (Courtesy L. R. Parsons)

39. Microsprinkler irrigation and soil banks for freeze protection. (Courtesy L. R. Parsons)

40. Trees protected by microsprinkler irrigation during a severe freeze (right), compared with trees that were not irrigated (left). (Courtesy L. R. Parsons)

41. Trunk damaged by equipment. (Courtesy D. P. H. Tucker)

42. Clean cultivation of row middles. (Courtesy D. P. H. Tucker)

43. Orchard with a typical weed management program of mowing the row middles and application of herbicides in the tree rows. (Courtesy D. P. H. Tucker)

44. Row middle before chemical mowing. (Courtesy D. P. H. Tucker)

45. Row middle after chemical mowing. (Courtesy D. P. H. Tucker)

46. Specialized equipment for chemical wiping in row middles. (Courtesy D. P. H. Tucker)

47. Typical herbicide application equipment. (Courtesy D. P. H. Tucker)

48. Milkweed vine climbing an irrigation riser through the tree canopy. (Courtesy D. P. H. Tucker)

49. Roots of milkweed vine infected with DeVine mycoherbicide, a fungal pathogen. (Courtesy D. P. H. Tucker)

50. Weed regrowth around irrigation emitters in a herbicide-treated area. (Courtesy D. P. H. Tucker)

51. Symptoms induced by systemic pathogens of citrus in the field. (Courtesy S. M. Garnsey)

A, Quick decline of 10-year-old sweet orange on sour orange rootstock recently infected with a decline isolate of citrus tristeza virus. Trees with quick decline often have a yellow-brown stain at the bud union (see Plate 51G).

B, Symptoms of greening in a mandarin tree. Yellow shoots appear on part of the tree, which soon declines to a worthless condition. Chronically affected trees die back, deficiency symptoms develop in the leaves, and the fruit is small and poorly colored.

C, Early symptoms of citrus blight. Infected trees flush more slowly than healthy trees, because blockage of the flow of water in the wood causes moisture stress in the canopy. Blight is rare in trees less than six years old and primarily affects trees on rough lemon, Rangpur lime, citrange, and trifoliate orange rootstocks.

D, Advanced symptoms of blight. Trees decline to an equilibrium condition, with twig dieback on the outside of the canopy and vigorous shoots (water sprouts) growing from the main trunk.

E, Stunting of a young tangelo tree infected with stubborn. Note the compact canopy and short shoots. Infected trees typically produce only sparse fruit, which is often misshapen. Stunting of young trees also occurs when budwood infected with citrus tristeza virus is propagated on sour orange or when viroid-infected budwood is propagated on trifoliate orange or citrange rootstock.

F, Chlorotic leaves and dieback of sweet orange with citrus variegated chlorosis. Affected trees produce small, abnormally hard fruit.

52. Symptoms induced by systemic pathogens of citrus in the field. (Courtesy S. M. Garnsey)

A, Bark scaling of trifoliate orange rootstock infected with citrus exocortis viroid (the scion is sweet orange). Bark scaling is confined to the rootstock. This symptom also occurs in citranges and Rangpur lime.

B, Bark scaling of the trunk of sweet orange infected with a severe form of psorosis–citrus ringspot. Scaling is confined to the scion. The bark remains alive under the scales, and little gum is exuded. The canopy is usually sparse on limbs with severe lesions.

C, Deformities of the trunk induced by the concave gum–blind pocket agent. Gum may be present in the wood under the deep cavities and can be observed when bark is removed. Leaf patterns occur in some young leaves during the spring flush (see **I**).

D, Gum pockets in the bark of a young Temple orange on Orlando tangelo rootstock, a symptom of cachexia, revealed when shallow tangential cuts are made in the bark. Pegs and depressions in the surface of the bark and wood can be seen when a bark patch is removed. Symptoms may occur on either side of the bud union, if the scion or the rootstock is susceptible.

E, Tristeza stem pitting in Star Ruby grapefruit on rough lemon rootstock. The trunk is bumpy on the exterior, and deep pitting in the wood of the scion is revealed when a bark patch is removed at the bud union. Trees severely affected by stem pitting rarely die, but yield and fruit size are reduced. Other diseases, such as cachexia and cristacortis, also cause stem pitting.

continued on next page

53. Undersurface of a leaf infested with nymphs of the citrus whitefly, *Dialeurodes citri.* The nymph in the center has been infected and killed by the fungus *Aschersonia aleyrodis,* spores of which are visible as orange areas in the white mycelial mats. (Courtesy H. W. Browning)

54. Citrus snow scale, *Unaspis citri,* and a parasitic wasp, *Aphytis lingnanensis,* in its adult stage on a citrus leaf. The adult wasp parasitizes immature female scales. (Courtesy H. W. Browning)

55. A citrus leaf infested with Florida red scale, *Chrysomphalus aonidum.* The scale cover has been flipped over to reveal a pupa of the parasitic wasp *Aphytis holoxanthus,* which has developed from an egg deposited on the body of the scale. (Courtesy H. W. Browning)

56. Late nymphal stages of the citrus whitefly, *Dialeurodes citri,* on the underside of a citrus leaf. The nymph on the right contains a pupa of the parasitic wasp *Encarsia lahorensis.* (Courtesy H. W. Browning)

57. A colony of the spirea aphid, *Aphis spiraecola,* being grazed by an adult coccinellid beetle. This predator consumes numerous aphids per day in such a colony. (Courtesy C. W. Melton)

52. *continued*

F, Stem pitting in a twig of Marsh grapefruit infected with a severe stem-pitting isolate of citrus tristeza virus. Lime, grapefruit, and sweet orange infected with severe stem-pitting isolates of the virus are often stunted and produce small fruit, and their twigs are brittle and can be easily broken.

G, Yellow-brown stain just below the bud union of sweet orange on sour orange rootstock infected with a quick decline isolate of citrus tristeza virus. In trees that decline slowly, many small pits, resembling a honeycomb when magnified, may form on the inner face of the bark.

H, Groove at the bud union of satsuma mandarin on trifoliate orange rootstock infected with citrus tatter leaf virus. The bud union incompatibility stunts and weakens the tree. Other incompatibility problems also occur at the bud union in citrus and can often be detected by removing a bark patch from that area.

I, Leaf fleck in a young sweet orange infected with the concave gum pathogen. Similar symptoms occur in trees with cristacortis and impietratura. Flecks may appear in only a few leaves on a tree and fade rapidly as the leaves mature.

J, Leaf mottling in pummelo infected with citrus greening. Leaves are often marked with a blotchy, mottled pattern when trees are first infected.

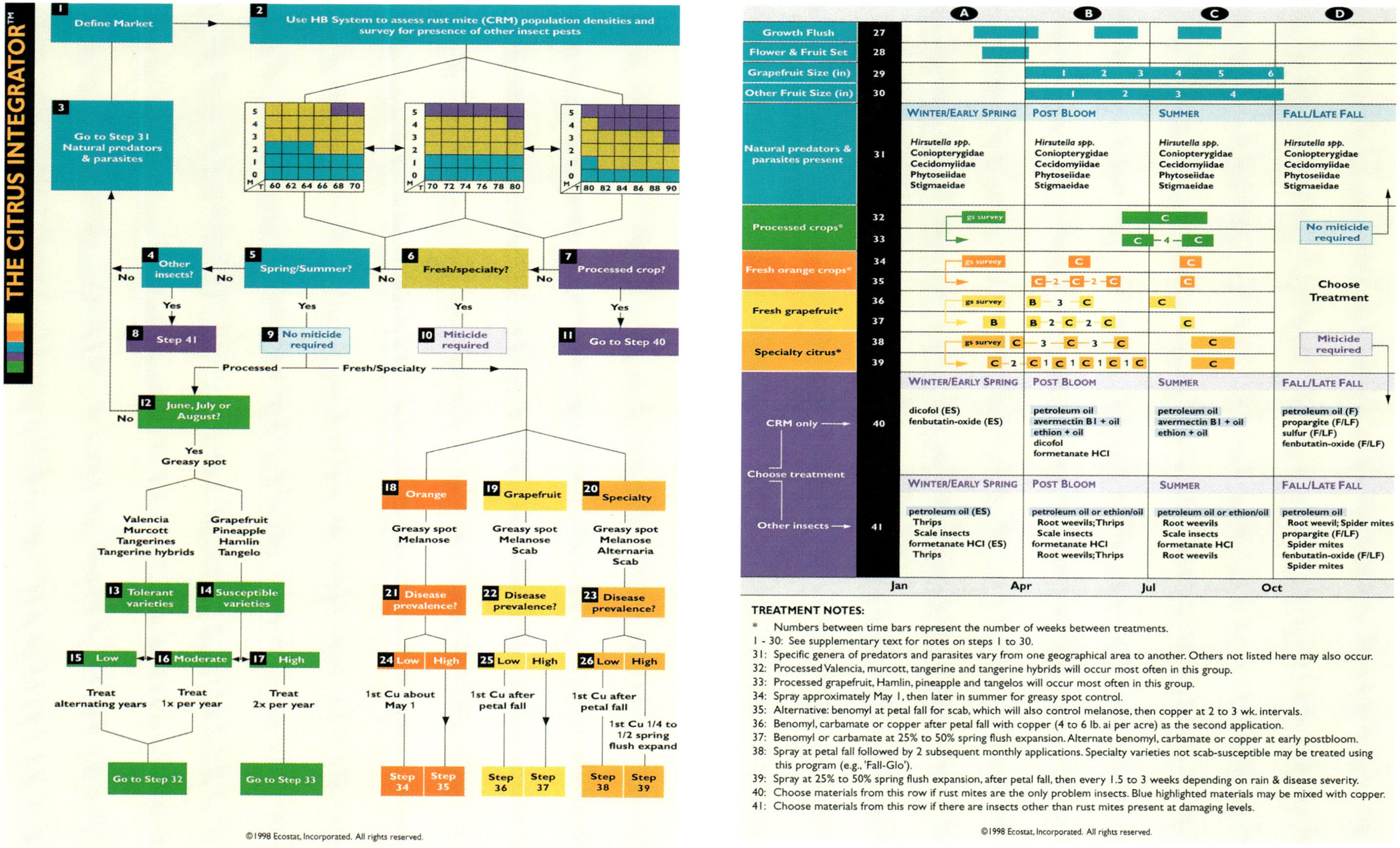

58. The Citrus Integrator. For an example illustrating the use of the flowchart, see the appendix to Chapter 11. (Reproduced, by permission, from Ecostat, Inc.)

59. A–D, Citrus root weevils in the adult stage: **A,** *Diaprepes abbreviatus.* **B,** *Pachnaeus litus.* **C,** *Asynonychus godmani.* **D,** An *Exophthalmus* species. **E,** Foliar feeding by adults. **F,** Root injury caused by larvae. **G,** Adult *D. abbreviatus* infected with the pathogenic fungus *Beauveria bassiana.* **H,** Infection of a larva by entomogenous nematodes. (Courtesy C. W. McCoy)

60. Drench application of thiabendazole to pallets of oranges on a truck bed before unloading and degreening. (Courtesy G. E. Brown)

61. A degreening room with horizontal air movement for effective removal of chlorophyll from citrus fruit. (Courtesy G. E. Brown)

62. A cleaning unit in which a sanitizer is applied to fruit conveyor rollers for daily sanitation of the packing line. (Courtesy G. E. Brown)

63. Stabilization of a unitized pallet of packed fruit with wooden pallet, corner boards, and strapping. (Courtesy G. E. Brown)

64. Bulk bin for domestic retail marketing of Florida citrus fruit. (Courtesy G. E. Brown)

Table 8.2. Susceptibility of weeds to herbicides[a]

Common name	Scientific name	Bromacil + diuron 1:1	Bromacil + diuron 2:1	Bromacil	Diuron	Simazine	Nor-flurazon	Oryzalin	Paraquat	Glypho-sate	Fluazifop-butyl	Sethoxy-dim
Broadleaf weeds												
Florida beggarweed	*Desmodium tortuosum*	S	S	S	U	S	PS	T	S	S	T	T
Black nightshade	*Solanum nigrum*	S	S	S	S	S	PS	T	I	S	T	T
Camphorweed	*Heterotheca subaxillaris*	S	S	S	S	U	I	T	PS	S	T	T
Common purslane	*Portulaca oleracea*	S	S	S	S	S	S	S	S	I	T	T
Common ragweed	*Ambrosia artemisiifolia*	S	S	S	PS	S	PS	T	S	S	T	T
Cudweed	*Gnaphalium* sp.	S	S	S	U	S	PS	T	S	S	T	T
Dayflower, tropical spiderwort	*Commelina benghalensis*	S	S	S	S	T	PS	T	S	PS	T	T
Dog fennel	*Eupatorium capillifolium*	PS	PS	PS	PS	PS	S	T	PS	S	T	T
Florida pusley	*Richardia scabra*	S	S	S	PS	I	PS	S	S	S	T	T
Goatweed	*Scoparia dulcis*	S	PS	T	PS	I	PS	T	U	I	T	T
Hairy beggarticks, Spanish-needles	*Bidens pilosa*	S	PS	T	S	S	I	T	S	S	T	T
Jerusalem oak goosefoot	*Chenopodium botrys*	PS	PS	PS	I	T	I	T	PS	PS	T	T
Lambsquarters	*Chenopodium album*	S	S	S	PS	S	PS	S	S	S	T	T
Lantana	*Lantana camara*	PS	PS	PS	PS	T	T	T	I	S	T	T
Pepperweed	*Lepidium virginicum*	S	S	S	S	I	PS	T	PS	S	T	T
Phaseybean	*Macroptilium lathyroides*											
Pigweed	*Amaranthus* spp.	PS	S	T	S	S	I	S	S	S	T	T
Spurge	*Chamaesyce hyssopifolia*	S	S	S	PS	S	S	S	S	S	T	T
Teaweed	*Sida acuta*	S	PS	PS	PS	I	PS	T	T	PS	T	T
Virginia pepperweed	*Lepidium virginicum*	S	S	S	U	I	PS	S	PS	S	T	T
Grasses												
Bahiagrass	*Paspalum notatum*	S	S	S	I	T	PS	PS	I	S	T	S
Bermudagrass	*Cynodon dactylon*	PS	PS	PS	T	T	PS	PS	I	S	S	S
Crabgrass	*Digitaria ciliaris*	S	S	S	S	S	S	S	PS	S	S	S
Crowfootgrass	*Dactyloctenium aegyptium*	S	S	S	S	S	PS	S	PS	S	PS	S
Goosegrass	*Eleusine indica*	S	S	S	S	S	PS	S	S	S	S	S
Guineagrass	*Panicum maximum*	PS	PS	PS	PS	PS	PS	PS	S	S	S	S
Johnsongrass	*Sorghum halepense*	PS	PS	PS	PS	PS	PS	PS	PS	S	S	S
Natalgrass	*Rhynchelytrum repens*	S	S	S	PS	I	PS	PS	U	S	U	S
Paragrass	*Brachiaria mutica*	S	S	S	I	T	PS	PS	T	S	S	S
Southern sandbur	*Cenchrus echinatus*	S	S	S	S	I	PS	S	S	S	S	S
Hairy signalgrass	*Brachiaria piligera*	PS	PS	PS	I	PS	PS	S	S	S	S	S
Texas panicum	*Panicum texanum*	S	S	S	U	I	PS	S	S	S	S	S
Torpedograss	*Panicum repens*	PS	PS	PS	T	T	S	PS	T	PS	I	PS
Vaseygrass	*Paspalum urvillei*	PS	PS	PS	T	T	PS	PS	PS	S	S	S
Sedges												
Purple nutsedge	*Cyperus rotundus*	I	I	I	T	T	I	T	T	PS	T	T
Yellow nutsedge	*Cyperus esculentus*	I	I	I	T	T	I	T	T	PS	T	T
Vines												
Balsamapple	*Momordica charantia*	S	S	PS	PS	S	PS	T	PS	S	T	T
Bigroot morningglory	*Ipomoea pandurata*	U	T	T	U	S	I	T	PS	PS	T	T
Cat's claw vine	*Bignonia unguis-cati*	T	T	T	T	T	U	T	T	PS	T	T
Maypop	*Passiflora incarnata*	PS	PS	PS	I	T	T	T	PS	S	T	T
Milkweed vine, strangler vine	*Morrenia odorata*	PS	PS	PS	PS	PS	PS	T	PS	I	T	T
Moonvine	*Ipomoea alba*	I	I	I	T	T	U	T	PS	U	T	T
Morningglory	*Ipomoea* sp.	PS	PS	PS	PS	I	PS	T	PS	S	T	T
Peppervine	*Ampelopsis arborea*	T	T	T	T	T	U	T	PS	I	T	T
Virginia creeper	*Parthenocissus quinquefolia*	T	T	T	T	T	U	T	PS	PS	T	T

[a] S = Susceptible at germination and early seedling stage of growth.
PS = Susceptible only at germination. Repeat applications may sometimes control established seedlings.
I = Intermediate susceptibility; partial control may be achieved but will be erratic, killing some members of a species but not others.
T = Tolerant, either showing no sign of injury or able to recover from injury.
U = Status unknown because of a lack of experimental data and reliable field observations.

Frequency of Application

Under tropical and subtropical conditions, two to four herbicide applications are required per year for weed control. This is particularly true at the initiation of a chemical control program, when difficult-to-control species, including vines, shrubs, and perennial grasses, are well established. As a satisfactory degree of weed control is established with time, residual herbicide rates may be reduced. Where trees are being continually replaced, it is prudent to lower the rate and increase the frequency of application, since lower rates are less likely to injure newly planted trees, particularly in sandy soils. An increase in the frequency of application means higher costs for equipment operation, but costs may be reduced by including split rates of herbicide in fluid fertilizer applications. With the current groundwater protection guidelines in Florida and other areas, the trend is toward more frequent application of soil-residual herbicides at lower rates. However, herbicide rotation is recommended, as the practice of repeated applications at low rates has the potential to lead to the development of herbicide resistance.

Biological Control

Classical biological control of weeds uses natural enemies, including parasites, predators, and pathogens, to reduce the vigor, growth, and reproduction of a target species over time. Introduced biocontrol agents generally affect only a single species and are slow to become established. One of the few shining examples of success with mycoherbicides in weed management is DeVine, *Phytophthora palmivora,* a fungal pathogen developed for the control of milkweed vine (strangler vine) (Plates 48 and 49).

The larvae of noctuids (armyworms, cutworms, loopers, and others) have been observed to effectively decimate stands of grass species in orchards. A number of leaf-mining beetles (in the family Chrysomelidae) have been used successfully in Australia and Hawaii for control of lantana, and they could be introduced into Florida. However, such introductions are frequently delayed or denied by federal or state agencies, which must conduct a risk-benefit analysis. There has been some opposition to the introduction of biological control agents, due to the fear that the introduced insects or pathogens will attack other plants once their primary food source has been consumed. Also, a weed of economic importance in one crop may be a desirable plant of some value in other situations. The economic impact of a weed in croplands must be compared with its positive aspects (e.g., its importance as an ornamental plant or as a host of beneficial organisms).

Keen observers in orchards frequently notice weeds under stress, perhaps with symptoms of foliar discoloration or distortion caused by one or more control agents. Such activity should be noted and exploited if possible, although in most cases it may not be of economic significance.

Cultural Practices

Cultural practices, including the establishment of cover crops, tree spacing, and water and nutrient management, can influence weed invasion. Closer tree spacing results in earlier suppression of vegetation through shading, while reduced rates and better placement of irrigation water and nutrients can eventually have similar desirable effects (Plate 50).

While citrus orchards have traditionally been managed for the benefit of the trees rather than the ground cover in row middles, there is an increased awareness of the benefits of more intensive ground cover management. A number of plant species can be grown as cover crops, in monoculture or in mixed stands. They may be used for one or more primary purposes, often being identified on the basis of their function. *Smother crops* are used to suppress weed growth, *catch crops* to reduce nutrient leaching, *green manures* for incorporation into the soil, and *insectary crops* for the purpose of attracting beneficial insects and mites.

The recognized advantages of cover crops in tree fruit such as citrus include the following:

- The addition or conservation of nitrogen
- Reduced soil erosion
- Improved soil structure in heavy soils
- Increased water penetration
- Increased organic matter content of soil and associated biological activity
- Suppression of weed species
- Attraction of beneficial insects and mites and suppression of soilborne pests, such as certain nematodes
- Reduced mowing and cultivation for the suppression of vegetation
- Improved traction for vehicles and equipment on wheels

Cover crops may also have some undesirable characteristics: they compete with the trees for water and nutrients; they may serve as reservoirs for viruses or other plant pathogens and may harbor rodents and insect pests; they have the potential to become economically important weeds in the tree rows if they not adequately controlled by available herbicides; and they can lower the air temperature during a freeze.

The perennial peanut, *Arachis glabrata,* now being evaluated by growers in Florida as a ground cover in row middles, is considered to provide some of the aforementioned benefits that are not offered by the existing perennial grass and indigenous vegetation species.

Weed Management Programs

Weed management programs must vary with location and individual orchard sites, depending on the climatic conditions, soil type, weed species, tree age, rooting depth, and planting system. Programs for areas with deep, well-drained sandy soil include herbicide applications in tree rows and mechanically cultivated or mowed row middles, and some chemical mowing is now being practiced. Programs for areas with shallow, inherently poorly drained soils which have been disturbed by bedding and ditching also include herbicides in tree rows and mechanically or chemically mowed middles; no cultivation is suggested in these areas, because of possible damage to the shallow root systems, although choppers (rotating bladed drums) are being reevaluated, because of their low cost of operation. In both growing areas and both planting systems, supplemental weed control includes the injection of herbicides into low-volume irrigation systems (herbigation) and localized spot spray treatments (applied by battery-powered sprayers on all-terrain vehicles) for weeds escaping the standard applications in the tree rows. Hand labor is still essential for the removal of some established woody shrubs and vines in Florida and elsewhere, although it is considered to be prohibitively costly in the United States.

Global Aspects of Weed Management

While general principles of vegetation management are similar for various citrus-growing areas, management strategies will vary with the climate, soils, terrain, weed species, and economic conditions.

Climate and Soils

Differences in temperature, humidity, and rainfall in tropical Central America and the Caribbean, humid subtropical Florida, and arid subtropical and desert growing areas in California and the Mediterranean region present very different growing conditions for citrus and native vegetation. In tropical areas there is a wider spectrum of vegetation with more aggressive growth habits, requiring greater weed management inputs year-round.

Soil types in citrus-growing areas vary considerably, ranging from the deep sandy soils of Florida to the loams of California and the heavier clay and clay loam soils of the tropics. In the more fertile soils of tropical areas, rapid tree growth is matched by greater competition from the aggressive growth of other vegetation. The selection of herbicides and application rates vary with soil type, tree age, climatic factors, and management programs.

Weed Species

Climate and soils will influence the nature and composition of indigenous vegetation in various growing areas, and the spectrum of weed species and their respective emergence periods influences weed management decisions. In tropical and humid subtropical areas, native vegetation has a greater impact on citrus production than in subtropical arid and desert areas. However, high nutrition and irrigation inputs in arid areas stimulate the growth of vegetation around trees.

Terrain

The terrain of the various growing regions varies from essentially flat in Florida to quite sloping in some tropical and arid areas, where citrus may be established on hillsides. This presents a problem for cultivation, mowing, and herbicide application equipment, especially in heavy soils during prolonged periods of high rainfall. As access is limited, the growth of vegetation goes unchecked. Soil erosion also becomes a factor if the soil surface is left bare of vegetation and exposed to heavy rainfall. Erosion from rainfall and wind can be a factor in bedded plantings in Florida and other areas, especially before native or introduced cover has been established during orchard development. Beds can be eroded, root systems exposed, and soil-residual herbicides washed from treated soil surfaces. In hillside plantings, particularly in tropical areas, herbicide strips in the tree rows should be kept narrow, and residual vegetation should be maintained to stabilize the soil surface. Low rates of application of soil-residual herbicides in combination with postemergence herbicides are preferred.

Economic Development

The local economic conditions and stage of development of citrus industries in various areas will determine to a large extent the sophistication of weed management operations. Generally, the more developed the area, the greater the mechanization and use of agrichemicals, including herbicides. In underdeveloped areas, labor is more readily used, as it is available at a lower cost, but it is generally less skilled. Weed control is practiced with hand-held equipment such as machetes and hoes and to some extent herbicides applied with backpack sprayers. However, machetes and other tools can produce trunk wounds that are susceptible to pathogen invasion. Calibration is a problem with backpack sprayers, and therefore postemergence contact and systemic herbicides are preferred. Poor access to equipment and difficult operating conditions lead to a greater reliance on hand labor. Limited availability of capital for investment in application equipment and agrichemicals is also a factor. In some cases growers have little access to equipment and chemical dealers, as they may not reside in the country or may be unable to commute long distances to remote farming areas to consult with growers.

The Role of Weed Management in the Crop Management System

Weed management influences and is influenced by other orchard practices. Many groves have converted from high-volume overhead to low-volume under-tree irrigation systems, in which less of the orchard floor is wetted, so that weed growth in the row middles and areas outside the emitter zones is reduced. In low-volume systems during extended dry weather, it is difficult to establish cover crops in unwetted zones in newly developed orchards, and wind erosion can become a problem. The intensive usage of low-volume irrigation has led to more rapid leaching of herbicides in the emitter zones, resulting in earlier resumption of weed growth. Fertilizer applied through these systems also contributes to greater weed growth. Lower rates of herbicides applied more frequently, herbicide applications through irrigation systems, and spot treatments with postemergence herbicides are being used to control return weed growth more effectively and minimize herbicide leaching. As planting densities increase, the orchard floor becomes shaded more rapidly by the tree canopy, particularly in the row, so that weed growth is more rapidly suppressed. Conversely, when there is tree loss due to tree declines, freezes, and pruning, the orchard floor is exposed to sunlight, and weed growth becomes reestablished.

Traditionally, weeds have been controlled culturally and chemically, with little attention paid to the effects of control practices on the environment, including the diversity and stability of insect and mite populations and disease organisms. While little is known about such interrelationships in citrus, they have been documented in other cropping systems. Studies are in progress to evaluate the role of weed management in integrated crop management (ICM) for citrus, including the role of weeds as hosts of various mite predators.

Weed management strategies are known to influence the environment, which in turn may increase or reduce the incidence of certain insects, mites, and diseases. For example, minimal control of vegetation between tree rows can result in dense growth of native or introduced species. This favors certain predator mite populations and can increase certain foliar fungal diseases and insects which can invade citrus canopies.

Weeds are often the only source of flowers (pollen and nectar), vital in maintaining high populations of beneficial insects in agro-ecosystems. Weeds provide important habitats for natural enemies, which may be more abundant and effective in weedy areas. Some weeds also maintain a certain level of insects that serve as alternative food for predators or parasites. There are no precise references confirming the influence of specific weeds on the microclimate of citrus. However, weedy agro-ecosystems do have a complex external appearance and associated pattern of microclimates which influence insect behavior. Vegetation modifies climatic conditions to a marked degree, with plant height and vegetation quality and density affecting temperature and humidity.

Extensive new growth through increased tree vigor resulting from the removal of the cover crop has on occasion resulted in increased red mite populations. Higher temperatures near the soil surface associated with bare-ground culture are known to shorten regeneration time in rust mites, so that they may produce more generations per year, especially in the canopies of young trees, which are close to the orchard floor. Lower levels of spider mite infestation have been found in orchards with cover crops. *Hirsutella,* a fungal pathogen of rust mites, is known to require high humidity and free water for maximum activity—conditions associated with dense weed cover rather than bare ground.

Grasshoppers, plant bugs, leafhoppers, and root weevils are often associated with weed cover. Therefore, cultivation operations such as tillage and mowing when a large insect population is present in the weed cover can cause the insects to move into the tree canopy. This is particularly true for plant bugs such as the citron bug, which pierces and injures fruit, causing them to drop prematurely. Cover crops should not be disturbed when they are harboring populations of these insects. However, cultivation prior to the emergence of grasshopper nymphs can disturb egg masses and reduce future populations of these insects. The flightless species of root weevils, such as the Fuller rose beetle, utilize weeds as a route into the tree canopy to lay eggs. Hence, areas near the tree skirt should be kept relatively weed-free if populations of these insects are known to be present.

Weed growth around tree trunks and canopies creates humid conditions favorable for the development of Phytophthora foot rot. The influence of weed cover as opposed to bare ground on the severity of attack by the greasy spot fungus, *Mycosphaerella citri,* is not clearly understood. However, under weed-free conditions the incidence may be greater, because of more air movement around the leaf litter and improved ascospore dispersal. The reduced rate of decomposition leaf litter around and under the trees in weed-free areas exposes the canopy to an extended period of infection by ascospores. Conversely, with heavy weed growth, the humidity is greater, and the decomposition of leaf litter is more rapid. Hence, there may be a shorter period of ascospore release from fallen leaves. This relationship is probably complicated by the use of microsprinkler irrigation, which enhances the decomposition of leaf litter by frequent wetting. Leaf litter can be buried by tillage to reduce the amount of inoculum, which would otherwise lie on the surface until the leaf litter rotted.

Cultivation during the dry months frequently produces heavy airborne dust, which settles on the tree canopy. This dust has been shown to interfere with the searching abilities of *Aphytis* species, parasitic wasps which attack armored scale insects. Armored scale often builds up in the lower canopy in frequently cultivated orchards. Dust can also reduce the effectiveness of foliar-applied pesticides.

Orchards infested with nematodes, particularly the burrowing nematode, *Radopholus similis,* and root weevils should be kept well irrigated and weed-free to minimize stress. Tillage should not be used for control in the tree rows, as it destroys surface feeder roots, on which the trees become more dependent when their deeper roots are decimated by the burrowing nematode. Mechanical mowing and chemical weed control are recommended in these situations. Cultivation or nonhost cover crops may be recommended in row middles in orchards to control host vegetation of certain nematode species.

Allelopathy in Citrus Weed Management

Allelopathy, the regulation of growth of one plant species by chemicals released by another, occurs widely in plant communities and is believed to regulate species density and distribution. Direct allelopathic effects of weeds on citrus trees have not been well documented, and therefore their contribution to tree and weed competition is not known. However, allelopathic effects between weed species has been documented in citrus orchards, and this may itself have a positive impact on tree growth.

Lantana, a rapidly growing perennial woody shrub, is an economic pest in Florida citrus orchards. Its proliferation and spread are attributed to several factors, including natural propagation, the reduced use of manual weeding, tolerance to some currently used herbicides, and little competition from other weed species. Interference with the growth of plants in the vicinity of lantana has been attributed, at least in part, to the allelopathic effects of the weed. The absence of milkweed vine infestations near lantana has been observed.

PART FOUR

Crop Health

Stephen M. Garnsey
U.S. Department of Agriculture, Agricultural Research Service
U.S. Horticultural Research Laboratory, Orlando, Florida

CHAPTER 9

Systemic Diseases

This chapter covers the major graft-transmissible diseases affecting citrus trees, here referred to collectively as systemic diseases. These diseases are caused by several types of pathogens that affect trees in similar ways, including viruses, viroids, uncharacterized viruslike agents, and prokaryotic (bacteria-like) pathogens (Table 9.1). Citrus production personnel often overlook important diseases caused by these pathogens or confuse them with other problems. The symptoms are often subtle and may easily escape notice. Some symptoms, such as stunting, decline, or yellowing of the canopy, can be mistaken for other problems. Trees may also be infected by several pathogens, each of which contributes to the symptoms observed. Many systemic pathogens cause disease only in a particular cultivar or a particular combination of scion and rootstock and infect other cultivars without causing symptoms. These symptomless carriers are important reservoirs of inoculum from which diseases can spread to susceptible cultivars. Infections can often be detected only by performing a specific testing procedure.

This chapter provides some basic information intended to help citrus producers recognize systemic disease problems and to understand how these diseases affect citrus trees and how they may be controlled. Methods are described for recognizing systemic disease problems, and management and control strategies for them are discussed. Suggestions for preventing damage to future plantings through use of disease-free budwood and disease-tolerant cultivars are also presented.

Basic Properties of Systemic Pathogens

A general understanding of the properties and the differences between systemic pathogens of citrus is useful, since these properties affect diagnosis and control. All viruses and viroids are obligate parasites (they multiply only in live tissue), cannot be cultured, and are so small that they cannot be seen without an electron microscope. Unlike prokaryotes, viruses and viroids do not have metabolic systems, and they do not respond to antibiotics. The bacteria-like pathogens can sometimes be seen with a light microscope, some can be cultured, and they respond to certain antibiotics.

True viruses, such as citrus tristeza virus, consist of particles of nucleoprotein of a specific size and shape. The core of the particle is composed of either ribonucleic acid (RNA) or deoxyribonucleic acid (DNA), which provides the essential genetic information that determines how the virus functions. A protein coat protects the RNA or DNA core and is involved in

All systemic pathogens can be transmitted by grafting and are commonly spread by the propagation of infected budwood.

transmission, infection, and movement within the plant. The coat protein of each virus is unique, and if it is injected into animals it will trigger the production of antibodies. The antibodies formed to a virus are specific to that virus only. Such antibodies can be purified from blood collected from injected animals and used to detect virus infections by serological methods (Box 9.1).

Viroids are much smaller than viruses and are composed of only a small piece of nucleic acid, which is structured so that it is very stable. Since viroids do not have a protein coat, they cannot be detected serologically.

At least four distinct types of prokaryotes cause systemic diseases in citrus. A *Xylella* species infects the xylem (wood) of the tree and causes citrus variegated chlorosis in South America. *Xylella* species have distinct cell walls and can be cultured. *Spiroplasma citri* infects sieve elements of the phloem (the food-conducting system in the plant) and causes stubborn disease. Spiroplasmas lack cell walls, have a distinctive helical shape, and can be cultured. A phytoplasma causes witches'-broom. Phytoplasmas, sometimes called mycoplasma-like organisms, have no cell walls and cannot be cultured. A *Liberobacter* species infects the phloem and causes greening (huanglongbing), one of the most deadly diseases of citrus.

These bacteria have a thin cell wall, which distinguishes them from spiroplasmas and phytoplasmas. They have never been cultured, despite many attempts.

A number of diseases are caused by systemic pathogens which have not been well characterized. Psorosis was the first viruslike disease of citrus recognized, yet the causal agents of this disease complex remain poorly characterized. Citrus blight is apparently caused by a pathogen which can be transmitted by root grafting, but the nature of the pathogen remains unknown. These diseases can be recognized primarily by the symptoms they cause in the field or in graft-inoculated indicator plants.

All systemic pathogens can be transmitted by grafting and are commonly spread by the propagation of infected budwood. Some systemic pathogens are also spread by insects, and these are the most difficult to control. Insect vectors include psyllids (which transmits citrus greening), leafhoppers (stubborn, citrus variegated chlorosis, and witches'-broom), aphids (tristeza), a whitefly (citrus chlorotic dwarf), and mites (leprosis). Citrus viroids and some viruses can be transmitted as contaminants on pruning tools. Seed transmission of systemic citrus pathogens is rare, but concave gum is seed-transmitted in some varieties.

Certain cultivars or combinations of scion and rootstock may be severely affected by a specific pathogen, but the same pathogen can be symptomless in others. Budwood selected from a vigorous tree carrying a symptomless infection on one rootstock may produce diseased trees when propagated on a sensitive rootstock. Similarly, in areas where many different varieties or scion-rootstock combinations are planted, symptomless trees may be the reservoir of infection for pathogens that are carried by insects, such as aphids.

Origin and Distribution of Systemic Diseases

Some systemic diseases of citrus are found nearly everywhere citrus is grown. Others have been found only in certain areas (Table 9.1). Commercially cultivated citrus varieties originated in various parts of Southeast Asia, where citrus has been propagated for thousands of years. Growers undoubtedly selected cultivars that were tolerant of the diseases prevalent in each location. The major citrus industries in Africa, Aus-

Table 9.1. Graft-transmissible pathogens of citrus

Disease	Type of pathogen[a]	Natural means of transmission[b]	Geographic distribution	Occurrence[c]
Virus diseases				
Tristeza	Closterovirus	Aphids	Worldwide	Widespread
Citrus variegation	Ilarvirus	Mechanical (sap); other (?)	Worldwide	Localized
Satsuma dwarf[d]	"Nepovirus"[e]	Soilborne	China, Japan	Localized
Tatter leaf	Capillovirus	Mechanical (sap); other (?)	Asia	Common
Ringspot-psorosis complex[f]	NC	Aphid (?)	North America, South America, Mediterranean Basin	Common
Citrus mosaic (India)	Badnavirus	Mealybug (?)	India	Common
Viruslike diseases				
Cristacortis	NC	ND	Mediterranean Basin	Common
Impietratura	NC	ND	Mediterranean Basin	Common
Concave gum[g]	NC	ND	Mediterranean Basin, North America	Common
Citrus chlorotic dwarf	NC	Whitefly	Turkey	Localized
Viroid diseases				
Exocortis	Exocortis viroid	Mechanical (sap)	Worldwide	Common
Cachexia	Citrus viroid IIb	Mechanical (sap)	Worldwide	Common
Dwarfing factor	Citrus viroid III	Mechanical (sap)	Worldwide	Common
Others	Citrus viroids I and IV	Mechanical (sap)	North America, Mediterranean Basin	Rare
Systemic prokaryote diseases				
Stubborn	*Spiroplasma citri*	Leafhoppers	North America, Mediterranean Basin, Near East	Common
Greening (huanglongbing)	*Liberobacter*	Psyllas	Asia, South Africa, Near East	Widespread
Witches'-broom	Phytoplasma	Leafhopper	Oman, United Arab Emirates	Localized
Citrus variegated chlorosis	*Xylella fastidiosa*	Sharpshooters (?)	South America	Common
Diseases caused by unknown agents				
Blight	NC	ND	North America, South America, South Africa, Australia	Common
Australian citrus dieback	NC	ND	Australia	Localized

[a] NC = not characterized.
[b] ND = not determined. All diseases are transmitted by grafting.
[c] Widespread = endemic, with a large portion of trees infected. Common = occurring in many areas, but not affecting all trees. Localized = occurring only in some areas, although possibly widespread in those areas.
[d] Includes Japanese citrus mosaic and navel orange infectious mottling. All may occur in other citrus-growing areas in cultivars propagated from infected budwood from Asia.
[e] Status uncertain.
[f] Includes citrus necrotic ringspot, naturally spread psorosis, psorosis A, and psorosis B isolates that induce bark scaling and necrotic shock reactions and are mechanically transmissible.
[g] Includes blind pocket.

tralia, Europe, and the Americas were developed by the importation of citrus from Asia. The earliest movement of citrus was by seed, and since only a few systemic pathogens are seed-transmitted, early plantings of seedling trees were free of systemic diseases. Epidemics of Phytophthora foot rot subsequently promoted interest in propagating citrus on foot rot–tolerant rootstocks, such as sour orange.

As transportation systems improved, live plants or budwood was frequently shipped between countries. Much movement of citrus germ plasm occurred before the existence of viruses and other systemic pathogens was even recognized. Tolerant cultivars that did not show symptoms of infection were undoubtedly more likely to be selected for propagation. Early accounts of failures in propagating some new citrus introductions on certain rootstocks indicate that some of these plants were carrying symptomless infections. For example, satsuma mandarins imported from Asia into the United States could not be grown on sour orange, a clear indication that they were infected with citrus tristeza virus. In retrospect, many of the systemic disease problems that have developed in various countries can be traced to the importation and propagation of infected budwood. In some cases, vectors of these pathogens were also moved with infested plants.

However, not all systemic diseases of citrus can be traced to Asia. Some of these diseases appeared only after citrus was introduced into a new area of cultivation. In these cases the pathogen apparently was already present in other plants and spread to citrus after citrus plantings were established in the new area. Stubborn (which first appeared in California), blight (which first appeared in Florida), citrus variegated chlorosis (which was first found in Brazil), lime witches'-broom (which was discovered in Oman), and citrus chlorotic dwarf (which was discovered in Turkey) all apparently originated in secondary sites of citriculture.

The complicated patterns of exchange of citrus germ plasm make it difficult to pinpoint the origin of some pathogens, such as citrus viroids and the causal agents of the psorosis complex. Some viruses and viruslike pathogens may have been introduced into citrus by chance feeding by vectors not normally associated with citrus or by mechanical inoculation. These events may be rare, but once an infection is established it is maintained for the life of the tree. If budwood is taken from that tree for propagation, the number of infected trees can increase rapidly, even if the original vector does not transmit the disease within the plantings.

It is natural for growers to focus their attention on the problems that cause damage in their own localities; however, the prudent grower is also aware that other citrus pathogens and their vectors elsewhere in the world can arrive suddenly in an age of extensive international travel and commerce.

Major Systemic Diseases of Citrus

Some of the most common and important systemic diseases of citrus are briefly described in this section. The diseases selected illustrate the major types of systemic pathogens affecting citrus and also illustrate the various types of effects and symptoms of systemic diseases. A more extensive listing of systemic citrus pathogens is presented in Table 9.1. Several recent references are also recommended for more detailed information on specific diseases or pathogens and methods of detection and diagnosis (see Sources of Additional Information).

Tristeza

Citrus tristeza virus (CTV) is one of the most important pathogens affecting citrus. There is a wide range of strains or isolates of CTV. Some isolates are very mild and cause little damage. Others may cause a severe decline of trees grafted on sour orange rootstock (Plate 51A). Some isolates also cause a serious stem-pitting disease in limes, grapefruit, and sweet orange (Plate 52E and F). Well over 50 million trees on sour orange have been destroyed to date, and stem pitting has affected the production of limes, grapefruit, and oranges in many areas.

CTV is spread by several species of aphids. The most efficient vector is the brown citrus aphid, which is now endemic in most major citrus production areas. The brown citrus aphid continues to spread to new areas, including the Caribbean Basin and Florida, and as of 1996 only the citrus regions of the Mediterranean Basin, Mexico, and the western United States remained free of the insect. CTV spreads more rapidly in the presence of the brown citrus aphid, but severe disease epidemics can occur where the only vectors are the melon and spirea aphids.

Several countries, such as Brazil, Venezuela, and Peru, have experienced a double CTV problem. An initial epidemic destroyed plantings on sour orange rootstocks and forced growers to replant on rootstocks which are less well adapted horticulturally or are susceptible to other diseases, such as blight, exocortis, and Phytophthora foot rot. Subsequently, stem-pitting isolates have caused production losses in new plantings on decline-tolerant rootstocks. Severe stem pitting limits commercial production of limes and grapefruit in many areas, and it affects sweet orange production in some locations. Many isolates of CTV are symptomless in sweet orange, mandarins, and many other cultivars, and infection can be detected only by testing trees for the virus. Graft inoculation of Mexican lime plants is the classic method, but some laboratory tests, especially enzyme-linked immunosorbent assay (ELISA), are now commonly used and are cheaper and faster.

Psorosis and Ringspot

Psorosis was the first viruslike disease recognized and has affected citrus production in many countries. The psorosis disease complex is still not well understood, but several different causal agents are probably involved. Isolates in one group cause an obvious bark scaling in sweet orange (Plate 52B) and may also cause leaf symptoms. Young shoots on newly infected plants frequently die back (a symptom known as shock reaction). This group includes pathogens associated with psorosis A, psorosis B, citrus necrotic ringspot, and naturally spread psorosis. Natural spread of these psorosis-ringspot isolates has been observed, but the means of spread has not been determined. Most members of this group are mechanically transmissible to herbaceous plants and are associated with the presence of a unique type of virus particle. Several laboratory assays are under development for rapid detection of isolates in the psorosis-ringspot group.

Concave Gum and Blind Pocket

Another subgroup of the psorosis complex includes isolates which cause abnormalities of the trunk or limbs but do not cause bark scaling. These isolates cause various chlorotic patterns in young leaves of the spring flush (Plate 52I), including a distinctive oak leaf pattern. Isolates in this group cause concave gum and blind pocket, and it is likely that they cause impietratura and cristacortis, which are common in Mediterranean countries. The fruit of trees affected by impietratura frequently has gum pockets in the albedo, and the rind over these areas does not color normally. The causal agent of this group of diseases has not been characterized. Seed transmission of concave gum has been reported in trifoliate orange and Troyer and Carrizo citranges. The concave gum diseases can be diagnosed only by graft inoculation of sensitive sweet orange or mandarin indicators, grown under controlled greenhouse conditions.

Diseases Caused by Viroids

The exocortis viroid was the first viroid discovered in citrus, and exocortis is the most well-known viroid disease. However, at least five different groups of viroids affect citrus, and three of these are very common. Furthermore, each viroid group includes several members. Citrus viroids that are now known to belong to the distinct viroid groups I, III, and IV were previously considered mild isolates of exocortis viroid, on the basis of symptoms in Etrog citron indicators. Exocortis viroid causes bark scaling in susceptible cultivars, such as trifoliate orange, some trifoliate orange hybrids (Carrizo and Troyer citranges), and Rangpur lime, used as rootstocks for exocortis-infected scions (Plate 52A). Infected trees are usually stunted. The stunting may also be caused, in part, by co-infection with other viroids, especially members of citrus viroid group III. The group III viroids cause stunting in trees on susceptible rootstocks, but they do not cause extensive bark scaling.

Cachexia affects some mandarin and mandarin hybrid scion cultivars and *Citrus macrophylla* rootstock. It is caused by some members of citrus viroid group II, also referred to as the hop stunt group. The symptoms include gum deposits in the bark (Plate 52D) and pits and pegs in the wood.

Viroids which affect citrus are closely related to viroids attacking other crops, such as hops, cucumber, grapes, and apples. Citrus viroids are not known to be transmitted by insects, but they are easily spread by contaminated pruning tools.

All citrus viroid groups, except group II, can be detected by graft inoculation of sensitive clones of Etrog citron. The standard laboratory procedure for viroid detection has been gel electrophoresis of partially purified nucleic acid extracts from graft-inoculated citron plants. Viroid identification by gel electrophoresis of extracts taken directly from field trees is usually difficult, because the viroid concentration is low. However, the molecular technique of RT-PCR (reverse transcription polymerase chain reaction) shows promise for direct detection of viroids in field trees. This technique produces copies of the viroid molecules until enough are present that they can easily be detected by gel electrophoresis (Box 9.1).

A beneficial use of select members of citrus viroid group III, to produce dwarfed trees, has been suggested and is under field trial in Australia and other countries. Trees on a susceptible rootstock infected with these dwarfing viroids are smaller than normal, but their fruit is the same size and quality as the fruit of uninfected trees. A greater number of the smaller trees are planted per acre to achieve normal levels of fruit production.

Box 9.1

Methods of Identifying Systemic Pathogens

Systemic pathogens in commercial cultivars often cannot be detected just by looking at the trees, and it is often desirable to confirm identification based on field symptoms. Identification of systemic pathogens, often referred to as indexing, is done by a variety of techniques. The choice of a technique will often depend on the pathogen to be detected and the equipment, personnel, and diagnostic reagents available. The ability to detect systemic pathogens of citrus is essential in a budwood certification program.

Changes in Plant Chemistry Caused by Pathogens

Some pathogens leave a specific "footprint" that can be used for diagnosis. Fruit from trees affected by greening contains fluorescing compounds which can be detected by chemical analysis. Blighted trees have a xylem dysfunction that can be detected by injecting water into the trunk, and proteins associated with blight infection also accumulate in the leaves. Depletion of starch in the root system of trees affected by tristeza can be detected with the iodine test for starch.

Biological Tests

The classic approach for detecting systemic citrus pathogens is to graft-inoculate varieties of citrus which show distinctive symptoms when infected with a particular pathogen (see C. N. Roistacher, 1992, *Graft-Transmissible Diseases of Citrus: Handbook for Detection and Diagnosis,* Food and Agriculture Organization, Rome). The minimum time required ranges from one to 18 months. Biological testing is the basic technique for most systemic pathogens except the blight agent.

Some citrus viruses can be transmitted experimentally to herbaceous plants such as beans or cucumber by inoculating the plants with extracts from young leaves of infected citrus trees. Diagnostic symptoms appear in three to 15 days. This technique can be used to detect citrus variegation, satsuma dwarf, tatter leaf, and ringspot viruses.

Serological Tests

Proteins extracted from pathogens can be injected into animals to produce antibodies. These antibodies can be used to detect pathogens in infected plants, since they react to the pathogen proteins but not to the host proteins. A label, such as an enzyme, is often chemically attached to an antibody to make it a more sensitive detector. Enzyme-linked immunosorbent assay (ELISA) and immunoblotting are serological tests commonly used for routine diagnosis. Serological tests are rapid and can be completed in 1 to 24 hours. They are used for numerous pathogens, including those causing tristeza, citrus variegation, satsuma dwarf, stubborn, greening, citrus variegated chlorosis, and blight.

Greening (Huanglongbing)

Greening is one of the most devastating diseases of citrus. No other disease can so completely limit citrus production or be so difficult to control once it is established. Greening is common throughout much of Asia and Africa. It has not been reported in the Americas, the Mediterranean Basin, Australia, and Japan.

Greening was once considered a virus disease, but the pathogen is actually a bacterium (*Liberobacter* species) which grows in the bark, leaves, and veins of infected citrus. *Liberobacter* was identified as the cause of greening when unique bacteria were discovered in phloem cells of infected plants examined by electron microscopy, and greening-affected trees were found to respond to treatment with antibiotics. Most plant-pathogenic bacteria can be cultured on artificial media, but all attempts to culture the greening liberobacter have failed. The bacterium is spread by two species of psyllids (insects similar to small leafhoppers).

The name *greening* was chosen because affected fruit fail to color properly. However, the Chinese names *yellow dragon* and *yellow shoot* are more descriptive, since one or more branches of the tree often turn yellow when first infected (Plate 51B). Leaf symptoms in chronically infected trees resemble various deficiency symptoms, and diagnosis can be difficult in citrus groves where trees are suffering from lack of appropriate care and fertilization. Affected trees soon die back and decline to an unproductive state (Plate 51B). Greening severely affects nearly all sweet oranges, mandarins, and grapefruit. Lemons and limes are somewhat more tolerant.

Asian and South African species of the greening (huanglongbing) pathogen have been described. The Asian species causes more severe symptoms and is more adapted to warm climates. In areas where Asian greening and its vectors are common, new citrus plantings are rapidly infected and become unproductive within five to 10 years.

The use of pathogen-free budwood and eradicating existing infected trees may delay disease development, but these steps have not been fully successful. The best control has been obtained on Réunion Island, where the psyllid population was reduced to a very low level after the introduction of a parasitic wasp. Unfortunately, biocontrol of the psyllids has been less successful elsewhere, because the parasitic wasps themselves are frequently parasitized by other wasps (hyperparasites).

Greening can be identified by graft inoculation of mandarin seedlings, which develop strong symptoms and are rarely affected by CTV. Severe symptoms appear in graft-inoculated grapefruit and sweet orange seedlings, but these varieties also develop similar symptoms from infection by severe isolates of CTV, which are common in areas where greening is endemic. Several laboratory procedures have been developed for the detection of the pathogen.

Stubborn

Stubborn is another disease that was once thought to be caused by a virus, until the pathogen was discovered to be *Spiroplasma citri*. The disease occurs in the southwestern United States, North African countries bordering the Mediterranean Basin, and parts of the Middle East.

Microscopy

Light microscopy. *Spiroplasma citri, Xylella fastidiosa,* and citrus tristeza virus inclusions can often be detected in stained sections of infected plants viewed under a good-quality compound light microscope. The minimum time required is 1 hour.

Electron microscopy. Individual virus particles can be detected in extracts from infected plants, and viruses and bacteria, including the greening pathogen, can be identified in thin sections cut from infected plants. Extracts can be prepared and examined rapidly (1 hour), but the preparation and examination of thin sections is a multistep process requiring at least one week. Electron microscopy is used to detect citrus tristeza virus, *S. citri,* and *Liberobacter.*

Culturing

S. citri and *X. fastidiosa* can be cultured from extracts of infected plants on appropriate media. The identity of these bacteria is then verified by microscopy or serology. The minimum time required is two to seven days. Culturing is used for diagnosis of citrus variegated chlorosis and stubborn.

Electrophoresis

Sequential gel electrophoresis is used to detect viroids in nucleic acids extracted from infected plants. When the extracts are placed on a gel and exposed to an electric current, viroids migrate in specific bands, which are made visible by staining the gel. Electrophoresis is also used in molecular detection of viruses and other pathogens. The minimum time required is one to two days.

Labeled Genomic Probes

Each pathogen has a unique genetic code, and once a portion of this code has been determined, a complementary sequence can be constructed and labeled with a marker. A complementary probe will hybridize with a nucleic acid of the same sequence but not with other nucleic acids. Probes are useful when the concentration of the pathogen nucleic acid in the extract is sufficient to be detected. The test can be completed in one to two days. Probes are used to identify viroids and the pathogens causing greening, tristeza, and citrus variegated chlorosis.

PCR

When the concentration of the pathogen nucleic acid in an extract is too low to detect directly with complementary probes, the polymerase chain reaction (PCR) technique can be used to amplify the RNA or DNA of the pathogen to a concentration that can be easily detected by electrophoresis or labeled probes. PCR can be done in one to two days but requires experienced personnel. This technique can be used to detect tristeza, tatter leaf, and ringspot viruses, viroids, and other pathogens whose genetic code is sufficiently known.

S. citri is distinct from the greening liberobacter, but it also infects the phloem of citrus plants and disrupts normal plant functions. Young trees infected by the pathogen are stunted and tend to have a bushy, upright growth habit and branches with shortened internodes. Trees which become infected after they are mature show less obvious effects, but yield is reduced, and the fruit is of poorer quality than normal. Infected trees are less prone to the extensive decline and dieback associated with greening.

Stubborn is spread by several species of leafhopper and is a major disease problem in arid areas where these insects are abundant. *S. citri* infects many plants, including several weeds, and leafhoppers become infected when feeding on these plants. Citrus is not the primary host of leafhoppers, but natural infections of mature citrus trees can occur in association with the migration of infected leafhoppers from other plants. Stubborn epidemics in young plantings are often associated with trees propagated from stubborn-infected budwood or nursery stock infected prior to planting.

Stubborn can be detected by graft inoculation of orange, grapefruit, or tangelo seedlings. The symptoms are not highly specific, and indicator plants must be grown under warm conditions with good care to develop clear reactions. Several laboratory tests can be done; they are frequently based on culture of *S. citri* from infected tissue on special media. Both light and electron microscopy are also used in diagnosis (Box 9.1).

Citrus Variegated Chlorosis

Citrus variegated chlorosis (CVC) is caused by a bacterium, *Xylella fastidiosa*. In contrast to the greening and stubborn bacteria, which attack the phloem, *X. fastidiosa* affects the xylem (the woody, water-conducting part of the plant). CVC was first discovered in Brazil and has become widespread there and in several other South American countries. Affected trees have reduced vigor, leaves may turn yellow in some parts of the canopy, and twig dieback may occur (Plate 51F). The fruit is reduced in size, and large brown spots may form on fruit and leaves.

Similar bacteria also cause diseases in coffee, grapes, oaks, plums, peaches, and several other woody plants. It is likely that the CVC bacterium was first transmitted to citrus from other hosts and has been spread in citrus by propagation and by vectors. The identity of the vectors of CVC has not been fully determined, but various sharpshooters transmit other strains of *Xylella,* and several have transmitted CVC under experimental conditions. *Xylella* can be cultured from diseased trees for identification and can also be detected by serological tests and molecular procedures (Box 9.1).

Blight

Citrus is affected by several decline diseases for which the causes remain unknown. Citrus blight is one of the most common and widespread of these. It has been found in a number of countries, including parts of the United States (Florida and Hawaii), most of South America, the Caribbean, South Africa, and Australia. Trees affected by blight grow normally until they are at least five or six years old and then suddenly wilt and begin to die back (Plate 51C and D). Blighted trees rarely die, but they remain in a chronic decline condition and are unproductive. When heavily pruned, they may recover temporarily but soon relapse. The xylem of infected trees is nonfunctional, except for the outer wood. Some people believe that blight is an abiotic disease (i.e., that no pathogen is involved). However, it has been transmitted experimentally by grafting roots from blighted trees to healthy trees. The bacterium *Xylella* has been suggested as a cause of blight, and it has been isolated from some blighted trees, but its ability to cause the disease has not been confirmed experimentally.

Blight affects both seedlings and grafted trees. Trees on all commercial rootstocks are affected, but the disease appears more quickly and spreads more rapidly in trees grafted on rough lemon, Rangpur lime, trifoliate orange, and citrange rootstocks. Disease incidence is lower in trees on sweet orange, sour orange, mandarin, and some citrumelo rootstocks, but eventually some trees on these stocks become affected and show symptoms similar to those of trees on other rootstocks.

Impairment of water uptake and accumulation of zinc in the tree are characteristics of blight infection. In diagnostic tests, water is injected into the trunk to detect impairment of water uptake, and wood samples collected from the tree trunk are analyzed for accumulation of zinc. Stress-related proteins also form in leaves of blighted trees and can be detected by serological tests (Box 9.1).

Other Virus and Viruslike Diseases

Growers must not only deal with major disease problems that are already present in their own region but also be aware of diseases that exist in other regions and may become problems in the future.

Satsuma dwarf virus (SDV) and several related viruses are present in China and Japan but are not yet widely distributed elsewhere. SDV has a significant effect on some mandarins and sweet oranges. It is soilborne and has hosts other than citrus, and thus it can be a serious problem. Once an area is contaminated, SDV remains in the soil, and new trees planted in the area become infected.

Citrus tatter leaf virus (CTLV) is widely distributed in commercial cultivars in China and some other areas of Asia. It is also present in many other countries, in trees propagated from budwood which originated in Asia. No vector of CTLV has been reported, but viruses closely related to it have been found in apples and lilies, and this suggests that some unrecognized type of natural spread occurs. CTLV does not cause symptoms in orange, grapefruit, lemon, or mandarin seedlings or in trees of these varieties grafted on tolerant rootstocks. However, when budwood of infected trees is grafted on trifoliate orange, citranges, or citrumelos, a serious bud union incompatibility occurs, and the trees are stunted and may die. Citranges, citrumelos, and trifoliate orange are now extensively used as rootstocks worldwide, and the millions of trees on these rootstocks are susceptible to injury should they become infected.

Some viruses are quite common but do not have a serious impact on production. Vein enation (woody gall), an aphid-transmitted viruslike disease, is common in many countries but does little damage. Galling can occur in rootstocks of susceptible cultivars (such as rough lemon) in the nursery, especially when infected plants are wounded, but the overall effects of the disease are not serious.

Other viruses have been detected in citrus trees indexed for virus infection. These viruses do not cause disease in existing citrus trees; however, they could affect new rootstocks and scions in the future. Some virus and viroid pathogens were first recognized when new cultivars were substituted for existing scions or rootstocks, and problems then appeared.

New disease problems may appear suddenly at any time. A recent example is citrus chlorotic dwarf, a new viruslike disease spreading in commercial citrus plantings in Turkey. It is transmitted by the bayberry whitefly and appeared a few years after the insect was introduced into Turkey. It is probable that the disease agent was transmitted to citrus from other plants in the area by the bayberry whitefly.

Field Diagnosis of Systemic Diseases

Prompt diagnosis is often fundamental to the management or control of systemic diseases. For example, if growers fail to recognize that poor growth of trees in a new planting is due to a virus or viroid infection, they may waste several years vainly applying extra fertilizer or other treatments to overcome the problem rather than quickly replacing the diseased trees and avoiding further loss. Prompt identification and removal of diseased trees in mature plantings also minimizes the period of crop loss and may reduce sources of inoculum. Accurate identification of diseases present in existing groves is important in selecting replacement trees which will tolerate those diseases if they become infected.

Astute citrus growers can frequently determine that a systemic pathogen is causing problems, through field observation and analysis of the scion-rootstock combination affected.

Precise diagnosis of systemic pathogens of citrus is often difficult without greenhouse or laboratory tests. However, astute citrus growers can frequently determine that a systemic pathogen is causing problems, through field observation and analysis of the scion-rootstock combination affected. When called to diagnose a problem, citrus pathologists and extension specialists often rely heavily on the observations and information that the grower has collected, and with a little practice the grower can often make a reasonable preliminary diagnosis using the same information. Some key steps in field diagnosis of diseases caused by systemic pathogens are described below, and a quick field guide to diagnosis of common systemic diseases is presented in Box 9.2.

Young Plantings

The presence of stunted, unthrifty trees (Plate 51E) in a young planting with no obvious cause of the disorder is often a clue that a systemic pathogen is involved, especially when some trees of the same scion-rootstock combination are growing normally. Infections in young trees usually originate from infected budwood, and trees grown from registered budwood by reputable nurseries are less likely to be infected with systemic pathogens than trees propagated from untested sources. Stunting is usually the primary effect noticed. Yellowing of leaves may also occur, especially in trees infected with greening or tristeza. Rapid decline and death resulting from infection by systemic pathogens are rare in young trees.

The first step in diagnosis is to eliminate other factors, such as improper planting, poor nutrition, herbicide damage, unfavorable soil conditions, and injury to the trunk, major root system, or major limbs, as causes of the symptoms. The failure of trees to respond to extra fertilizer or other enhanced cultural practices is often a clue that a virus or viroid is involved.

The second step is to look for an association between the disease and a particular scion or rootstock cultivar or source of budwood. For example, if trees are growing vigorously on mandarin, rough lemon, or citrange rootstocks but are stunted on sour orange, tristeza may be the problem (Table 9.2). Stunting of trees on citrange or trifoliate orange rootstocks often indicates exocortis or other viroids. It is often difficult to pinpoint the effects of a systemic pathogen when all trees in a planting are uniformly affected. A field diagnostic test for some diseases can be built into new plantings by propagating the same scion on several different rootstocks.

The third step is to look for specific symptoms. For example, exocortis viroid infection often causes bark scaling of trifoliate orange, Rangpur lime, and some citrange rootstocks after the trees are three or four years old (Plate 52A). Removal of a patch of bark across the bud union and examination of the

Table 9.2. Susceptibility to systemic diseases in citrus trees on different rootstocks[a]

Rootstock	Tristeza	Tatter leaf	Exocortis[b]	Cachexia	Blight
Sour orange	×	○	○	○	○
Cleopatra mandarin	○	○	○	?	×/○
Rough lemon[c]	○[d]	○	○	?	×
Rangpur lime	○[d]	○	×	?	×
Trifoliate orange	○	×	×	○	×
Carrizo citrange	○	×	×	○	×
Swingle citrumelo	○	×	?	○	(○)
Citrus macrophylla	×[e]	○	○	×	×
Sunki mandarin	○	○	○	○	ND
Gou tou	○[d]	(○)	(○)	(○)	ND

[a] × = Susceptible. ○ = Tolerant or resistant. ? = Some disease reaction is suspected but is not well documented. (○) = Tolerant response is expected but has not been fully tested. ND = not determined.

[b] The pattern of response to viroid-induced dwarfing (citrus viroids group III) is similar.

[c] Rough lemon, Volkamer lemon, and Milam lemon have similar reactions.

[d] Some isolates may cause stem pitting in the rootstock.

[e] Severe stem pitting in rootstock.

bark and the surface of the trunk frequently provide other clues. A groove in the wood at the bud union (Plate 52H), gum impregnation of the wood at the bud union (Plate 52D), pitting in the wood of the rootstock or the scion (Plate 52D), and gum pockets in the bark (Plate 52E and F) are all indicators of a possible virus or viroid infection.

The fourth step is to consider the probability of encountering a specific disease that may be suspected. For example, greening is automatically suspected as a cause of tree decline in most of Asia and parts of Africa, but it would be unlikely in the Americas or the Mediterranean Basin unless the budwood used to propagate the tree came from a country where the disease is common. Stubborn is common in arid areas, such as the southwestern United States, northern Africa, and the Near East, but is uncommon elsewhere. Blight is not associated with stunting in young plantings and does not affect trees less than four years old.

Diagnostic testing may be needed to confirm a field diagnosis or to determine the cause of a problem where there are no obvious clues (Box 9.1).

Mature Trees

In mature trees, symptoms of systemic disease are usually of two general types: stunting and decline.

Some diseases cause trees to be stunted, although they may have a healthy-looking canopy. Systemic disease in such trees can usually be traced to infected budwood, and the steps in diagnosis are the same as those described for young trees.

The second general symptom is a decline of mature trees

Box 9.2

Quick Field Guide to Diagnosis of Common Systemic Diseases of Citrus

Young trees in well-maintained new plantings are stunted and do not grow properly. No obvious injuries are present.

- Stunted trees are on sour orange rootstock. Leaves are often yellowish. Flowering is abnormal. .... **Suspect decline isolates of citrus tristeza virus**
- Stunted trees are mandarin or mandarin hybrid (e.g., tangelo) scions or are on *Citrus macrophylla* rootstock. Pockets of gum in the bark are visible when the surface of the trunk is scraped, and the wood appears bumpy when the bark is removed (Plate 52D). .... **Suspect cachexia**
- Stunted trees are on trifoliate orange, citrange, or Rangpur lime rootstock, and the budwood source is uncertified. The bud union is normal, but scaling may be present on the bark of the rootstock (Plate 52A). .... **Suspect exocortis or dwarfing viroids**
- Stunted trees are on trifoliate orange, citrange, or citrumelo rootstock. A deep crease at the bud union is visible when the bark is removed. .... **Suspect tatter leaf**
- Stunted trees are on tristeza-tolerant rootstock. Foliage is chlorotic. Bunchy growth and twig dieback are common.

  Twigs are brittle. Stems are pitted (Plate 52F). Primarily limes and grapefruit are affected. .... **Suspect tristeza stem pitting** (Asia, South America, Africa)

  Mandarins are affected. Pitting is mild or absent. Affected trees are growing in a region where greening occurs. .... **Suspect greening (huanglongbing)** (Asia, Africa)

  Affected trees are bushy (Plate 51E). Leaves are small and upturned. Fruiting is sparse. No stem pitting is present. .... **Suspect stubborn** (arid regions, western United States, North Africa, Middle East)

Mature trees are wilted and in decline.

- Declining trees are on sour orange rootstock. Numerous fruit remain on the tree. A brown stain just below the bud union is visible when the bark is removed (Plate 52G). Most affected trees die, with fruit remaining on the tree. .... **Suspect tristeza quick decline**
- Declining trees are on tristeza-tolerant rootstocks. Spring flush is delayed (Plate 51C). The bud union and feeder roots look normal when the tree is in the early stages of decline. Water cannot be injected into the trunk. .... **Suspect blight**

Mature trees are unthrifty and lack vigor, new growth is sparse and may be somewhat yellow, and fruit is undersized.

- Affected trees are on sour orange rootstock. There are no visible lesions or cankers on the trunk or major limbs. The bark is thicker just below the bud union, and pinholes are visible on the inner bark face just below the bud union when a patch of bark is removed. .... **Suspect tristeza decline**
- Bark scaling (Plate 52B) occurs on the trunk or major limbs. Tissue under the scaling areas is alive. ..... **Suspect ringspot or bark-scaling forms of psorosis**
- Affected trees are on trifoliate orange, citrange, or Rangpur lime rootstock and are somewhat stunted. Bark scaling often occurs on the trunk below the bud union (Plate 52A). .... **Suspect exocortis or dwarfing viroids**
- Affected trees are mandarin or mandarin hybrid scions or are on *C. macrophylla* rootstock. Pockets of gum are present in the bark of mandarin or mandarin hybrid scions or *C. macrophylla* rootstock. The wood appears bumpy when the bark is removed. .... **Suspect cachexia**

that previously grew well and were of normal size and vigor. Decline is normally associated with infection that occurs after the tree has been planted, and the pattern of declining trees is often random. Trees often decline rapidly, as in the case of blight or tristeza, or may deteriorate more slowly.

The first step in diagnosis is again to eliminate other factors, such as soil conditions, horticultural factors, and injury due to water damage, herbicide, and spray damage. The major roots, the trunk, and the major scaffold limbs should be examined for wounds and injuries to eliminate mechanical injury or other disease problems, such as Phytophthora foot rot, as the cause of decline.

The next step is to look for an association of symptoms with a specific scion cultivar or scion-rootstock combination. For example, if rapid decline is occurring only on sour orange, and trees on tristeza-tolerant rootstocks remain healthy in the same area, tristeza may be the problem. Blight is a prime suspect in rapid decline of trees on Rangpur lime, rough lemon, and citrange rootstocks. Stunting of trees on trifoliate orange or citrange rootstocks often indicates a viroid problem.

Abnormal flowering and flushing habits are also a clue. Trees affected by phloem pathogens, such as those causing tristeza, greening, and stubborn, tend to bloom earlier than normal. Trees affected by blight bloom and flush later than healthy trees and are easy to spot at the end of the spring flush.

Mature trees may also present specific disease symptoms. Bark scaling (Plate 52A) of trifoliate orange, Rangpur lime, and citrange rootstocks often indicates exocortis infection. Bark scaling of sweet orange and grapefruit scions (Plate 52B) often indicates ringspot psorosis. Deep depressions in the trunk and

Leaves turn yellow on one or more limbs. Chlorosis spreads, and dieback occurs.

- Sweet oranges, mandarins, pummelos, and grapefruit are affected. The fruit is small and does not color properly. Seed abortion is common. Citrus psyllas are present. .... **Suspect greening (huanglongbing)** (Asia, Africa)
- Sweet orange is affected. Yellow to necrotic spots form on leaves. The fruit is small and hard. Trees may recover if pruned. .... **Suspect citrus variegated chlorosis** (Brazil, Argentina)

Trunks or major limbs of mature trees are marked by irregular growth and creases or sunken pockets.

- Grapefruit and lime are affected. Deep, sharp pits in the wood are visible when the bark removed (Plate 52E). .... **Suspect tristeza stem pitting**
- Sour orange, Orlando tangelo, and some mandarins are affected. Stem pitting occurs, but affected trees are growing in an area where tristeza stem pitting is rare. .... **Suspect cristacortis**
- Gum deposits are present under deep depressions in the wood (Plate 52C). Patterns form in young leaves (Plate 52I) and fade as the leaves mature. .... **Suspect concave gum**

Some limbs have twigs with short internodes and small, upturned leaves. Flowering is abnormal. Fruit is small or misshapen.

- Affected trees are growing in a hot, arid area where citrus tristeza virus is rare. Stem pitting is absent. Seeds are aborted. .... **Suspect stubborn** (arid regions, western United States, North Africa, Middle East)
- Twigs are brittle. Bark is thick. Stem pitting is visible in twigs when the bark is removed (Plate 52F). .... **Suspect tristeza stem pitting** (Asia, South America, Africa)

Leaves on some trees have abnormal chlorotic patterns or distortion not associated with spray damage or fungal diseases.

- Lemons and grapefruit are affected (worldwide). .... **Suspect citrus variegation virus**
- Mandarins and some oranges in Japan and China are affected. .... **Suspect satsuma dwarf**
- Sweet oranges and mandarins in India are affected. .... **Suspect citrus mosaic**
- Grapefruit, tangelos, lemons, and some mandarins in Turkey are affected. .... **Suspect citrus chlorotic dwarf**

Fruit has ringlike patterns or areas that do not color properly.

- Sweet orange and grapefruit are affected with bark scaling. .... **Suspect ringspot (psorosis)**
- Sweet orange and mandarin are affected. Chlorotic patterns (Plate 52I) form in young leaves in the spring and later disappear. Gum pockets are present in the albedo under green areas of the rind. .... **Suspect impietratura** (Mediterranean Basin)

main limbs may indicate stem pitting, concave gum, or blind pocket (Plate 52C). Shallow depressions, 0.5–1 inch long (1–3 centimeters), with a crest in the inner face of the bark are associated with cristacortis. Twigs on trees with a stunted, bushy appearance and abnormal-looking fruit should be checked for stem pitting. Trees should be carefully examined for evidence of a virus-induced incompatibility by removing a patch of bark at the bud union and looking for a stain or groove at the bud union (Plate 52G and H) or stem pitting in the rootstock or scion (Plate 52D). Genetic incompatibility between a scion and a rootstock may occur even when no pathogen is present.

Distinctive patterns may form on young leaves of trees infected with concave gum, cristacortis, or impietratura (Plate 52I), and a blotchy pattern may appear in young leaves of trees infected with greening (Plate 52J). Veinclearing is frequently a symptom in lime trees infected with tristeza.

Several systemic pathogens cause fruit symptoms. Trees infected with stubborn frequently have only a few fruit, most of which are misshapen and do not color normally at maturity. Trees infected with greening have small, misshapen fruit that remains green, and seed abortion is common. Trees infected with severe stem-pitting isolates of CTV also have small fruit of poor internal quality. Trees infected with impietratura frequently have fruit with gum pockets in the albedo, and the rind does not break color in these areas.

Some diagnostic tests can be performed in the field that may help identify the cause of disease. For example, if less water is taken up by declining trees than healthy trees when it is injected into a small hole in the trunk, blight may be involved. The root systems of healthy trees normally contain abundant starch. Depletion of starch reserves can indicate a bud union problem, such as tristeza. To test for starch, a solution of iodine and potassium iodide is applied to a bark patch cut across the bud union or to the ends of finger-size roots. A blue-black color indicates presence of starch.

Further identification of specific viruses and other systemic pathogens frequently requires laboratory or greenhouse tests (Box 9.1).

Control Strategies for Systemic Pathogens

There are no practical treatments to cure citrus trees once they become infected by systemic pathogens. Viruses and viroids do not respond to antibiotics, and antibiotics are not practical for field use in controlling systemic prokaryote diseases, even if they work experimentally. The basic control options are to prevent disease from occurring by using pathogen-free budwood, to use disease-tolerant or resistant cultivars, and to reduce or delay losses by certain disease management strategies.

Prevention

Quarantine

Many systemic disease problems begin with the importation of citrus plants or budwood infected with systemic pathogens. Many systemic pathogens are symptomless in common cultivars, and presence of these pathogens is not easily recognized. The epidemics of tristeza quick decline which decimated citrus production in Brazil and Argentina in the 1930s and 1940s began with the importation of infected trees or budwood from South Africa, where trees were grown on tolerant rootstocks. Greening was also introduced into numerous countries in Asia by the movement of infected plants. As a consequence of these and other diseases, such as citrus canker, many citrus-growing countries have adopted strict regulations to prevent the introduction of live citrus plants or plant parts.

While quarantines are important for preventing long-range movement of systemic pathogens to new areas, it is difficult to thoroughly check the large volume of baggage and cargo transported in international travel and commerce. Citrus growers and nurserymen are often tempted to smuggle promising new cultivars from other countries for testing and may not recognize the risks involved. While it is difficult to prevent all introductions of plant material, the hazards of illegal introductions can be minimized by effective nursery certification to prevent the multiplication and propagation of infected budwood (see Chapter 4). Regulations that allow safe and rapid importation of promising new cultivars which can be tested by growers after screening for systemic pathogens reduce the temptations of smuggling.

Budwood Certification

The most important step in preventing problems with many systemic pathogens of citrus is the use of pathogen-free rootstocks and budwood. Many countries have implemented budwood certification programs to produce pathogen-free budwood for new plantings. Valuable cultivars are freed of graft-transmissible pathogens by shoot tip grafting and heat treatment. Budwood sources are indexed periodically to verify that they are free of virus infections, and in some cases they are housed in screenhouses to protect them from infection. Budwood certification is described in more detail in Chapter 4.

> **The most important step in preventing problems with many systemic pathogens of citrus is the use of pathogen-free rootstocks and budwood.**

The full-fledged certification programs operated by major citrus-growing countries may not be practical in smaller production areas. However, growers in countries without budwood certification programs can often avoid disease problems by propagating only rootstocks and scions in combinations which have already proved successful in the area and by avoiding the importation of diseased cultivars from other countries. Careful importation of well-tested, pathogen-free budwood from a well-established program can be used to establish budwood sources for varieties that are not available locally.

The presence of many viruses and viruslike pathogens can be detected by propagating scion selections on a few well-selected rootstocks (such as sour orange, Carrizo citrange, trifoliate orange, *C. macrophylla,* and Cleopatra mandarin). Scions that grow well on all of these, as well as the other rootstocks that will be used commercially, are likely to be free of harmful pathogens. Marked differences in the performance of trees on different rootstocks often indicates the presence of a virus or viroid pathogen.

Eradication

In some cases, damage from inadvertent introduction of a systemic pathogen can be averted by removing infected plants before extensive spread occurs. In general, eradication is effective only when 1) only a few plants are infected, 2) the infected plants can be identified accurately and rapidly, and 3) no vector is present and significant natural spread has not occurred. For example, if greening were found in a small number of plants imported from another country and no psyllids were present, then rapid destruction of the infected plants would eliminate a future hazard.

Eradication is difficult if a vector is present and secondary spread of the pathogen has occurred. The removal of infected trees may suppress further movement if all new infections can be promptly identified and removed. Suppression or eradication works best when the infection level is still very low, but finding a few infected trees scattered among thousands of healthy trees is often difficult. After the infection level exceeds 1%, eradication may not be feasible, because of the costs of sampling and tree removal and the progressively more rapid increase in the number of infected trees. Eradication and suppression programs should be considered only when it is clear that a significant risk exists if the pathogen is not eradicated and that the methods proposed for eradication can accomplish the job at a cost the industry can afford.

Use of Resistant or Tolerant Varieties

New Plantings

A key step in avoiding future problems with systemic diseases is to select disease-tolerant rootstocks and scions for new plantings. This is especially important in controlling diseases that are naturally spread to trees after they are planted. For example, in regions where decline-inducing isolates of CTV are present or a threat, a tristeza-tolerant rootstock should be used. In locations where blight is a major problem, blight-susceptible rootstocks, such as rough lemon and Rangpur lime, should be avoided.

When a combination of disease factors is present, finding scion-rootstock combinations tolerant to all problems is often difficult, and growers have to accept the risk of one or more diseases. In some cases, no adequate resistance or tolerance is available. Lime and grapefruit scions are highly susceptible to stem-pitting isolates of CTV and often cannot be grown in areas where these isolates are common and their vectors are abundant. Greening and stubborn also pose significant problems, since nearly all orange, grapefruit, mandarin, and mandarin hybrid scions are susceptible to injury.

If the pathogen is not spread by a vector, susceptible scion-rootstock combinations can be planted if pathogen-free budwood is used and appropriate sanitation is maintained. For example, viroid-susceptible rootstocks such as trifoliate orange and citranges can be used with viroid-free budwood.

Existing Plantings

Diseased trees in existing plantings should be replaced with tolerant or resistant trees. Replacement trees on tolerant rootstocks must be free of pathogens to avoid creating an additional source of inoculum for vectors to spread to the remaining susceptible trees.

Disease-free plantings may be damaged by a pathogen carried from nearby new plantings by insect vectors. Extensive losses to tristeza and greening have occurred in existing productive groves following the establishment of new groves nearby containing infected trees. The use of pathogen-free trees in new plantings in areas where citrus is already being grown minimizes the risk of spreading disease to existing plantings.

Sanitation

Sanitation is important for controlling some systemic diseases. Two basic elements of sanitation are the removal of sources of inoculum and the reduction of accidental spread by contamination.

Removal of Infected Trees

Removing infected trees from plantings or nearby areas may reduce hazards to new plantings. Tree removal is often a part of the management of citrus greening in areas where the vector is present and new plantings of disease-free trees are to be made. New trees will become infected unless existing infected trees are removed. Removal of infected trees near nursery sites or budwood blocks may also be useful. Removal of viroid-infected trees reduces the chances of spreading the viroids on cutting tools in propagating areas where budwood source trees and nursery stocks are maintained (see the following section). Pruning infected branches has been recommended for controlling CVC in Brazil, since the disease spreads slowly within the tree.

Decontamination of Cutting Tools

Most citrus viroids and some viruses can be transmitted by pruning tools that are used on infected trees and then used on healthy ones. Infection by contamination of tools is a special concern for budwood sources, nurseries, and young plantings. Viroids pose the greatest threat as contaminants, because they are highly infectious, very stable, and difficult to inactivate. Fortunately, tristeza, greening, and stubborn are not easily transmitted this way. Wiping or dipping cutting tools in a dilute solution of sodium hypochlorite (the active ingredient in household bleach) is highly effective in preventing the spread of viruses and viroids. A solution containing 0.5% sodium hypochlorite is adequate (about 1 pint of bleach per gallon of water). A fresh solution should be made daily. Bleach is corrosive, and treated tools should be washed immediately.

Contamination is of less concern in mature plantings, unless the trees will be used as budwood sources. The effects of viroids on mature trees are less severe if infection occurs once the trees are well established. Viroids can be spread by hedging or topping operations, but normally only to trees near those that are infected. Continual decontamination of hedging equipment is not practical, but decontamination of cutting blades may be warranted when equipment used in an infected block is moved to a viroid-free planting.

Mild-Strain Cross-Protection

Many years ago plant pathologists observed that plants infected with some mild strains of a virus did not develop typical severe symptoms when inoculated with severe strains of the same virus. This effect, known as mild-strain cross-protection, has been proposed as a means of controlling certain virus diseases when tolerant or resistant cultivars are not available. Exactly how cross-protection works is unknown. Protected plants are sometimes described as being immunized, but this is not really accurate, since plants do not have an immune system. Cross-protection remains a tantalizing strategy for controlling certain systemic pathogens, but it is often difficult to find isolates that are both mild and protective. Cross-protection is rarely a permanent solution for long-lived crops, such as citrus. There are certain hazards associated with the deliberate use of infectious pathogens, and this strategy is not recommended when other options are available to control systemic diseases.

The first step in cross-protection is to locate an isolate of the pathogen mild enough that infected trees will continue to produce adequate crops. Mildness must be verified for all cultivars subject to infection by vectors which have previously fed on protected plants. For example, tristeza virus isolates that protect against decline in orange trees grafted on sour orange but cause severe stem pitting in grapefruit cannot be used in sweet oranges near grapefruit plantings.

Once a mild isolate is found, its protective ability must be confirmed. If the isolate is not closely related to the severe isolates against which protection is desired, cross-protection may not occur. Testing cross-protection in citrus is often a major problem, since field evaluation is a long-term proposition, and it is difficult to develop advance information for strains of pathogens that exist only in foreign countries. Cross-protecting isolates of CTV have generally been selected from vigorous trees that survived a natural epidemic and have then been tested to confirm their protective effects experimentally.

Mild-strain cross-protection has undoubtedly been practiced in an unrecognized way by growers and nurserymen in some regions where severe pathogens are endemic. If trees infected with mild, protecting isolates are the most vigorous and productive trees, they will be selected as bud sources for propagation of new trees. Continued selection for vigor and productivity over time would theoretically yield trees infected with mild strains provided a high degree of protection against the severe strains in the area.

Cross-protection is currently used to reduce the effects of tristeza stem pitting in grapefruit in South Africa and Australia and Pera sweet orange in Brazil. However, mild isolates of CTV found in a number of countries have frequently failed to provide protection against various severe isolates. Cross-protection has also been observed for some isolates of the psorosis agent.

Cross-protection occurs between citrus viroids in the same group, but not between citrus viroids from different groups. Infection with citrus viroid IIa, which does not cause cachexia symptoms in tangelos and other susceptible hosts, protects trees against subsequent infection with the closely related citrus viroid IIb, the causal agent of cachexia, but does not protect against exocortis.

Therapy

There are essentially no practical chemical controls for citrus diseases caused by systemic pathogens. Trees infected with systemic diseases caused by bacteria-like pathogens, such as greening, stubborn, and CVC, may improve temporarily when treated with selected antibiotics. At one time trees in South Africa were injected with tetracycline compounds to obtain remission of greening. However, antibiotic treatments do not cure infected trees, and antibiotics are not practical as a long-term control measure. Viruses and viroids do not respond to antibiotics, but they may respond to certain antiviral compounds similar to those used for cancer therapy. These antiviral agents, however, are toxic to plants and are expensive. They are also, in general, somewhat hazardous to use.

Replication of some systemic pathogens is reduced by high temperatures. New growth of systemically infected trees produced under high temperatures may be pathogen-free or contain a reduced level of pathogens and show milder symptoms. Heat therapy is sometimes used to free budwood sources of virus infection. However, the effects of heat are not consistent, and viroids and *S. citri* are quite tolerant of high temperatures.

Most systemic pathogens of citrus do not invade the meristematic cells in the tips of new buds. Pathogen-free plants can be obtained by removing the tip of an expanding bud and micrografting it to a virus-free seedling that has been propagated under sterile conditions and reduced light. This procedure, called shoot tip grafting, is widely used to obtain virus-free sources of important cultivars (see Chapter 4).

Vector Management and Integrated Disease Management

A suggestion sometimes made for the control of systemic diseases spread by vectors is to eliminate the vector and thereby prevent further spread of the pathogen. Preventing the spread of vector-borne virus and viruslike diseases solely by chemical control of vectors is unlikely to work in commercial plantings. Insecticides usually do not act quickly enough to kill a vector before it can inoculate a previously sprayed plant. An insecticide may make vectors restless and cause them to visit numerous plants before they die, thereby actually increasing transmission. Effective chemical suppression of insects would require treatment of each flush of growth over the life of the citrus trees. This is generally not economically feasible and also raises health and environmental safety concerns. Furthermore, the primary reservoir of the vector may be in another crop.

Chemical control of vectors may be effective in certain situations, however. For example, chemical treatments may be useful to protect budwood source plantings and nursery stock in small areas. Treatment of a small number of infected trees or a small area may be useful to prevent secondary spread until infected trees can be eradicated. It can also be a component in an integrated approach to management of vector-borne virus and viruslike pathogens.

Biological control of vectors can eliminate many of the problems associated with the use of chemicals. Biocontrol agents include insect parasitoids, pathogens, and predators. Pathogens and predators may not be very effective for preventing virus spread, since these agents normally become effective only after sizable populations of the insect prey are available, and at this point the pathogen probably has already been transmitted to new trees. Parasitoids can effectively maintain a target insect at a low level but do not eliminate it. The level of control needed to reduce the spread of a disease may vary with the disease and the efficiency of the vector. A parasitic wasp has provided effective control of citrus psyllas in Réunion and has greatly reduced the natural spread of greening there.

Integrated Disease Management

Some systemic diseases may be controlled by a combination of practices that are not fully effective when used alone. For example, to grow citrus successfully where the citrus greening pathogen and its psylla vectors are present may require a combination of suppressing inoculum by removing infected trees in remnants of old plantings, the use of greening-free nursery stock, plus some vector control. Mild-strain cross-protection to control tristeza stem pitting may be enhanced by reducing sources of severe isolates and vector management. Increased application of integrated management will require more information on critical aspects of host, vector, and pathogen biology and disease epidemiology.

L. W. Timmer
University of Florida
Citrus Research and Education Center, Lake Alfred

CHAPTER 10

Diseases of Fruit and Foliage

Important Diseases of Fruit and Foliage

Several significant diseases affect the fruit and foliage of citrus trees. Some cause fruit losses, whereas others affect only the external quality of the fruit. One major disease of citrus fruit and foliage, citrus canker, is caused by a bacterium; the other diseases described below are caused by fungi.

Greasy Spot

Greasy spot is caused by *Mycosphaerella citri,* which produces yellow-brown to black lesions on the underside of leaves and minute black lesions on fruit rind. Pseudothecia of the fungus on decaying leaves on the orchard floor discharge wind-dispersed ascospores in response to wetting (Fig. 10.1 and Table 10.1). Ascospores germinate on the undersurface of leaves and grow epiphytically prior to penetration of stomata. The fungus grows slowly in the leaf, and often several months pass before symptoms appear.

Postbloom Fruit Drop

Postbloom fruit drop is caused by *Colletotrichum acutatum.* Reddish orange to orange-brown lesions form on the petals of infected flowers, and subsequently the fruitlets drop off. The calyces and floral disks (buttons) remain attached to the stems; this characteristic is diagnostic of the disease. The fungus produces abundant conidia on diseased petals, and these spores are dispersed to healthy flowers by rain splash (Fig. 10.2 and Table 10.1). The fungus survives between flowering periods as resistant appressoria on leaves, twigs, and buttons.

Melanose

Melanose is caused by *Diaporthe citri,* which produces raised, brick red to black lesions on fruit, leaves, and twigs. In its asexual stage, *Phomopsis citri,* the fungus forms pycnidia in

Table 10.1. Inoculum production, means of spread, conditions for infection, and overwintering of important foliar diseases of citrus

Disease	Inoculum source	Type of inoculum	Means of spread	Optimum conditions for infection: Temperature (°F)	Optimum conditions for infection: Wetting period	Symptom development period	Overwintering substrate
Greasy spot	Decomposing leaves on orchard floor	Ascospores	Wind	75–80	Several nights	4–6 months	Infected leaves
Postbloom fruit drop	Infected flowers	Conidia	Water splash	77–83	10–12 hours	4–5 days	Resistant structures, buttons, leaves
Melanose	Recently dead twigs	Conidia	Water splash	77–85	10–12 hours	5–7 days	Dead twigs
Citrus scab	Pustules on leaves and fruit	Conidia	Water splash	70–85	5–6 hours	4–6 days	Infected tissue
Citrus canker	Lesions on twigs and leaves	Bacterial cells	Windblown rain	77–83	4–6 hours	3–4 days	Infected tissue
Black spot	Decomposing leaves on orchard floor	Ascospores	Wind	70–90	1–2 days	4–6 months	Infected tissue
Alternaria brown spot	Infected leaves on orchard floor	Conidia	Wind	70–80	12–14 hours	1–2 days	Infected tissue

abundance on infected dead twigs (Fig. 10.3 and Table 10.1). Symptomatic leaves and fruit do not serve as a source of inoculum. Conidia from dead twigs in the tree canopy are dispersed by rain and spread to young fruit, leaves, and twigs. Ascospores are also produced in perithecia embedded in dead twigs, but in much lower numbers. Ascospores have little importance in disease development, but since they are airborne, they serve to disseminate the fungus to young plantings nearby.

Citrus Scab and Sweet Orange Scab

Citrus scab and sweet orange scab are caused by *Elsinoe fawcetti* and *E. australis,* respectively. Raised, warty pustules appear on infected leaves and fruit and become flattened as the fruit matures. In their asexual stages, *Sphaceloma* species, these fungi produce conidia directly on the surface of scab pustules on leaves or fruit (Fig. 10.4 and Table 10.1). Conidia are dispersed primarily by rain splash, but one type of spore can be dispersed by wind for short distances. Ascospores of *E. fawcetti* are rare; the windborne ascospores of *E. australis* may be more important in the disease cycle. There are several pathotypes of *E. fawcetti,* which attack citrus species and cultivars differentially.

Citrus Canker

Citrus canker is caused by the bacterium *Xanthomonas axonopodis* pv. *citri,* which produces lesions on young fruit and leaves of citrus. Bacteria ooze from lesions under moist conditions and are dispersed by windblown rains (Table 10.1). High winds with rain are needed to spread the bacteria and force them into stomates on leaves or wounds on leaves, twigs, or fruit. Several strains exist and differentially attack species and cultivars of citrus.

Citrus bacterial spot, a disease of minor importance known only in Florida, is caused by other *Xanthomonas* species.

Black Spot

Black spot is caused by *Guignardia citricarpa,* which produces black lesions of various sizes on fruit and leaves of susceptible species. The fungus forms ascospores in perithecia on decaying leaves on the orchard floor (Fig. 10.5 and Table 10.1). These spores are wind-dispersed. They germinate on the fruit or leaf surface and cause quiescent infections. Symptoms develop several months later, when the tissues mature, near or even after harvest.

Alternaria Brown Spot

Alternaria brown spot is caused by a host-specialized *Alternaria* species, which produces dark lesions on fruit, stems, and leaves of susceptible cultivars. The fungus produces a toxin, which often rapidly kills large portions of infected leaves and occasionally causes dieback of entire shoots. It forms conidia on lesions on mature leaves remaining in the canopy or leaves

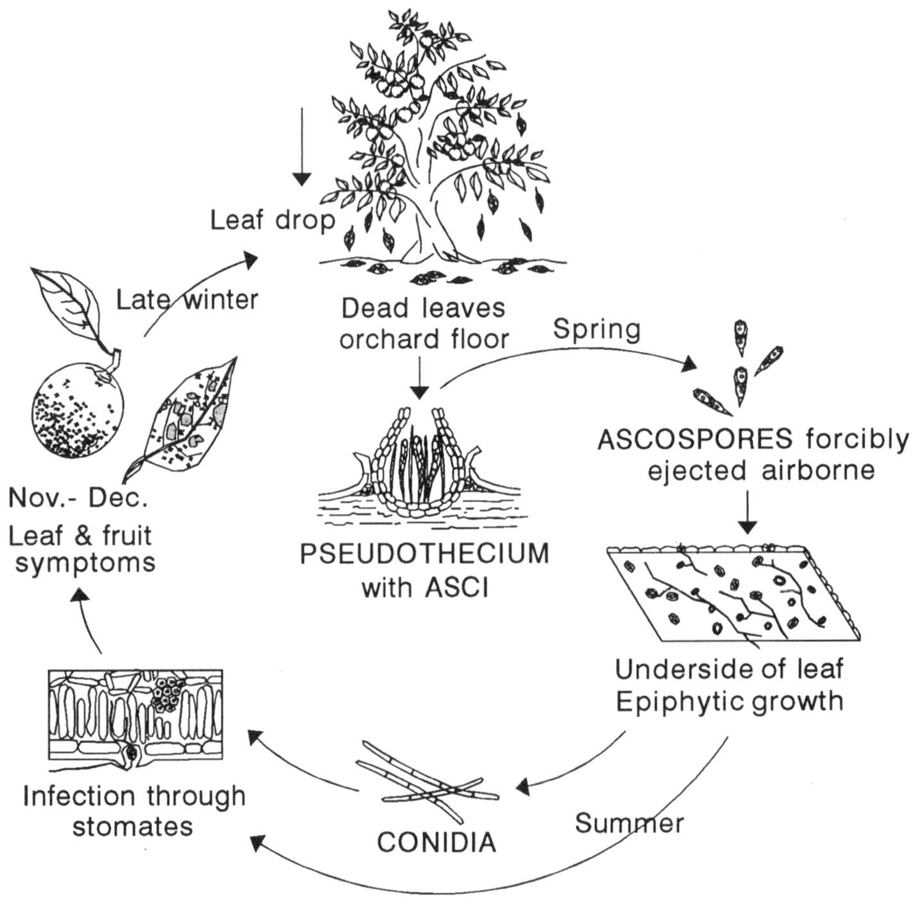

Fig. 10.1. Disease cycle of greasy spot, caused by *Mycosphaerella citri.* (Drawing by Diana Drouillard)

fallen to the grove floor (Fig. 10.6 and Table 10.1). Spore germination, infection, and symptom development may occur in as little as 24 to 36 hours if conditions are favorable.

Other Diseases of Fruit and Foliage

Many other diseases affect aboveground parts of citrus trees, but most are only of local importance or affect only some cultivars.

Mal secco is a systemic disease of lemons and other citrus species caused by the fungus *Phoma tracheiphila,* which is common in the Mediterranean area. It infects leaves and penetrates vascular tissues, causing a wilt and dieback of branches.

Rio Grande gummosis results in a tree decline and is common in Florida and Texas. The cause is still uncertain, but fungal colonization of heartwood through wounds and branch stubs may be involved.

Pink disease and thread blight, caused by *Corticium* species of fungi, result in bark decay and twig dieback. These diseases are important only in humid tropical areas.

Lime anthracnose, caused by *Colletotrichum acutatum,* is a serious blight of young leaves, flowers, and twigs. It affects only Mexican lime and is important in the humid tropics.

Economic Significance

Most fungal and bacterial diseases of citrus are important only in humid areas. Even pathogens whose spores are windborne still depend on moisture for infection and symptom development. Most truly arid citrus areas are free of serious foliage and fruit diseases caused by fungi and bacteria. However, Alternaria brown spot causes significant damage in many semiarid regions, including Australia, South Africa, Israel, and Turkey. *Alternaria* spores are pigmented and can survive harsh conditions. They also germinate and infect very rapidly, so that dew can provide the moisture needed for infection. Thus, economic losses due to *Alternaria* may occur in regions which are rain-free during the infection period. Likewise, black spot may cause significant problems even in semiarid regions. Substantial infection may occur if sufficient moisture is present at the time of spore release.

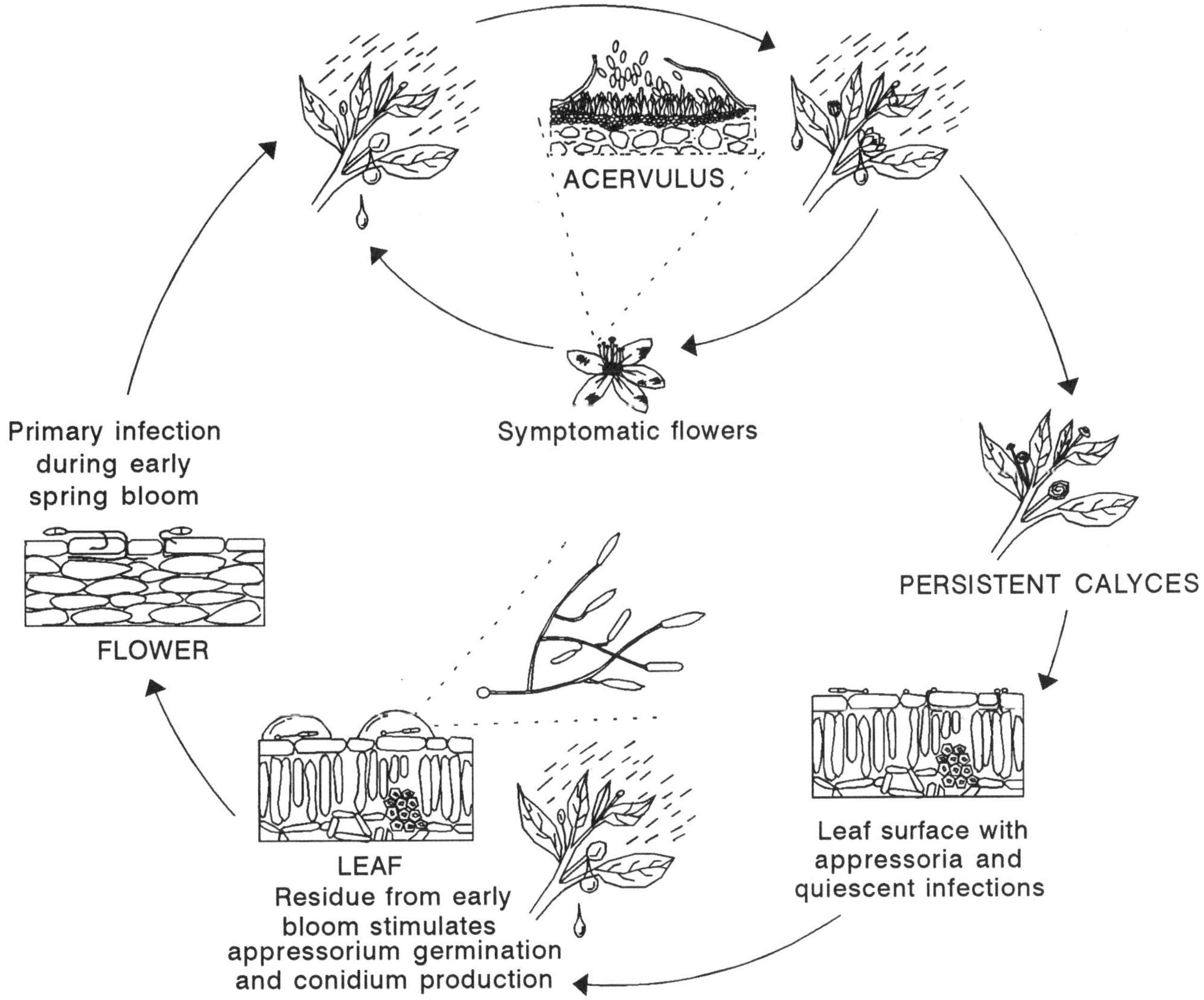

Fig. 10.2. Disease cycle of postbloom fruit drop, caused by *Colletotrichum acutatum.* (Drawing by Diana Drouillard)

The destination of the fruit affects the need for control of many of these diseases. There is seldom a need to control melanose or scab on fruit intended for processing. Black spot and canker cause some premature fruit drop when infection is severe, but under most circumstances there is little need to control these diseases on processing fruit. Greasy spot causes significant premature leaf drop, which may reduce yield and thus must be controlled whether the fruit is intended for the fresh market or for processing. Postbloom fruit drop causes yield losses and must be controlled regardless of the desti-

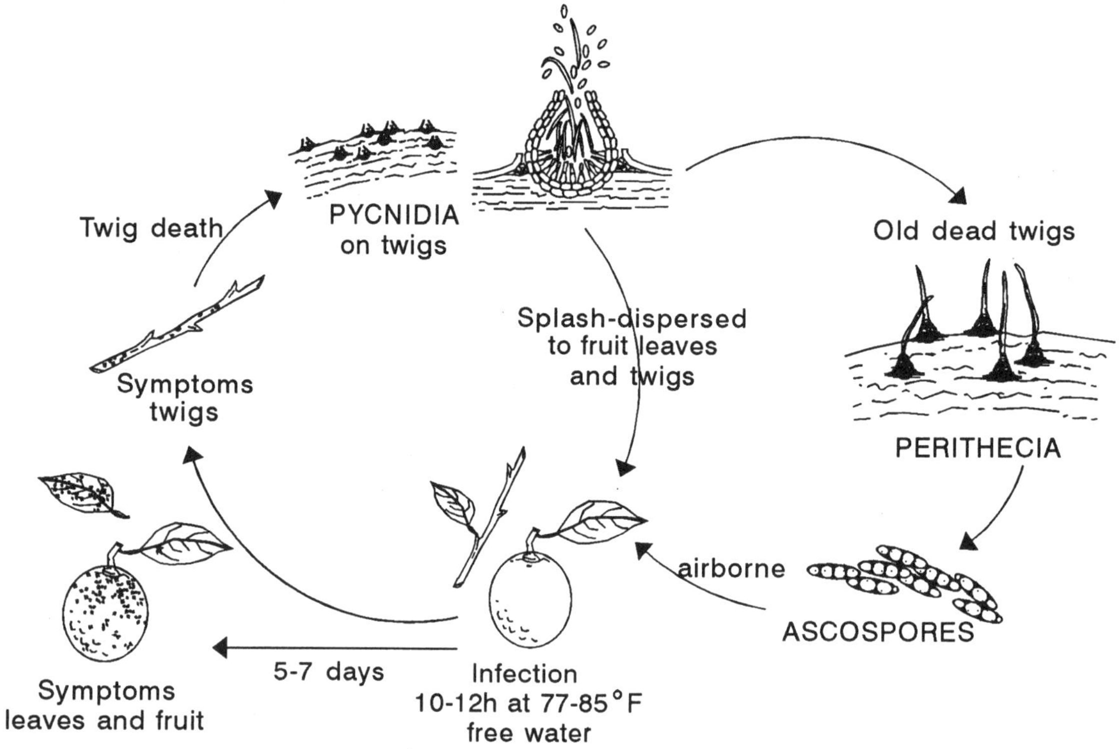

Fig. 10.3. Disease cycle of melanose, caused by *Diaporthe citri.* (Drawing by Diana Drouillard)

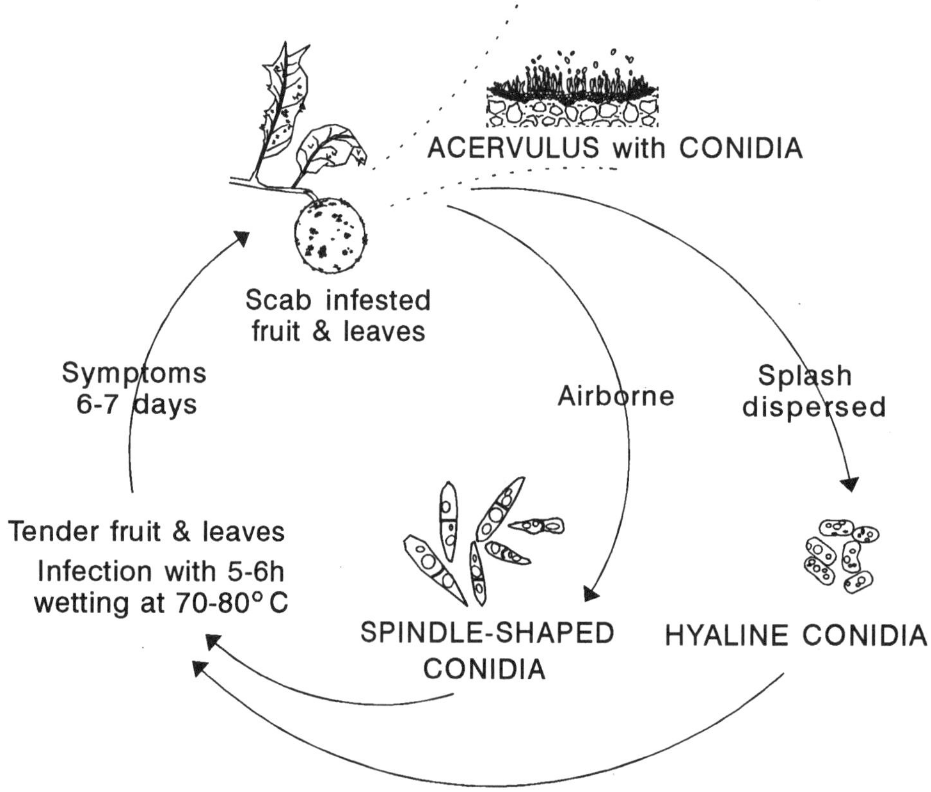

Fig. 10.4. Disease cycle of citrus scab, caused by *Elsinoe fawcetti.* (Drawing by Diana Drouillard)

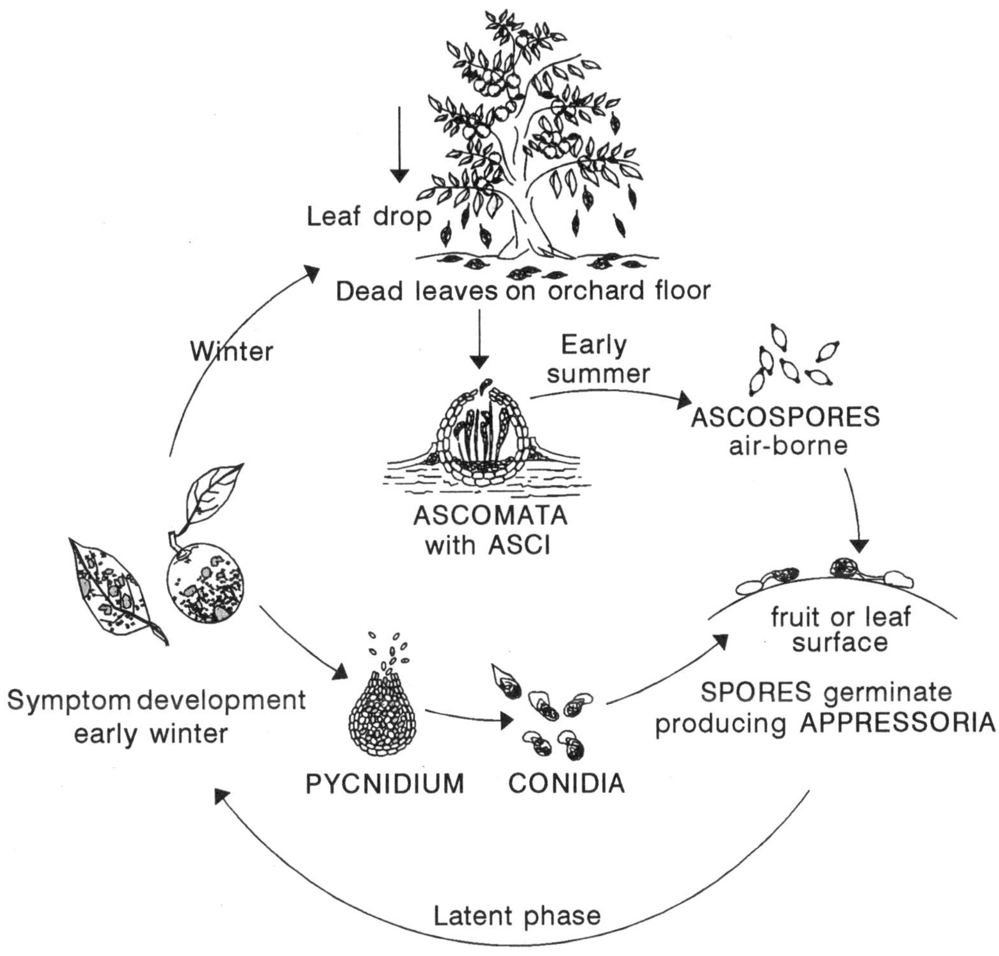

Fig. 10.5. Disease cycle of citrus black spot, caused by *Guignardia citricarpa*. (Drawing by Diana Drouillard)

nation of the fruit. Alternaria brown spot occurs primarily in varieties grown for the fresh market, but it must be controlled in both processing and fresh-market fruit, since it may result in considerable premature fruit drop.

Close attention to weather conditions and well-timed fungicide applications are needed in order to produce high-quality fruit in humid areas. The loss of fungicides through regulatory

The destination of the fruit affects the need for control of many foliar and fruit diseases.

action and the development of resistance by fungi has made the task even more difficult.

Foliar diseases can cause damage in seedbeds and nurseries as well as in groves. Citrus scab affects many rootstock varieties, including rough lemon, Rangpur lime, sour orange, and Carrizo citrange. Severe infection may cause stunting and excessive branching of seedlings. Seedlings infected with scab serve as sources of inoculum for susceptible scion varieties. Most other diseases have relatively minor effects on the production of nursery trees. However, the potential for dissemination of diseases in nursery trees far outweighs any direct effects. Citrus canker is often spread by nursery trees. Several attempts to eradicate this disease in groves have failed after it

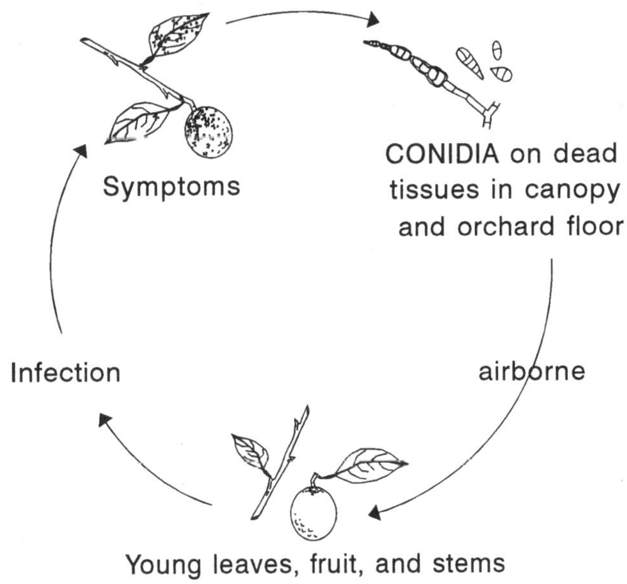

Fig. 10.6. Disease cycle of Alternaria brown spot, caused by a host-specialized *Alternaria* species. (Drawing by Diana Drouillard)

became established in nurseries and subsequently was widely disseminated in nursery stock. If nursery trees are already severely affected by scab or Alternaria brown spot when planted, disease control becomes much more difficult in the grove. However, disease buildup can be delayed significantly if disease-free trees are planted. In Florida, a benomyl-resistant strain of the scab fungus has been spread by replanting groves with infested trees. Increased production of seedlings and budded trees in greenhouses where foliage is not wetted should help reduce these problems in the future.

Epidemiology and Environmental Effects

Wind is important in the dispersal of ascospores of the fungi causing greasy spot, melanose, and black spot, conidia of *Alternaria,* and inoculum of the citrus canker bacterium. Ascospores are actively discharged following rainfall. Relatively light winds can carry the spores and distribute them widely within a grove and to nearby groves (Figs. 10.1, 10.3, 10.5, and 10.6 and Table 10.1).

Insects are relatively unimportant in the dissemination of these pathogens. Spores and bacteria produced in a slimy matrix may be carried short distances by any insect coming in contact with a lesion. For example, bees and other insects may pick up spores of *Colletotrichum* from flowers and carry them to other trees or groves. Transmission of disease by insects is seldom an important factor in disease severity, but it can introduce pathogens to healthy trees.

Conidia of the fungi causing postbloom fruit drop, melanose, and scab as well as bacteria are produced in a slimy matrix. They cannot be distributed by wind except in droplets of water splashed into the air by the force of rainfall (Figs. 10.2–10.4 and Table 10.1). Rain washes spores down through trees and splashes them to healthy tissues. High-intensity rain accompanied by wind will disperse pathogens over much longer distances than gentle rain. Rain may also be important in triggering the release of spores which are primarily windborne. For example, wetting of decomposing leaves will bring about the release of sexual spores of *Mycosphaerella,* which are then carried by the wind. Conidia of *Alternaria* are also dislodged by rain droplets.

Inoculum for infection of new healthy tissue is often present in the grove on diseased tissue from earlier infections (Table 10.1). Some fungal pathogens can germinate and infect citrus tissues in a very short time. The scab fungi produce spores on pustules within 1 to 2 hours after wetting. Once the inoculum reaches susceptible tissue, a brief moist period (5 to 6 hours) is all that is required for germination and infection, although symptoms do not appear for five to seven days. Spores of *Alternaria* germinate and infect with 12 to 14 hours of wetting, and symptoms appear in 24 to 36 hours. In extreme cases, such as greasy spot, spores of the pathogen germinate rapidly but require up to several weeks of intermittent moisture in order to grow on the leaf surface prior to infection.

The period of wetting required is affected greatly by temperature. At optimum temperature (about 83°F, or 25°C), spores of *Phomopsis* germinate and infect in 10 to 12 hours, whereas at 60°F (15°C) germination and infection may require 18 to 24 hours. Thus, in Florida, melanose is much more common in late spring and early summer than immediately after petal fall.

Temperature also affects the time required for symptoms to appear. For example, citrus areas with cold winters seldom have problems with greasy spot. Symptom development is delayed until spring, and leaf drop does not occur until after the spring flush, thus diminishing the impact of the disease on the carbohydrate reserves of the tree.

Disease Management

Regulatory Measures

Disease avoidance has been widely practiced to limit problems with citrus pathogens. These pathogens are readily spread on vegetative material, such as nursery trees and budwood. Most important citrus-growing areas have strict regulations on movement of vegetative material, and long periods of quarantine are required to ensure the health of introduced material. Diseases of fruit and foliage can also be transported in shipments of commercial fruit, but this is a much less likely avenue of introduction. Movement of fresh fruit from areas with canker, sweet orange scab, and black spot is often restricted or prohibited.

The relative merit of attempting eradication of introduced pathogens is the subject of great controversy. Citrus canker eradication has been attempted in a number of citrus-growing areas. Much time, money, and effort has been expended in some areas to try to eradicate canker when it was much more widespread than originally thought. On the other hand, eradication efforts have delayed spread of the disease within countries and delayed disease losses or the need to implement control measures. The key to eradication seems to be the availability of accurate survey information on disease distribution prior to the initiation of a vigorous eradication campaign. Eradication of localized outbreaks of diseases which have water-borne inoculum, such as canker, scab, and postbloom fruit drop, is probably feasible. However, diseases with windborne inoculum, such as black spot and Alternaria brown spot, would be difficult to contain.

Certification of nursery stock to control the spread of systemic pathogens of citrus is widespread, but virtually no regulations exist for nursery trees with foliar pathogens. While freedom from pathogens is unlikely to persist for long periods once plants are moved to the field, serious foliar problems can be delayed by setting out healthy trees. Citrus canker, scab, and Alternaria brown spot are examples of diseases which are much more severe when trees are infected at an early age. In areas where these diseases occur, seedlings and budded trees of susceptible cultivars should be grown in greenhouses where they are protected from rain.

Disease Resistance

None of the currently used citrus cultivars was selected specifically for resistance to foliar pathogens. Many important diseases, including greasy spot, postbloom fruit drop, citrus canker, and melanose, affect most commercial citrus cultivars, although with some differences in disease severity (Table 10.2). Thus, it is unlikely that cultivars can be selected with specific resistance to these diseases. For instance, in some areas it may be possible to replace early-maturing varieties with Valencia oranges, which tolerate greasy spot better. However, economic considerations are often more important than disease tolerance in cultivar selection. Given the difficulty of breeding citrus, it

Table 10.2. Relative susceptibility of important commercial citrus species to foliar and fruit diseases[a]

Disease	Sweet oranges: Early-season and midseason varieties	Sweet oranges: Valencia	Grapefruit	Tangerines and hybrids	Lemons and limes
Greasy spot	MS	MT	S	T[b]	S
Postbloom fruit drop	MT	S	MT	T[c]	S
Melanose	MS	MS	S	MT	S
Citrus scab	I[d]	I[d]	S[e]	V[f]	V[g]
Citrus canker	MS	MT	S	T	S
Black spot	MT	S	S	MT	S
Alternaria brown spot	I	I	T	V[h]	T

[a] I = immune; MS = moderately susceptible; MT = moderately tolerant; S = susceptible; T = tolerant; V = variable.
[b] Except tangelos.
[c] Most members of this group are tolerant.
[d] Immune to most pathotypes.
[e] Susceptible to many pathotypes.
[f] Satsumas, Murcott, Temple, and many tangelos are susceptible; Clementines, Dancy, and Sunburst are immune to most biotypes.
[g] Lemons are susceptible; limes are tolerant.
[h] Most tangelos and Dancy, Emperor, and Murcott are susceptible; most others are tolerant or immune.

is unlikely that cultivars having a tolerance for these diseases will be developed soon.

On the other hand, specific resistance exists to certain pathotypes of the fungi causing scab and Alternaria brown spot (Table 10.2). With the progress in genetic engineering in recent years, it is conceivable that genes for resistance for these diseases could be introduced into susceptible cultivars or that resistant lines could be selected or bred. The development of a Minneola tangelo resistant to Alternaria brown spot or a Murcott resistant to scab seems within the realm of possibility.

Biological Control

There are currently no known biological controls of foliar diseases of citrus, and little research has been conducted in this area. There is some possibility of the development of organisms which are antagonistic to citrus pathogens or compete with them on the leaf or fruit surface. In other crops, avirulent strains and unrelated organisms have been used to induce resistance in the host. With the long period of susceptibility of citrus to foliar and fruit diseases, the development of commercially useful biocontrol agents would require an extensive research effort.

Cultural Practices

Management practices have been manipulated to reduce the impact of disease for many years. While these measures are seldom effective enough to eliminate the need for chemical control, they can make diseases more manageable.

In Florida, the shift from overhead irrigation to under-tree microsprinklers has reduced the severity of scab and postbloom fruit drop. In some cases, irrigation at night when trees are usually wet with dew allows the plants to dry during the day, thus minimizing the wetting period. Timing of irrigation so that it does not coincide with peak bloom has been helpful in reducing the severity of postbloom fruit drop. Restricting the movement of equipment and personnel to periods when the foliage is dry can reduce the spread of disease in the grove and nursery.

Windbreaks have been highly effective in controlling citrus canker. Canker severity is usually reduced to such an extent, even under environmental conditions favorable for disease development, that bactericide applications can be reduced or eliminated. Alternaria brown spot is favored by luxuriant foliar growth. Thus, practices such as avoiding excessive irrigation and fertilization with nitrogen and avoiding pruning during rainy seasons can reduce the impact of the disease. With varieties susceptible to *Alternaria,* planting groves in locations with good air drainage reduces the length of periods of leaf wetness and thus disease severity. Other measures, such as increasing planting distances and pruning tree skirts, increase air flow and decrease disease.

Other changes in horticultural practices may increase disease severity. The switch from disking and clean cultivation to weed control by herbicides in Florida has had many benefits but has probably aggravated greasy spot problems. Disking buries leaf debris on the orchard floor and thus eliminates much of the available inoculum. Other changes have had beneficial effects. Leaf litter is wet more frequently by microsprinkler systems than by overhead irrigation; with more frequent wetting, ascospores mature more rapidly and, in Florida, are released during dry weather in the spring instead of during the summer rainy season, so that disease severity is reduced.

Fungicides and Bactericides

Products Available

Whether desirable or not, chemical control is still the mainstay of management of foliar and fruit diseases. However, in recent years, the number of products available has been reduced by regulatory action or the development of resistance by pathogens.

Copper products are the primary materials used for control of melanose, greasy spot, and canker. They are less effective for control of scab and Alternaria brown spot, but still useful. They are not effective against postbloom fruit drop.

Benomyl has been widely used for control of greasy spot, scab, black spot, and postbloom fruit drop, but in many areas the development of resistance by the pathogens causing these diseases has limited its usefulness.

Dithiocarbamates are still used to some extent.

Ferbam is effective against scab and maneb against black spot, but the short residual activity of these products often makes frequent applications necessary.

Iprodione is highly effective against Alternaria brown spot, but the pathogen has developed resistance to it in some areas.

Whether desirable or not, chemical control is still the mainstay of management of foliar and fruit diseases.

Citrus spray oil is effective for control of greasy spot but not other diseases.

Captafol is highly effective for control of greasy spot, melanose, scab, and Alternaria brown spot, but it is no longer legal in most citrus areas.

Sterol-inhibiting fungicides, which are effective against some diseases, have not yet been registered for use on citrus in most areas. Difenoconazole is very active against scab and is now registered in some countries.

Programs for Specific Diseases

In Florida, greasy spot is usually controlled by a single midsummer application of citrus spray oil or spray oil plus copper. In warmer, more humid areas in the Caribbean, Mexico, and Central America, two or three applications may be needed to protect growth flushes, which may occur at any time of year. Spray oil alone is adequate on the more tolerant varieties, such as Valencia oranges and mandarins. Copper fungicides are usually needed on grapefruit and tangelos, which are more susceptible, and must be used where greasy spot rind blemish is a problem. On oranges used for processing, a single fungicide application for greasy spot may suffice for the entire year.

Copper products are generally the only materials used for melanose control. No sprays are needed for processing fruit. For less susceptible species, such as oranges and mandarins, a single application three to four weeks after petal fall may be sufficient. For fresh-market grapefruit grown in humid areas, several applications may be needed from petal fall till the fruit becomes resistant in midsummer.

For control of citrus scab, fungicides such as benomyl, difenoconazole, or ferbam may need to be applied at the initiation of spring flush to reduce inoculum if old leaves are heavily infected. Protectant materials, such as copper fungicides or ferbam, can be used after petal fall to prevent fruit infection. One to three applications may be needed from petal fall until the fruit becomes resistant to infection, about when it attains one-third of its final diameter. Sweet orange scab does not affect leaves, and thus one or two applications of copper fungicides, benomyl, or other materials after petal fall are sufficient to provide protection.

Copper products are the only materials effective against citrus canker. Depending on the amount of inoculum, the susceptibility of the species, and the conditions in the area, one to several sprays may be needed to produce high-quality fruit. Fruit usually becomes resistant by midsummer, but older fruit may be infected if wounded.

Black spot was initially controlled by several applications of protectant fungicides, such as maneb or copper, during the growth of the fruit. When benomyl became available, a single late-season application was sufficient to eliminate latent infections. Currently, with the predominance of resistant strains in many areas, growers have returned to applying multiple protectant sprays.

Alternaria brown spot is usually controlled by one to several applications of protectant materials, such as iprodione or copper. In humid areas, epidemics of this disease are often so severe that control is not possible. The production of highly susceptible cultivars, such as Minneola tangelo, Emperor mandarin, and Dancy tangerine, has been abandoned in these areas.

Postbloom fruit drop is controlled by one to three applications of benomyl during the bloom period. The number of applications needed depends on the length of the bloom period and the weather conditions during that time. Other fungicides, such as captafol, captan, or ferbam, may be used but are not legal in some areas or are less effective.

Problems Associated with Fungicide Use

The development of resistance in pathogens to fungicides has been a frequent concern. In disease control programs for other crops, fungicides have been alternated or mixed to delay or avoid the development of resistance. However, with so few products available for use on citrus, there may be only one product which is highly effective for a given disease. In Florida, only one application of benomyl per year is usually recommended. However, benomyl is the only product effective against postbloom fruit drop, and sometimes more applications are needed to control this disease. Ferbam is often mixed with benomyl to help avoid the development of resistance.

Fungicide applications are sometimes detrimental to biological control of insects and mites. In Florida, *Hirsutella* species, fungi which parasitize the citrus rust mite, may be affected by fungicide applications. For processing fruit, fungicide applications are limited and probably not detrimental. For fresh fruit, miticides may have to be applied with fungicides to avoid stimulating mite populations in certain cases.

Citrus disease control has been heavily dependent on copper fungicides and is even more so today than in the past. They are quite effective against some diseases, but many problems are associated with their use. Copper sprays may cause fruit blemishes if applied at low pH or in mixtures with other pesticides or nutritional materials. They also may blacken or enhance existing blemishes caused by other agents, particularly if applied at high rates or at elevated temperatures.

Copper accumulates in soil and may reach a level at which it is toxic to plants. The phytotoxicity problem has been largely controlled by liming the soil until it reaches pH 6.5 or above, since copper is less available under these conditions. However, continued application of high annual rates will eventually have negative effects on plant growth and may cause other environmental problems. Current copper fungicides are more effective than older products per unit of metallic copper and can be used at lower rates. However, no highly effective copper fungicide has been developed with a low metallic copper content.

Box 10.1

Predictive Model for Postbloom Fruit Drop

A model has been developed to assist growers in determining the need for fungicide applications and the timing of applications to control postbloom fruit drop. The model is based on the amount of inoculum of the fungus present (i.e., the number of diseased flowers per tree) and rainfall over the preceding five days, and it predicts the percentage of diseased flowers that will be present four days in the future. The formula for predicting this percentage is

$$y = -7.15 + 1.28\sqrt{\text{TD}} + 0.44\sqrt{2{,}500R}$$

where y = the percentage of diseased blooms four days in the future, TD = the total number of diseased blossoms on 20 trees, and R = the cumulative rainfall (in inches) over the preceding five days. The values in the following table are approximate percentages of diseased flowers predicted by the equation for different levels of disease and amounts of rainfall.

	R (inches)				
TD	**0.00**	**0.25**	**0.50**	**1.00**	**2.00**
20	0.0	9.6	14.1	20.6	29.7
100	5.7	17.7	21.2	27.7	36.8
200	11.0	22.0	26.5	33.0	42.0
400	18.5	29.5	34.0	40.5	49.6
800	29.1	40.1	44.6	51.1	60.2

Fungicide application is indicated if three criteria are met:

1. Sufficient bloom is present or developing to represent a significant portion of the total crop.
2. The predicted disease incidence is greater than 20%.
3. No fungicide has been applied in the last 10 to 14 days.

The values in the outlined portion of the table indicate the need for fungicide application if a sufficient number of flowers are present.

Groves with buttons from the previous year should be closely examined once the bloom begins. If infected flowers are found on scattered early bloom, the recommendations of the model should be followed once sufficient bloom is present. Groves with a history of postbloom fruit drop should be checked twice weekly during the bloom period.

Ground and aerial applications are effective in controlling postbloom fruit drop. The combination of benomyl and ferbam has been effective and should lessen the risk of selecting strains resistant to benomyl.

Timing and Integration of Pesticide Applications

If the numbers of applications listed above are totaled, it would seem that many sprays are needed each year to control diseases. In reality, relatively few are needed, compared to the number needed for most other fruit crops. Firstly, not all diseases are present in all areas. Secondly, not all varieties of citrus are susceptible to all diseases. Thirdly, a single application may serve to control more than one disease. For more specific information, see the Citrus Integrator, in the appendix to Chapter 11.

For processing oranges, the only sprays needed are those to control postbloom fruit drop in areas where it is a problem and a single summer application of oil to control greasy spot. Citrus spray oil is a multipurpose spray, also controlling scale insects, suppressing mites, and removing sooty mold. For fresh-market oranges, an additional application of copper in the spring is sufficient to control melanose. Many applications are needed only in some areas, such as Argentina, where sweet orange scab, black spot, and canker may occur. Even there, a single copper spray may serve several purposes.

Frequent fungicide applications are often needed for the production of fresh-market grapefruit, tangerines, and lemons. Grapefruit is highly susceptible to greasy spot, melanose, and canker and is affected by some strains of citrus scab. However, these diseases do not all occur in the same area. Usually, three to five applications are adequate to protect the fruit during the susceptible growth period. Citrus fruit becomes resistant to all diseases once it attains nearly its final size, and thus no applications are needed prior to harvest. Many tangerines are susceptible to Alternaria brown spot or citrus scab and may require several applications to produce high-quality fresh-market fruit. Lemons are susceptible to melanose, scab, and some strains of the citrus canker bacterium, and multiple applications are usually needed in humid areas.

The timing of applications is usually based on fruit phenology. A protective coat of fungicide must be maintained during the period of active fruit growth. Adjustments are often made during exceptionally wet or dry periods. A predictive model has been developed for postbloom fruit drop (Box 10.1), but such models have not been developed for most citrus diseases. Models for timing fungicide applications to control greasy spot would not be especially useful where only one or two applications are needed and the timing is not critical. Predictions of exceptionally severe or mild years would be helpful for management of greasy spot and other diseases. Currently, growers in most citrus areas have insufficient equipment to respond quickly to predictions of disease outbreaks and thus must depend on preventive applications. However, with improvements in spray equipment, response time may be reduced. Means are needed to estimate inoculum levels for citrus diseases and to predict more accurately the effects of weather conditions. With greater knowledge in these areas, fungicide applications can be reduced, and disease control improved.

CHAPTER 11

Harold W. Browning
University of Florida
Citrus Research and Education Center, Lake Alfred

Arthropod Pests of Fruit and Foliage

One of the significant contributors to annual variable costs of citrus production is investment in efforts to reduce or prevent damage to aboveground portions of the tree caused by arthropod populations. Economic damage occurs when populations of one or more species of insects and mites build up on leaves, flowers, fruit, or branches of the tree until they affect fruit yield and quality and, in some cases, tree health and survival. Most damage is associated with an accumulation of generations of arthropods over time, and this period of buildup provides an opportunity for monitoring populations and making informed decisions about when it is appropriate to use various tactics of integrated pest management (IPM). These decisions involve numerous additional factors, such as grove location, tree age, cultivar, cultural practices in effect, seasonal weather patterns, grove management structure, and fruit destination at harvest. This chapter will discuss some major groups of canopy-inhabiting arthropod pests and management of these pests in the context of IPM. Pest species constitute only a small portion of the total arthropod community, which includes many species that are unimportant in citrus production and some species that are beneficial to production. The beneficial organisms include pollinators, such as honeybees, and natural enemies that feed on other arthropods in the community.

Effects of Arthropods on the Citrus Canopy

There are probably as many means by which arthropods affect tree health and productivity as there are pest groups. This chapter describes the more common forms of arthropod damage and illustrates species and groups which can cause economic damage if left unchecked. For the sake of brevity and generality, we have highlighted the most injurious species and groups, those that are most widespread in citrus around the world, and those that are the most representative of their type (Table 11.1).

Several arthropod species or groups are associated with each of the aboveground parts of the tree—the flowers, fruit, foliage, twigs and branches, and trunk. Many arthropods inhabit more than one part of the tree, and thus their population dynamics and impact may be multidimensional. We shall discuss the most important aboveground tree parts utilized by the highlighted species and other tree parts that are seasonally important (for example, the citrus rust mite is most important as a fruit pest, but it utilizes leaves when fruit is not available). The discussion focuses primarily on pests found in humid subtropical growing regions, particularly Florida, but pests that may be more abundant or more important in arid production areas are also described. Thus, the group of pests discussed does not represent a complex found in any one area but is a selection of examples to illustrate principles representative of broader groups of pests.

Citrus blooms (flowers) provide a short-term substratum for colonization by insects. Feeding on petals and on developing fruitlets can result in fruit scarring and drop. After flowering there is a protracted period of development and growth, during which mites and insects colonize and feed on and within the fruit. The period of susceptibility to fruit feeding varies greatly with different pests. Some species are able to utilize fruit in early stages, while others attack only mature fruit.

Leaves are another highly suitable substrate for insects and mites on citrus, and a host of species utilize them throughout their growth and protracted life span on the tree. Twigs and branches are host to several groups of insects which feed on or in the wood. Finally, the trunk, the main structural component of the tree, is susceptible to attack by various insects, including those that feed on the outer bark cells and those that penetrate the trunk to feed on the wood.

External fruit quality is a critical characteristic affected by infestations of insects and mites and attracts the most attention in systems of arthropod management. Rind cell punctures by mites, malformation due to insect feeding, and physical contamination with insect bodies demand attention in fruit grown for fresh consumption. The duration and type of feeding and the stage of fruit development determine the extent and type of damage to the rind. Contamination by honeydew and sooty mold are secondary effects associated with feeding by homopterans on citrus foliage, and the deposition of honeydew may be sufficient to foster a layer of black fungal growth on fruit hanging beneath infested leaves.

Internal fruit quality can likewise be compromised by infestation by certain insect pests, which probe into fruit and damage tissue or deposit eggs, from which larvae hatch to feed on internal tissues. The damage may result in spoilage, a decrease in the soluble solids content, premature fruit drop, and altered juice content, color, and flavor.

Fruit yield is a primary consideration where pests develop season-long populations and thereby reduce the photosynthetic capability of trees by damaging their leaves or cause premature fruit drop and yield loss as a result of feeding.

Economic Significance of Damage to the Canopy

General tree health is affected by the cumulative impacts of arthropods and diseases, through reduced photosynthetic capability, reduced foliage production, and structural damage to twigs, limbs, and trunks. Tree health is also affected by many other canopy-related factors, including climate, fertility, water relations, and competition with other plant species, and by factors influencing root growth (refer to Chapters 12–14, on root problems). Thus, the economic impact of canopy-inhabiting arthropods may be difficult to assess and often quite difficult to discriminate from other factors affecting tree health.

Marketing standards are important in assessing the potential economic damage due to foliar arthropods. Standards for external appearance are low for fruit destined for processing into juice and other citrus products, while fruit destined for the fresh market must meet standards established by local consumer aesthetics. In some citrus markets, these standards are stringent, resulting in arthropod and disease control expectations that are difficult and expensive to achieve. However, in some domestic or local markets, standards for external fruit appearance are less rigorous, and the goals of pest management are more easily achieved with less expense.

Since the potential market for a citrus crop dramatically influences the goals of pest management, it is desirable to make marketing decisions early in the season and plan an IPM program accordingly. Many producers define marketing goals at planting, while others allow fluctuating prices, seasonal weather, and other events during the season to influence their marketing decisions. For many growers, the goal is to produce as high a percentage packout as possible, with the fruit not making fresh grade going to juice processing. The importance of the market destination cannot be overemphasized in setting goals for IPM of foliage-inhabiting arthropods, since most of the pest pressure either directly or indirectly influences external fruit quality.

A final consideration that strongly influences pest management of foliar arthropods in citrus is the age and productivity of the trees. Pest management for trees that are not yet producing commercial volumes of fruit may be very different from that for commercially producing trees. Different practices are required to facilitate vegetative growth in nonproducing trees and reproductive growth in producing trees.

Pest management for nonproducing trees should promote foliar growth and photosynthetic productivity, and thus damage assessment is concerned with pests that affect vegetative growth, such as the citrus leafminer, *Phyllocnistis citrella,* a recent introduction into many citrus production regions.

Pest management for bearing groves must also protect against economic loss of fruit yield and aesthetic quality, in addition to promoting foliar growth and tree health. For example, Florida red scale, *Chrysomphalus aonidum,* is an important pest in mature trees; it colonizes developing fruit and thus detracts from external fruit quality, and it can also cause injury to leaves and twigs. Foliar growth and tree health become particularly important as trees age and production declines through the cumulative impacts of weather, disease, and pests. Pest management goals for aging groves may need to be adjusted in accordance with their economics and site-specific considerations. Attention may need to be given to the economic benefit of intensive or expensive treatments required to maintain fruit quality standards for fresh marketing, such as season-long control of citrus rust mites or mealybugs, in groves with declining productivity or tree health. The difference in the market prices of fresh fruit and juice may not warrant the additional cost required to protect fruit in a grove in which trees are declining.

Pest management for trees that are not yet producing commercial volumes of fruit may be very different from that for commercially producing trees.

Major Arthropod Pests

Rust Mites and Bud Mites

Among the most important arthropod pests of citrus worldwide are mites in the family Eriophyidae, of which the dominant species is the citrus rust mite, *Phyllocoptruta oleivora.* Other species, including the citrus bud mite, *Eriophyes shel-*

Table 11.1. Generation times and primary (1) and secondary (2) substrates infested by canopy-inhabiting insects and mites

		Substrate infested				
Insect or mite group	Minimum generation time	Flowers	Fruit	Leaves	Twigs and branches	Trunk
Rust mites, bud mites	10 days	. . .	1	1	. . .	. . .
Spider mites	10 days	. . .	2	1	. . .	. . .
Scale, mealybugs	30 days	. . .	1	1	2	2
Whiteflies	45 days	. . .	. . .	. . .	. . .	. . .
Fruit flies	25 days	. . .	1	. . .	. . .	. . .
Aphids	10 days	. . .	. . .	1	. . .	. . .
Foliage feeders	30 days	. . .	. . .	1	. . .	. . .
Thrips	15 days	1	2	. . .	. . .	. . .
Woodborers	Months	. . .	. . .	. . .	1	2
Leafminers	13 days	. . .	2	1	2	. . .

doni, and the pink citrus mite, *Aculops pelekassi,* are pests in some parts of the world.

The citrus rust mite owes its reputation to its year-round survival in citrus, its ability to build up a tremendous population during the major growing season, and the extensive rind injury that it induces at high population density. Rust mites feed and reproduce on citrus foliage and move to young fruit as they become available. The mite population builds throughout the period of fruit development. The mites puncture rind cells and extract the cell contents. Cells damaged by repeated puncturing discolor, and the result is a russeting of the fruit surface. This injury can contribute to accelerated water loss from the fruit, but its primary effect is to reduce the aesthetic quality of the rind, and thus reduced fresh fruit value. Only when rust mite populations persist and attack fruit over extended periods does the accumulated injury result in a noticeable reduction in fruit size and, in some cases, premature fruit drop. Leaf symptoms of rust mite feeding generally go unnoticed. Certain citrus cultivars with moderate to high susceptibility to the mite may be subject to visible chlorosis and, in some cases, leaf drop.

During the peak period of rust mite development, when temperatures are high and humidities are moderate, the mite can achieve tremendous rates of increase, completing a generation in less than 10 days. Their minute size (less than 1 millimeter, or 1/32 inch, in length) makes them invisible to the unaided eye and allows populations to develop unnoticed.

Spider Mites

Spider mites also are important pests of citrus worldwide, including widely distributed species such as the citrus red mite, *Panonychus citri.* Unlike the rust mites, spider mites feed primarily on foliage, and only occasionally on fruit cells. Injury from leaf feeding by spider mites can result in leaf loss under dry conditions, and terminal dieback occurs in heavy infestations. Spider mites are visible to the unaided eye, forming colonies on the upper and lower surfaces of leaves. Some species produce silken webbing on leaves, and colonies develop in the protected environment created beneath this silk layer. Spider mite populations may be present at any time during the growing season, increasing rapidly during periods when conditions are favorable. Species vary in their preference for the upper leaf surface as opposed to the lower and their preference for portions of the canopy for colonization. Generation times are similar to those reported for the citrus rust mite, with development from egg to adult being completed within 10 days during summer.

Scale Insects and Mealybugs

Among insect pests of citrus are a group of related species in several families in the order Homoptera, and numerous species in this group are present in most citrus production areas. Most insects in this group have several features in common:

- They are generally sessile, being more or less sedentary in most life stages.
- Adult females retain the immature form, and males take on a winged adult form.
- All species except armored scale insects excrete honeydew as a by-product of feeding on plant sap.
- They utilize a mix of substrata, including fruit, foliage, twigs, and the bark of the trunk.

Among the more injurious of these insects are the California red scale, *Aonidiella aurantii,* a serious pest of fruit, foliage, and twigs in arid citrus-growing areas, and black scale, *Saissetia* species, a group of soft scale species that produce generous amounts of honeydew on fruit and foliage. The citrus mealybug, *Planococcus citri,* is cosmopolitan and damages fresh fruit in many parts of the world; it causes rind blemish and produces extensive honeydew as it colonizes fruit clusters.

Whiteflies

Whiteflies are another family of homopterous pests well represented in citrus. Infestations by species in this family are limited to foliage, but economic damage can occur as a result of leaf injury and also contamination of fruit with sooty mold. Like scale insects, whiteflies are sessile when immature, occurring primarily on the lower side of the leaf. However, both sexes develop into winged adults. Whiteflies transmit plant-pathogenic viruses in other cropping systems, and they act as vectors of citrus chlorotic dwarf in Turkey.

Among the more common species are the citrus whitefly, *Dialeurodes citri* (Plate 53); the woolly whitefly, *Aleurothrixus floccosus;* and the citrus blackfly, *Aleurocanthus woglumi.* Whitefly species vary in their preference for leaves on which to lay eggs; some are restricted to newly emerging leaves (e.g., the bayberry whitefly, *Parabemisia myricae*), and others able to oviposit and develop on mature leaves (e.g., the citrus whitefly and citrus blackfly).

Fruit Flies

Fruit flies, in the family Tephritidae, cause some concern in citrus production and marketing. These insects oviposit in developing fruit, and hatching larvae burrow within the fruit to feed. The larvae in fruit cause direct injury resulting in local decay and often premature fruit drop. However, of even greater concern is the presence of fruit fly larvae in harvested fruit, because of the threat of introduction into uninfested areas, including Florida. This is of particular concern with the Mediterranean fruit fly (Medfly), *Ceratitis capitata,* which is widely distributed but still not present in many key citrus-producing areas. The broad host range of this insect makes it a tremendous threat to many important agricultural crops, including fruits and vegetables.

Other fruit flies of importance are more limited in distribution and in host range (e.g., the Caribbean fruit fly, *Anastrepha suspensa,* in Florida), causing less economic damage and posing a lesser threat than the Medfly.

Aphids, Sharpshooters, and Psyllids

Aphids, sharpshooters, and psyllids not only inflict direct damage by feeding on citrus leaves or fruit but also have the ability to transmit numerous plant pathogens. Most of these pest species seek newly emerging foliage as a primary feeding site, although populations of some species can be sustained on mature leaves and twigs. Aphid species on citrus are numerous, with seasonal populations of most species closely tied to the production of new flush. Colonizing individuals (most often gravid females) are attracted to new growth and form new colonies as the trees enter a new flush cycle. Colonizers that were previously feeding on trees infected with certain pathogens can transmit those pathogens to healthy trees. The details, efficiency, and exact mechanisms of transmission vary with different vectors and pathogens, but the outcome—disease transmission—is generally more serious than the feeding injury to the tree. Further details on the diseases transmitted by these agents are presented in Chapter 9.

Populations of most species in this vector group vary tremendously over time, fluctuating seasonally and with citrus tree phenology within a grove. Variable flushing patterns can cause major site-specific differences in vector populations. However, the transmission of diseases often occurs during colonization, and the risk of transmission increases during periods when vectors are moving among trees or from grove to grove. Thus, population monitoring and management of this pest group must take into consideration colonization and transmission characteristics.

Among the well-known vectors in this group are the brown citrus aphid, *Toxoptera citricida,* and the Asian citrus psyllid, *Diaphorina citri.*

The brown citrus aphid is widespread in citrus-producing countries and is the most effective vector of citrus tristeza virus. This species is capable of phenomenal population explosions during a brief period of favorable conditions, completing a generation in less than two weeks. Winged females develop as the leaves mature. They then colonize new terminals. After relocating to a new leaf, females reproduce without mating and give birth to live young.

The citrus psyllid has a more conventional biology, laying eggs singly or in small groups on emerging leaves. The hatching nymphs feed and develop on leaves and succulent stems. Adults of both sexes are winged and move from one feeding site to another, and this movement of adult psyllids is responsible for the transmission of greening disease.

Foliage Feeders

Diverse insects feed on citrus foliage by consuming leaf material partially or wholly with chewing mouthparts. These foliage feeders include crickets, grasshoppers, katydids, cockroaches, leaf-cutting ants, and lepidopterans. They are not key pests under most circumstances, but they are particularly injurious in citrus nurseries and young groves, where injury from leaf feeding has more impact on tree growth. A single individual can consume large pieces of a leaf, and when numerous individuals are present in a tree, they can damage a significant portion of the canopy. This group of insects includes species belonging to different families in several different orders, and thus their life histories vary. In most cases, they are most injurious in their immature stages, and the adults may or may not also feed on citrus foliage.

The orange-dog caterpillar, *Papilio cresphontes,* is a common foliage feeder in Florida citrus and can cause extensive injury to young citrus plants in the nursery and in young groves. Its larvae hatch from eggs laid upon new terminals, and as the larvae develop, their rate of foliage consumption increases.

Other foliage-feeding species common in Florida include the lubber grasshopper, *Romalea microptera;* the American grasshopper, *Schistocerca americana;* the long-tailed katydid, *Microcentrum rhombifolium;* and the recently introduced Asian cockroach, *Blattella asahinai.*

Thrips

Thrips can cause direct damage to citrus fruit. The citrus thrips, *Scirtothrips citri,* is particularly damaging. It is one of several species that feed within citrus blossoms, causing fruitlet drop in some cases or scarring the bloom end of the tiny fruitlet. As the fruit expands and develops toward maturity, the scarring expands into a ring around the shoulder of the fruit. Thrips injury may result in reduced quality, especially in fruit sold for fresh consumption.

The life cycle of the citrus thrips and other blossom-feeding thrips complicates the management of these pests. They insert eggs in fruit tissues, and the hatching larvae feed during several stages in citrus blooms. Once feeding is completed, the mature larvae drop to the soil to pupate in surface litter. This characteristic, with the overlapping of generations of thrips, limits the effectiveness of on-plant suppression techniques and illustrates the need for effective population monitoring.

Wood Borers

Citrus groves in some tropical locations are susceptible to wood-boring beetles, which attack intact, active wood within twigs and stems. As adults, these insects may feed on citrus bark, and they lay eggs on the wood. The hatching larvae tunnel within the limbs or trunk, often beginning in dead or dying wood and subsequently moving into live tissue. The injury results in twig or branch dieback and may also provide an entrance for pathogenic organisms.

Leafminers and Peelminers

A few species of leaf-mining insects are pests of citrus, of which the citrus leafminer, *Phyllocnistis citrella,* is the most important in many areas of the world. In their immature stages, leafminers feed on epidermal tissue of leaves, and peelminers on the epidermis of the fruit rind. These lepidopteran miners tend to occur sporadically within seasons, generally synchronous with periods of increasing temperature and active flushing. Under tropical conditions, where flushing occurs throughout the year, leafminers can reach extremely high populations on foliage, and the resulting injury to leaves may reduce growth and yield. In combination with water stress, citrus leafminer injury can result in severe defoliation.

Leafminers have relatively short generation times, completing development in as little as 13 days under optimal conditions. Adults are nocturnal and oviposit on tiny, newly emerging leaves. The hatching larvae tunnel through the epidermis and feed through three instars, each lasting little more than one day. Only a brief portion of the life cycle involves feeding, and this fact has implications for remedial measures. Pupation occurs beneath the epicuticle and is usually accompanied by curling of the leaf margin in the vicinity of the silken pupation chamber. Populations are visible when leaf infestation levels exceed about 25% of expanding new flush leaves. As the percentage of leaves with miners increases, the number of mines per leaf also increases.

The life history and population development of peelminers is similar to that of leafminers. The period of population development is dictated by availability of fruit of the appropriate size for infestation.

Influence of Environmental Conditions

The wide range of environmental conditions under which citrus is propagated in different parts of the world allows for considerable variation in the response of arthropods which inhabit the canopy. Short-term patterns of temperature, humidity, wind, and rainfall influence the developmental rates and survival of arthropods. Interactions of environmental patterns with pest biologies provide a framework for predicting the potential impact of a pest species in terms of the population levels achieved and the damage due to feeding injury.

For example, the life cycle time for arthropods (the duration of the period of development from egg to reproductive adult) is temperature-dependent. In summer, with daily maximum temperatures of 85–95°F, the citrus blackfly, *Aleurocanthus woglumi,* completes a generation in about 60 days. During winter, with maximum temperatures in the 50s, the generation time stretches to more than 100 days, severely limiting any increase in the blackfly population.

Environmental conditions other than temperature also influence pest dynamics. Wind plays a role in disseminating foliar arthropods between and within trees, and it may be a mortality factor when the air is dry. Furthermore, humidity or rainfall interact with temperature to affect rates of arthropod development and survival. Some pest species thrive at low humidity, while others increase at higher rates under humid conditions, even at comparable temperatures.

An additional element in the influence of the weather on arthropod populations is the effect of temperature and moisture on biotic agents (predators, parasites, and pathogens) that attack citrus pests. Whiteflies, mites, and scale insects, among other pest groups, are highly susceptible to infection by fungal pathogens (Plate 53). Under moisture conditions that favor fungal sporulation, germination, and infection, some natural suppression of susceptible pests may occur. However, under dry conditions, the same pests may be relatively unaffected by disease, even though the fungal pathogens may be present in the orchard.

From these generalizations, it is easy to recognize the importance of weather to arthropod populations on citrus foliage, their pest status, damage, and seasonality of occurrence.

In arid and temperate or humid and subtropical citrus-growing areas, a combination of temperature, humidity, and rainfall allowing optimal development of some pest complexes can be expected during discrete periods. The typical pattern in temperate areas in the Northern Hemisphere is an optimal period of pest increase in late spring and early summer, following cold winter and cool spring conditions. Populations of pests such as leafminers, aphids, and scale insects that have survived with low rates of growth and reduced activity during winter become active and redistributed within trees and within groves. As summer progresses, populations peak, and as fall approaches, they are again limited by weather and its effects on tree growth. Winter pest populations in temperate or subtropical climates are often severely limited by weather, which provides a bottleneck in annual population dynamics. The severity of episodic as well as general winter temperatures can severely limit the potential for a pest species to build up a sufficient population over the growing season to cause economic injury.

In humid tropical citrus-growing areas, conditions favoring pest development and injury are sustained for a longer period than in temperate and subtropical areas. Temperatures in tropical settings do not drop below thresholds for pest development, and thus foliar pests are able to thrive year-round. Shorter generation times and increased numbers of generations per year lead to higher pest pressure. Because of the absence of a cold period, annual cycles are more predictable, and the complex of pest species tends to be broader.

Environmental factors which influence pest populations also affect citrus trees and define periods of active growth, flowering, fruit set, and other phenological events. The periodicity of tree growth defines the availability of new succulent leaves or twigs on which certain pests thrive. Citrus leafminers oviposit almost exclusively on newly emerging leaves less than half an inch long, and thus this pest is closely tied to seasonal periods of tree flushing. Climatic factors define the rates of growth of foliar arthropod pests, drive plant phenological events on which the pests rely for substrate and nutrition, and ultimately interplay with the pathogens, predators, and parasites which constitute the third trophic layer in the citrus ecosystem. Understanding these relationships allows one to develop the initial steps of an IPM program, wherein key pests and disease organisms are characterized by their periods of occurrence, ecology, and damage potential.

Finally, the relative damage resulting from feeding by foliar arthropods can be influenced by prevailing environmental conditions. Trees suffering from environmental stresses are most susceptible to injury from arthropod feeding. For example, water-stressed trees are subject to greater damage from spider mite feeding, and spider mite injury together with dry winds often results in defoliation. Similarly, cumulative injury due to the citrus leafminer can lead to defoliation under dry conditions, whereas little defoliation occurs with heavy mining of trees with good water status.

Management of Canopy-Inhabiting Arthropods

Pest suppression tactics and IPM strategies for foliage-inhabiting arthropods and pathogens focus on suppressing infestation or infection, to keep it below a level at which economic damage is inflicted, rather than complete, permanent control or elimination. In some specific cases, measures are implemented to exclude the introduction of serious pest species from citrus areas where they currently do not exist, but in most cases the intent is to limit the injury associated with sustained populations of pest species already present in the system. It has been repeatedly demonstrated that it is virtually impossible to eliminate an arthropod pest population once it has become established in a hospitable citrus production environment. Thus, elimination should not be an overriding goal of management of most pest species.

There is considerable geographical overlap in the complex of major arthropod pests of citrus in different parts of the world. This is a result of the widespread dissemination of infested plant material between countries and continents prior to the implementation of regulatory measures and the accidental introduction of species since these measures have been in place. Not all of these pests occur in any one citrus production area, as a result of regulatory efforts to prevent the introduction of arthropod pests. The recent introduction of the citrus leafminer, *Phyllocnistis citrella,* into many citrus production regions demonstrates the capacity of arthropods, like pathogens, to invade broad geographic areas and disrupt pest management programs. Thus, exclusion, while not usually a grower function, is an important feature of citrus pest management programs.

Year-round pest infestations in tropical citrus-growing areas complicate an economically based and ecologically sound management program, since the period of concern often spans eight or 10 months, if not the entire year. The extended vulnerability of fruit to infestation by the citrus rust mite and the year-round susceptibility of leaves to mites, scale insects, whiteflies, and other pests further illustrate the difficulties of pest management.

With these considerations in mind, it becomes obvious that monitoring the population of arthropod pests in groves is crucial. The dynamics of injurious populations must be characterized sufficiently well that management options can be selected that best fit the situation at a specific point during the season. Knowledge of the pest species involved, estimates of current and projected populations under prevailing conditions, and economic considerations revolving around projected yields and fruit utilization are essential to establishing action thresholds. An action threshold is the point in the development of a pest population where the implementation of a management practice will prevent the population from reaching a level at which it would inflict economic injury. Action thresholds have replaced economic thresholds in recent pest management vernacular, although they represent similar concepts.

Regulatory Measures

Regulatory measures have been invoked to protect citrus industries worldwide from invasion by exotic arthropod pests. Efforts to exclude plant material serving as hosts of exotic arthropod pests are focused on ports of entry in citrus production areas, to provide oversight of entry of potential pests via commercial shipping, recreational travel, and movement of infested fruit through all channels. Most foliage-inhabiting insects and mites can be transported on citrus vegetative material or fruit, but certain groups are of particular concern. Fruit flies pose a significant threat, because they are harbored within the fruit. Rates of fruit fly interception at ports of entry into the United States indicate the threat that infested fruit pose internationally.

Certain pest species have limited distribution in Florida, and efforts to maintain pest-free areas and to limit areas of infestation are ongoing. In areas designated "pest-free" through exclusion or regulatory treatment, harvesting and marketing of fresh fruit may be allowed. Such protocols are commonly utilized in the export of fresh fruit from "fly-free" areas within an industry. Furthermore, exclusion or pest-free designation is also used within the Florida citrus nursery industry to reduce the risk of movement of stock infested with such pests as the brown citrus aphid and root weevils. The success of regulatory efforts to restrict the movement of pests with limited distributions hinges on an awareness of the risks and compliance with voluntary or mandated control measures.

Arthropod Resistance and Susceptibility

Approaches to management which rely on differences in susceptibility of citrus cultivars have not been widely used to manage canopy-inhabiting insects and mites, although there is evidence of some variation in suitability among sources of citrus germ plasm. Arthropod injury varies in different germ plasm groups (oranges versus grapefruit, etc.), but these cases involve injury thresholds and market considerations, rather than suitability of the germ plasm for development of the pest. Differences in the duration of flushing periods, the succulence of new foliage, and rind characteristics of citrus varieties are some of the factors that determine their suitability for arthropod feeding and thus injury to the foliage or fruit. At present, host plant resistance is not a primary strategy in IPM of foliage pests. However, cultivars also differ in susceptibility to viruses transmitted by insect pests, and host plant resistance is an important consideration in this respect (see Chapter 9).

Biological Control

Citrus pest management is known worldwide as a testing ground for the development and implementation of biological control using natural enemies to reduce populations of foliage pests or their impact on the tree. Several features of the citrus production system contribute to this well-deserved reputation. The perennial nature of citrus and year-round populations of most pest species that persist in this system provide a stable resource on which parasites, predators, and arthropod pathogens can coexist. Furthermore, the sessile nature and highly contagious distribution of many insects on citrus, particularly the homopterans, make them highly susceptible to population reduction by biotic agents (Plates 53–57).

Biological control in citrus takes advantage of several approaches commonly employed in managing arthropods, including 1) conservation measures to foster the actions of existing natural enemy species, 2) the introduction of exotic species of natural enemies, and 3) periodic supplementation of existing natural enemies by the release of large numbers of parasites or predators at an appropriate time.

A variation on these approaches is the field application of insect or mite pathogens using technology similar to that employed for the delivery of conventional pesticides. These "biological pesticides" are used to produce field infection in pest populations sooner than it would naturally occur or to inoculate foliage and fruit so that natural epizootics can be induced when conditions are favorable. In practice, their effect on the pest is similar to that of a conventional pesticide; the major reduction

> **Citrus pest management is known worldwide as a testing ground for the development and implementation of biological control.**

in the pest population comes immediately after the application. Examples of this interface of biological control and pesticide technology include the use of various products containing Bt (the bacterium *Bacillus thuringiensis*) for the suppression of caterpillars and Mycotrol (the fungus *Beauveria bassiana*) for the suppression of leafminers and other foliar pests.

Biological control tactics tend to concentrate on one or a few species of related pests and, therefore, can be employed on a case-by-case basis. The biology of organisms which prey on mites or insect pests are closely tied to the biology of their prey, and this close relationship is responsible for a sustained interaction below the level of economic injury when biological control is obtained.

Conservation of Indigenous Natural Enemies

The conservation of indigenous natural enemies is probably one of the most important facets of biological control in citrus IPM today, since so many former key pests are now under successful biological control through the introduction of exotic agents. Florida red scale, purple scale, citrus blackfly, and brown soft scale are but a few examples of the many pests under partial or complete biological control in Florida citrus through the introduction of exotic natural enemies. The protection of these successful agents, as well as other natural enemies which are long-term residents of citrus ecosystems, requires skillful integration with judicious use of pesticides for management of other species. Choosing among available pesticides for mite control, sporadic insect suppression, and disease management must take into account the selectivity of the pesticide and its effects on parasites and predators of the target pest.

Classical Biological Control

Classical biological control, involving the introduction, release, and establishment of exotic natural enemies, has been extremely successful, in part because the majority of citrus pests are exotic in origin and have over time been reunited with citrus trees in various parts of the world. Upon establishment of a new pest organism, it often is desirable to determine which parasites, predators, or pathogens are responsible for maintenance of its host at low levels in the area of origin or in other areas with similar climatic profiles. Some well-known examples of classical biological control using introduced predators and parasites are described in Table 11.2.

Augmentative Biological Control

Management of some citrus canopy pests has been improved by augmentative biological control, the repeated use of applied biological control agents. Of particular note is the seasonal introduction of the parasitic wasps *Aphytis lingnanensis* (Plate 54) and *A. melinus,* which attack the California red scale, in the Central Valley of California. This tactic is most useful when there is a seasonal or spatial asynchrony between pest populations and existing species of natural enemies. Winter disruptions of their life cycles or other pest population bottlenecks can challenge the ability of natural enemies to stay linked with pest numbers. In these cases, the timely release of parasites or predators reared in insectaries or field nurseries can give a boost to reestablish the proper balance. This tactic has proved to be practical and economical in the case cited above, and it is a perennial component of citrus IPM for some producers.

Cultural Practices

Horticultural practices which promote tree health and productivity also play a role in determining the impact of foliage- and fruit-inhabiting arthropod pests. Fertilizer and irrigation promote uniform tree growth, for example, which in turn is favorable for feeding and reproduction by many citrus pests. The production of extensive flushes over prolonged periods favors increases in populations of pests which rely on new leaves for development. Citrus leafminers, aphids, and many whitefly species are limited by the availability of new flush, and thus horticultural practices which influence the availability of flush may also affect pest populations. Hedging and topping stimulate flushing, and trees that have been hedged may become heavily infested as unseasonal new flush emerges. Other cultural practices, such as pruning of lower branches (skirting), support other citrus operations such as subcanopy irrigation and weed control. Weed control programs variously affect the plant diversity within an orchard, and there is some evidence that plants other than citrus in and around citrus groves may provide alternate prey, pollen and nectar sources, and refugia for natural enemies of citrus pests. Skirting of trees can influence pest management by limiting the movement of some pests from the soil into the canopy. In California citrus, skirting to reduce contact between the canopy and the soil is a component of the management program for snail control and for restricting the movement of ants into the canopy, since ants interfere with biological control agents.

Cultural techniques may be employed specifically to reduce or eliminate the effects of canopy pests, as has been discussed in Chapter 10 for diseases. Under certain circumstances, selective pruning is used to eliminate either the damaged portion of the canopy (e.g., twigs killed by scale insects) or to limit further damage by removing the insect itself (e.g., limbs or twigs infested with boring insects). Thinning of canopy interiors and pruning of water shoots can alter the microclimate within the tree and change the penetration of light so that conditions are less favorable for pest development. Likewise, cultural decisions made at the time of planting (row and tree spacing, irrigation methods) influence the microclimate within the foliage and ultimately determine the suitability of the trees for pest attack and injury. While pest management may not be the driving force in the selection of horticultural practices implemented in a grove, it is certainly worthwhile to understand the impacts of such practices on canopy pests.

Population Suppression with Pesticides

The use of pesticides to reduce populations of pests and their impacts is very important in citrus IPM. Pest management can account for a significant portion of the total cost of production in areas where season-long pesticide applications are required. Application of pesticides is unparalleled in the immediacy of its impact, compared with other methods of pest control, and thus has been widely accepted by citrus growers as a management option. For several reasons, including the perennial nature of citrus, the presence of arthropod pests year-round, and the presence of beneficial arthropods (pollinators and natural enemies) as well as pests in orchards, the use of pesticides must be viewed broadly, within the context of IPM (see also the Citrus Integrator, described in the appendix to this chapter).

The goal of pesticide use in the citrus canopy is to selectively reduce populations of a target pest or pests at a critical time during their seasonal dynamics, so that injurious population levels do not develop. The commonly perceived short-term goal is to eliminate or dramatically reduce the present population and thus halt injury which is already occurring. However,

Table 11.2. Examples of successful biological control of canopy-inhabiting insect and mite pests of citrus

Target pest	Biological control agent	Type of agent	Approach	Results
Cottony cushion scale	*Rodolia cardinalis*	Predator	Classical	Complete success
Florida red scale	*Aphytis holoxanthus*	Parasite	Classical	Complete success
California red scale	*Aphytis* spp.	Parasite	Classical, augmentative	Partial success
Citrus whitefly	*Aschersonia aleyrodis*	Pathogen	Conservation	Complete success
Woolly whitefly	*Encarsia, Calles* spp.	Parasite	Classical	Complete success
Purple scale	*Aphytis lepidosaphes*	Parasite	Classical	Complete success
Citrus mealybug	*Leptomastix dactylopii*	Parasite	Classical	Partial success
	Cryptochaetum	Predator	Classical	Partial success

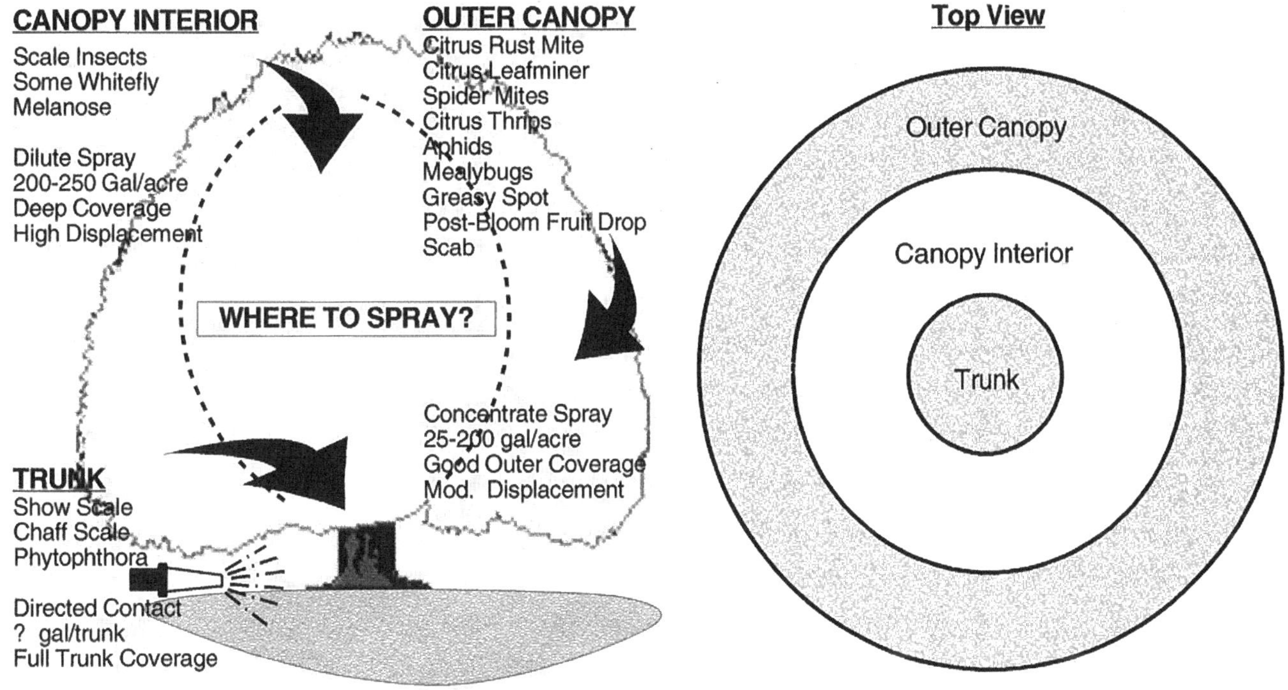

Fig. 11.1. Locations of arthropods and diseases within the canopy and sites of foliar application of pesticides.

pesticides rarely eliminate a pest population in its entirety, so the use of pesticides must be regarded as a method of reducing pest populations rather than eliminating them. Therefore, pesticide application must be implemented at the most opportune time to achieve success. Injury to leaves or fruit often has already occurred when the need to apply foliar pesticides is recognized, so the expectation of preventing injury is not realistic unless the applications are planned through close monitoring.

To be most effective, pesticide applications must be targeted in time and space. Air-blast applications to foliage of full-canopied trees is inefficient in providing full coverage, and contact with the spray directly determines the success of the effort in suppressing pests. Sessile pests, such as scale insects and whiteflies, do not actively contact residues of pesticides deposited on foliage, and thus they will only be affected by a pesticide if it is directly in contact with them at the time of application. The location of pest populations within the canopy (outer portions versus canopy interior, upper versus lower leaf surfaces, inside fruit clusters, etc.) should be a consideration in the selection of spray equipment, the volume applied, and the spray distribution, controlled by nozzle configuration, tractor speed, and other factors (Fig. 11.1). The details relevant to application targeting are as important as the selection of pesticide material and the timing of application.

The timing of pesticide applications should be determined primarily by current population levels of the major arthropod pests, but decisions must be made through consideration of other elements of IPM, such as disease management in the canopy. In-field monitoring or scouting should be conducted to determine if conditions for pesticide application are met. The economics of citrus marketing within the season, the crop load, the level of pest injury to date, and projected returns from suppression of target pests all play a role in the decision of whether to spray and the decision of when to spray.

Numerous pesticides are available for use in citrus, with wide variation in modes of action, specificity to target groups, efficacy, residual activity, and cost. Many attributes should be considered in choosing among the available pesticides, and the choices should reflect not just the immediate decision to spray but also the IPM program for the season.

Of primary concern is that the materials chosen for a specific target or targets should be efficacious against them, with the expectation that treatment will provide an economically justifiable return. The use of the material must also be consistent with its registration status and conditions of use stated on the product label. The labels of many pesticides used in Florida citrus specify citrus varieties, locations, periods of time, and other conditions of use. In addition, some pesticides authorized for use in bearing citrus are not approved for use on nursery trees or young, nonbearing trees in the grove, and some that are authorized for use on young trees are not approved for use on bearing trees.

Another consideration in the selection of a pesticide is its effect on other pests and diseases in the grove. Citrus spray oil is active against a range of mites, insects, and foliar diseases. Thus, spray oil treatments targeting the citrus rust mite may have numerous auxiliary benefits. Alternatively, several specific acaricides are useful for rust mite suppression, and these materials have limited activity against other species, including other mite pests.

Finally, a serious consideration in the use of pesticides for population suppression is the potential for the development of reduced susceptibility (resistance) in target pests through the repeated use of pesticides with the same or related modes of action. Selection for pesticide resistance is an increasing problem in many agricultural systems. In citrus, the need to address year-round problems involving a large complex of arthropod pests and diseases provides additional opportunity for pesticide resistance to develop. Pesticide resistance should be a consideration in the choice of pesticides for single applications and, more importantly, season-long pesticide strategies as part of IPM for citrus.

Steven Rogers
Ecostat, Inc.
Highland City, Florida

APPENDIX

The Citrus Integrator: Integrating Insect and Disease Management

Decision models for integrated crop management range from the simplistic to the highly complex. Most useful models fall somewhere between the two extremes. The needs of teachers, researchers, and growers should also be met, so that continuity can be maintained from the classroom to the field. Unfortunately, the decision methods of many growers are unresponsive to process improvement, since there is no common language by which new technologies are easily integrated into commercial practice. If significant advances are to occur (e.g., expert systems development), then an integrated template must be developed. One such model—the Citrus Integrator—is a step in that direction.

Abbreviations Used in the Citrus Integrator

- B benomyl
- C copper
- CRM citrus rust mite
- gs greasy spot
- M surveyed mite density (mites per square centimeter)
- T average daily temperature (°F)

How to Use the Citrus Integrator

The Citrus Integrator (Plate 57) is a decision flowchart for integrating citrus insect and disease management. It consists of the Decision Guide, on the first page, and the Management Guide, on the second page. The upper half of the Decision Guide is the insect management model, and the lower half contains disease management information. The upper portion of the Management Guide outlines crop phenology and biological agents associated with changes in pest populations, and the lower portion details the integrated management timeline.

The Citrus Integrator is an effective teaching guide, but some field management details have necessarily been omitted because of space constraints. The model involves five major pests and diseases—citrus rust mite, greasy spot, melanose, Alternaria brown rot, and scab—because the management actions and materials used to control these problems overlap.

The use of the Citrus Integrator is best illustrated by example. Following the arrows and responding to the queries at each junction box in the Decision Guide will lead to a timeline in the Management Guide. The numbered steps in the Citrus Integrator are not followed in sequence; the numbers are given as positional references only. All pest management decisions involve market considerations, so it is appropriate to start with an economic evaluation.

Beginning at Step 1, suppose that economic factors have been evaluated and it is determined that a Valencia crop will be intended for the processed market. In Step 2, suppose that surveys have established that the citrus rust mite density is about three mites per square centimeter and that populations of other economically important insects are below damaging levels. Weather data suggest that average daily temperatures of about 76°F can be expected for the next several weeks. Referring to the middle of the three charts in Step 2, find the square at the intersection of 76 on the T axis and 3 on the M axis and, noting that the square is olive green, move down to the query box of the same color (Step 6). Since the crop is not intended for the fresh market, follow the No arrow left to the Spring/Summer query box (Step 5). Then track the Yes arrow down through the Processed option and then down to the June/July/August query box (Step 12). If the current date is in one of these months, follow the Yes arrow down and to the left, where Valencia is listed (Step 13). An evaluation of disease severity is made, based on disease history. If disease severity is moderate (Step 16), follow the arrow to the go-to box at the bottom of the page. This box refers to the integrated management protocol detailed in Timeline 32 on the following page of the flowchart. Since a specific treatment for rust mites was not diagnosed, Timeline 32 shows that these processed Valencia oranges need only a single copper spray to control greasy spot, applied in June to early July. If a miticide were required, however, the grower need only to refer to Timeline 40 to find a list of materials that could be included in the June treatment.

Other management scenarios are handled in similar fashion. For example, had the mite density or average daily temperature been much lower, the path would have tracked through the blue option in Step 5, resulting in a recommendation of no miticide. If it were April instead of June, the arrows would have led to the Natural Predators go-to box (Step 3). Over time, recognizing natural predators by name can help growers better judge when pesticide applications can be delayed or avoided.

The modular framework of the Citrus Integrator is readily modified and expandable as new technologies are developed. For example, the predictive model for postbloom fruit drop (Box 10.1) could reasonably be integrated into the path before the economic evaluation of Step 1. Similarly, an economic model could replace the Step 1 junction box, to be invoked whenever diagnostics are initiated. Neither addition would disrupt the flowchart logic.

Management cells are the boxes in the Management Guide defined by the intersection of a row and a column in the chart. They are related to crop phenology. For example, management cell 35-B describes the development and protection of the spring growth flush from greasy spot disease on fresh orange crops. Thus, in addition to helping growers understand how management decisions are interrelated, the Citrus Integrator can facilitate the development of expert systems for pest management.

Notes for the Example

Step 2. Other insects are surveyed on the basis of presence or absence; however, other rating systems may be used without disrupting the flowchart logic. The divisions between the colored areas in the charts in Step 2 may vary, depending on the geographical location.

Step 15. Disease severity is estimated from foliar symptoms and winter inoculum surveys.

Timeline 31. Predators and parasites may vary from country to country.

Timeline 32. Processed Valencia, Murcott, tangerine, and tangerine hybrids are usually in this group.

Timeline 33. Processed grapefruit, Hamlin and Pineapple oranges, and tangelos are usually in this group.

Timeline 36. The numbers between the time bars represent the number of weeks between treatments.

J. H. Graham
University of Florida
Citrus Research and Education Center, Lake Alfred

J. A. Menge
Plant Pathology Department
University of California, Riverside

CHAPTER 12

Root Diseases

The fungi that attack the root system and trunk of the citrus tree are few in number, but their damage impairs major physiological functions, resulting in yellowing of foliage, poor growth and dieback of terminal shoots, leaf drop, branch dieback, and ultimately decline and death of the tree. Primary damage occurs to the fibrous roots, which take up nutrients and water from soil; to major roots and trunk tissues, which deliver water, mineral nutrients, and photosynthates between the root and shoot systems; and to fruit, causing preharvest and postharvest decay. Thus, these fungi directly or indirectly affect fruit size, yield, postharvest quality, and even tree survival under conditions favorable for disease. *Phytophthora* species are the most damaging soilborne fungi that attack citrus, because they affect virtually any part of the tree at any age. Soilborne diseases associated with other fungi are much less important. Nevertheless, diseases such as black root rot, mushroom root rot, and dry root rot can be locally significant, because they destroy fibrous roots or result in girdling of the large structural roots and crown of the tree.

Important Soilborne Diseases

Diseases Caused by *Phytophthora* Species

Damping-Off

Damping-off can affect newly germinated seed and young seedlings of any citrus cultivar. The typical symptom is the collapse of seedlings, which occurs when the pathogen infects and girdles the stem just above the soil line. *Phytophthora* species are soilborne and seedborne and also cause seed rot or preemergence rot. Infected seedlings are killed rapidly when soil moisture is abundant and temperatures are favorable for fungal growth. Plants usually become resistant to damping-off once the true leaves have emerged and the stem tissue has lignified. Symptoms of damping-off caused by *Phytophthora* species are similar to those caused by *Rhizoctonia* and *Pythium* species.

Foot Rot and Gummosis

The most serious diseases caused by *Phytophthora* species are foot rot and gummosis. Foot rot results from an infection of the scion near the ground level, producing lesions which extend down to the bud union on resistant rootstocks. In susceptible rootstocks, "frogeye" lesions on structural roots or crown rot may occur. Gummosis occurs where the inner bark of the scion is damaged. Infected bark remains firm, but small cracks form in it, through which abundant gum is exuded. Citrus gum is water-soluble and disappears after heavy rains but persists on the trunk under dry conditions. Lesions spread around the trunk, slowly girdling the tree. Severely affected trees have pale green leaves with yellow veins, a typical symptom of girdling. If the lesions cease to expand or the fungus dies, callus tissue forms around the affected area, and the lesions begin to heal over.

***Phytophthora* species are the most damaging soilborne fungi that attack citrus, because they affect virtually any part of the tree at any age.**

Nursery trees and young orchard trees of small trunk circumference can be rapidly girdled and killed. Large trees may also be killed, but typically their trunks are only partially girdled, and the tree canopy shows symptoms of defoliation, twig dieback, and stunted growth flushes. In susceptible rootstocks, lesions may occur on the structural roots and crown below the soil line, and symptoms in the canopy develop prior to obvious damage to the trunk aboveground. This form of the disease is referred to as crown rot, to indicate that the damage occurs on the rootstock, below the soil line.

Fibrous Root Rot

Phytophthora species infect the cortex of fibrous roots and cause a decay of these roots. Later, the cortex turns soft, becomes somewhat discolored, and appears water-soaked. The fibrous roots slough their cortex, leaving only the white, thread-like stele (vascular tissue) protruding at their ends.

Root rot can be especially severe in susceptible rootstocks in infested nursery soil or young trees in infested soil. In small trees, the loss of a significant number of roots can result in the death of the tree. Root rot also occurs in susceptible rootstocks in bearing orchards; the damage rarely kills the tree, but the tree declines in vigor and fruit production. In advanced stages of decline, the production of new fibrous roots cannot keep pace with the rate of root death. The tree is unable to maintain adequate uptake of water and mineral nutrients, and carbohydrate reserves in the root are depleted by repeated fungal attacks. This results in the reduction of fruit size and yield, loss of leaves, and twig dieback in the canopy.

Brown Rot of Fruit

Infection of fruit by *Phytophthora* species produces a decay in which the affected area is light brown, leathery, and not sunken below the adjacent rind. White mycelium forms on the rind surface under humid conditions. In the orchard, fruit on or near the ground becomes infected when it touches soil containing the fungus or is splashed with infested soil. If conditions are favorable for fungal growth, the fungus produces sporangia on the fruit surface, and splashing or windblown rain spread the sporangia to fruit throughout the canopy. Most infected fruit soon abscise, but those that are harvested may not show symptoms until after they have been held in storage a few days. If infected fruit is packed, brown rot may spread to adjacent fruit in the container. In storage, infected fruit has a characteristic pungent, rancid odor. Brown rot epidemics are usually restricted to areas where rainfall coincides with the early stages of fruit maturity. All cultivars are affected, especially winter crops of lemons in California and early-season oranges in Florida.

Crop Losses Due to *Phytophthora* Diseases

Phytophthora-induced diseases are economically important in all citrus-growing regions, but worldwide losses due to *Phytophthora* species are difficult to assess accurately. These pathogens are present in nearly all groves in California and Florida and cause substantial root damage in an estimated 8 to 20% of the orchards in these areas. However, disease losses due to root rot are unclear, because the relationship between root damage and yield loss is not directly proportional. Nevertheless, yield losses from fibrous root rot, foot rot, and gummosis in the United States have been estimated to range from about 3 to 6% per year, or $76 million per year, without fungicide control treatments. These losses do not include yield losses due to brown rot, which varies widely with climatic conditions from year to year. Overall, losses due to *Phytophthora* species are much more prevalent in some years in certain locations, because these diseases are particularly damaging under wet or flooded conditions.

Black Root Rot

Thielaviopsis basicola causes a widespread root rot in contaminated peat-based media in Florida greenhouses. The fungus forms small, well-defined, brown to black lesions on fibrous roots, where it produces black chlamydospores (long-lived resting spores) in the cortical tissue. Lesions at the root tip cause sloughing, which exposes the vascular cylinder and gives the root system a dry, stringy appearance. Veinal chlorosis of leaves is another symptom, typical of nonspecific nutritional deficiency due to root loss.

Black root rot affects seedlings of all commercial rootstocks. The most severe epidemics occur under cool, moist conditions in the winter. When conditions are more favorable for root growth in the spring, root rot decreases, but not before delays of up to two months until seedlings reach a suitable size for transplanting and budding. *T. basicola* is not considered a serious root rot pathogen in field soils in Florida, because fungal populations drop to nondamaging levels after affected seedlings are transplanted to the field nursery.

Mushroom Root Rot

Mushroom root rot, also known as Armillaria root rot and Clitocybe root rot, is caused by a group of *Armillaria* species, formerly described as *Armillaria mellea* and *Clitocybe tabescens*. These mushroom-forming fungi have a wide host range and are widely distributed in temperate and tropical regions of the world, including Australia, the Americas, and South Africa.

Mushroom root rot most often induces a slow decline of the whole tree canopy when part of the structural root system has been killed or a major portion of the trunk or crown roots is girdled. Positive identification of the disease is based on the presence of clusters of mushrooms at the base of an infected tree and white, fan-shaped mycelial mats between the rotting bark and the wood. Infected roots have a characteristic mushroom odor.

Unlike other declines, mushroom root rot tends to affect localized groups of trees, but it does not affect all of them at the same time. Limiting the spread of root rot to adjacent trees is difficult once the disease has appeared. Unless a complete but costly tree removal and replanting program is implemented, replanted trees within the existing orchard continue to die over an extended period.

Dry Root Rot

Dry root rot is a destructive but localized disease caused by the soilborne fungus *Fusarium solani*. It occurs in California, Arizona, Australia, and South Africa. The disease develops only when trees have been weakened by some other stress factor and are incapable of mounting a wound defense. Trifoliate orange, citranges, citrumelos, and rough lemon are more susceptible than other rootstocks. Trees develop a moist, dark decay of the bark which overlays the wood infected by the pathogen. The bark eventually dries and adheres to the wood, which is hard and dry and stained brown or gray where the fungus has colonized. Dry root rot cankers are confined to larger roots, the crown, and the trunk below the bud union and, unlike *Phytophthora*-induced lesions, do not ooze gum. Dry root rot may progress through the wood for years with the tree displaying only a slight wilt during periods of drought as an indication of the disease. Once the crown region of the tree is girdled, there is a sudden terminal wilt and collapse of tree canopy with the leaves still attached.

Trees are predisposed to dry root rot by root injuries due to waterlogged soil, poor drainage, gophers, root weevils, excess fertilizer, salinity, and excessive damage to roots by plowing, herbicides, and other pesticides. Thus, dry root rot is often a locally chronic disease. Tree decline may continue if the primary stresses are not relieved.

Epidemiology

Diseases Caused by *Phytophthora* Species

Causal Species

The most widespread and important *Phytophthora* species are *P. nicotianae* (synonymous with *P. parasitica*) and *P. citrophthora*. *P. nicotianae* is more common in subtropical areas of the world. It causes foot rot, gummosis, and root rot but usually does not infect far above the ground. *P. citrophthora* causes gummosis and root rot in Mediterranean climates, where seasonal rainfall occurs during the cooler winter months. It also attacks aerial parts of the trunk and major limbs and is most commonly the cause of brown rot in Mediterranean climates.

P. palmivora is highly pathogenic on roots and stems of citrus seedlings. It is commonly the cause of brown rot epidemics in Florida and probably the cause of brown rot and foliage blights in other subtropical and tropical areas of the world. *P. hibernalis* and *P. syringae* cause brown rot to a limited extent in areas of the world with cool, wet winters.

Disease Cycles

Phytophthora species are commonly present in citrus orchards throughout the world. Thus, in many situations, they cause endemic disease rather than epidemics. The disease cycle as it occurs under most conditions is portrayed in Figure 12.1.

Fungal populations in the soil are maintained by repeated infection of fibrous roots. Under favorable conditions of high moisture and temperature, infected roots produce sporangia, which release motile zoospores. Zoospores are probably attracted to the zone of elongation of new roots by nutrients which are naturally exuded from this root zone. Upon contact with a root, zoospores encyst, germinate, and then infect in the area of the zone of elongation. Once the fungus has entered the root tip, the infection may advance in the cortex, resulting in rot of the entire rootlet. The cycle can repeat itself as long as conditions are favorable and susceptible tissue is available.

P. nicotianae most likely survives unfavorable periods in root debris. The rotted cortex is sloughed, and the fungus produces chlamydospores, which persist in soil for long periods. When favorable conditions return, chlamydospores germinate indirectly to produce sporangia and zoospores or directly to produce mycelium. When both mating types are present, oo-

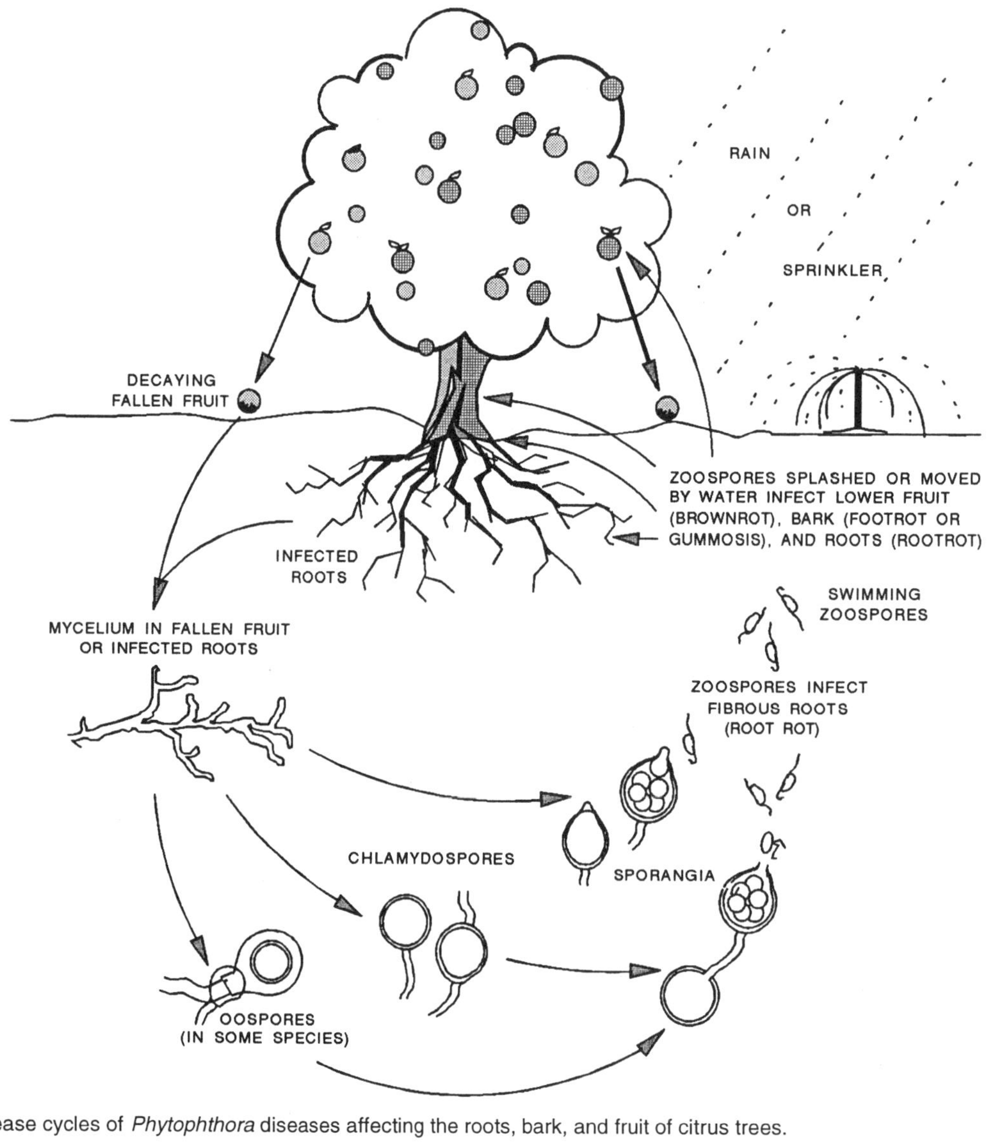

Fig. 12.1. Disease cycles of *Phytophthora* diseases affecting the roots, bark, and fruit of citrus trees.

spores may also be produced, and they aid in the survival of the fungus. *P. citrophthora* usually does not produce chlamydospores and never produces oospores, and its mechanism of survival is uncertain. However, both species can persist as mycelium as well as chlamydospores in infected, living roots or as sporangia.

Foot rot or gummosis occurs when zoospores or other propagules are splashed onto the trunk above the bud union. Infection occurs through wounds or natural cracks in the bark when moisture is present on or around the base of the trunk. Exposed cambium under the bark is susceptible to infection for up to 14 days. Foot rot lesions usually do not produce inoculum for subsequent infections, and thus they are of no epidemiologic significance.

Even though *Phytophthora* species commonly occur where citrus is grown, epidemics can develop in certain cases. When *Phytophthora* species are first introduced into an orchard or nursery, spread to other plants may be epidemic. However, epidemics of foliage blight and brown rot are much more common, especially in highly susceptible hosts, such as lemons. When conditions are favorable, fruit and young foliage can become infected by propagules splashed from the soil. Secondary infections are then caused by inoculum from the aboveground parts of the plant, which is dispersed by splashing or windblown rain. *P. citrophthora* and *P. palmivora,* which produce abundant sporangia on fruit and leaf surfaces, are commonly dispersed in this manner. Brown rot is most common in areas where these species are present and where citrus is grown in Mediterranean climates with prolonged winter rains. *P. nicotianae* is seldom dispersed in this manner, however, because it does not readily produce aerial sporangia.

Means of Spread

The primary means by which *Phytophthora* species are spread through citrus orchards are infested nursery stock and irrigation water. These pathogens may be present in soil or infected roots even though disease symptoms are not readily apparent. They can be dispersed within an orchard by irrigation, especially furrow or flood irrigation. Surface water following heavy rains may carry the fungus as it drains from the orchard. More serious problems can arise in irrigated citrus areas where runoff water carries the pathogen into canals, streams, or rivers. Water from these sources may then contaminate previously uninfested areas. The fungus is also carried in soil on farm equipment when vehicles move from infested to uninfested orchards and nurseries. The propagule density of *Phytophthora* species in soil declines sharply when the soil is air-dried, and the probability of spread is thereby reduced.

Wind is not a major factor in the dispersal of *Phytophthora* species. However, windborne soil can carry these fungi and recontaminate fumigated soils. Windblown rain also disseminates sporangia produced on the surface of aboveground plant parts.

Black Root Rot

Thielaviopsis basicola is often introduced into nurseries in contaminated peat-based media. In moistened media, resistant chlamydospores germinate and infect fibrous roots, forming lesions in the cortical tissue where black chlamydospores form. Once established, the fungus produces phialospores on the surface of the infested potting medium. Spores of the fungus may be splashed by irrigation water or become airborne and dispersed through the entire greenhouse. Black root rot is favored in media at low temperatures and by the high soil moisture prevalent in winter crops of seedlings. The population of the pathogen drops to a low level in the summer, but it can oversummer in the greenhouse and infest subsequent crops.

Mushroom Root Rot

Buried, infected roots or stumps left behind during land clearing are the primary source of mushroom root rot fungi for infection of healthy citrus roots. Several species of native trees may act as sources of inoculum: sycamores and oaks in California, oaks and other hardwoods in Florida, and eucalyptus in Australia. When native trees are felled, the fungi grow more actively in the remaining stumps and roots, increasing the inoculum potential. Once citrus trees are infected, the pathogen moves from tree to tree through root contacts. Aerial dispersal of basidiospores formed on the mushrooms presents no threat of spreading the fungi.

The mushroom root rot fungi rely on food reserves in the wood of the host, and their survival depends on how long these reserves last. These fungi do not spread much in the soil. They may produce infective rhizomorphs ("shoestrings"), but the rhizomorphs do not grow far from the infected root. Citrus trees weakened by drought or other stress factors are more susceptible to infection.

Dry Root Rot

Fusarium solani is commonly associated, as a saprophyte, with the roots, stems, and bark of citrus trees. It produces abundant macroconidia and microconidia on dead roots, leaves, bark, and organic matter in the soil. Primary infection of living tissue may occur from surrounding dead tissue where the fungus remains in a quiescent state. When citrus tissues are chronically stressed, their starch levels drop, and normal resistance is broken down. The fungus infects wood and cambial tissues, particularly during cool, wet seasons, when tissues are dormant and slow to heal.

Interactions of *Phytophthora* Species with the Environment, the Host, and Other Pests

Environmental Conditions

Phytophthora species are excellent parasites but are poor saprophytes in soil. They grow well on nutrients obtained from a living host and, under favorable conditions, pass through repeated disease cycles, from mycelium to sporangia, to zoospores, to mycelium again (Fig. 12.1). *P. nicotianae* is most active during the warmer seasons; it forms chlamydospores and oospores for survival during unfavorable periods. *P. citrophthora* is active during cooler periods and is believed to survive as resistant sporangia and possibly chlamydospores. Under well-aerated, moist conditions, chlamydospores can germinate immediately to form a sporangium, which liberates zoospores. The release of zoospores by *Phytophthora* species is optimal in saturated soils.

Host susceptibility is increased when roots are stressed or damaged in very wet or dry soil. Root exudates released by living roots attract zoospores. Soils with drainage restricted by hardpans or clay layers or those with shallow water tables that temporarily rise into the root zone provide ideal conditions for infection of fibrous roots and rapid buildup of pathogen inoculum. The frequency and duration of irrigation also influences the activity of the fungus and the predisposition of roots to rot. If soils are saturated during irrigation, zoospores are released and can infect roots to form more sporangia. When soils do not dry sufficiently between irrigations, sporangia can survive until the next irrigation, and zoospores will be released again. Citrus roots do not grow in saturated soil, because of the restricted availability of oxygen. Thus, new roots will not develop if the soil is not allowed to dry adequately between irrigation cycles.

The availability of oxygen in the soil atmosphere is closely related to soil moisture, because the amount of pore space occupied by water directly determines the remaining pore space occupied by air. Roots subjected to anoxia are damaged by reduced forms of minerals and by toxic metabolites produced by microorganisms living in the rhizosphere. Root regeneration is restricted and root exudation increases under flooded conditions. Prolonged oxygen deprivation makes roots more attractive to zoospores and increases infection. Moreover, chronic anoxia due to flooding of the root system damages roots even in the absence of *Phytophthora* species

Populations of *P. nicotianae* decline in the winter in Mediterranean climates, where cool weather (temperatures below 59°F) induces dormancy in chlamydospores. Populations of *P. nicotianae* do not drop significantly in Florida, because the soil temperature rarely falls below 59°F. In California, the increase in *P. nicotianae* populations coincides with a root flush in May, which occurs after the spring shoot flush in March or April. Similarly, a spring flush of roots may occur in Florida when soil temperatures rise above 68 to 74°F, followed by an increase in *P. nicotianae* populations.

P. citrophthora grows best at lower temperatures (75 to 82°F) than *P. nicotianae* (82 to 89°F) and is probably responsible for much of the damage to roots during the winter in Mediterranean climates. *P. citrophthora* may be more damaging, because it attacks dormant roots, which are incapable of regenerating during the winter months.

Interactions with the Host

Young fibrous roots of citrus are much more susceptible to root rot than older roots. The activity of *P. nicotianae* is strongly correlated with active root flushes and concentrations of sucrose, fructose, and fructans in the roots, whereas the activity of *P. citrophthora* coincides with root dormancy, when levels of storage starch are maximum. The level of carbohydrates in roots varies seasonally and annually, and the activity of *Phytophthora* species varies accordingly. Gummosis and foot rot lesions spread most rapidly in the spring and fall, when sugars and nitrogenous compounds are most actively moving in the tree. Under tropical conditions, where seasonality is lacking, *P. nicotianae* is probably active and host tissues are susceptible throughout the year.

Fibrous roots damaged by *Phytophthora* species may only live four to six weeks, but not every episode of root rot damage reduces tree productivity. Citrus trees periodically produce fibrous roots in excess of their needs for water and nutrient uptake. Excessive root growth probably competes directly with fruit growth for available photosynthates. The greatest damage from Phytophthora root rot is chronic annual root loss resulting in a deficiency of carbohydrates stored in the roots for fruit production in the following season. This occurs when trees are continually stressed as a result of poor water and fertilizer management.

Resistance to root rot and foot rot varies with the rootstock. Populations of *P. nicotianae* are lower on resistant rootstocks than on those susceptible to root rot. The resistance of fibrous roots to infection is not well understood, but the resistance of bark tissue appears to be related to the production of inhibitory compounds in affected tissues. Rapid regeneration of roots after damage is an expression of tolerance for root rot. Production of fewer propagules from infected fibrous roots may be an expression of root resistance.

Interactions with Other Pests

Fibrous root systems are commonly co-infected with the citrus nematode (*Tylenchulus semipenetrans*), which causes a slow decline of citrus and canopy symptoms similar to those of Phytophthora root rot. Both *Phytophthora* species and the citrus nematode feed on the cortex of fibrous roots and reduce the density of the fibrous root system. The citrus nematode increases the concentration of saline ions in the leaves and thus may exacerbate salinity stress in the orchard. Salinity stress may increase infection by *Phytophthora* species and resultant root damage. However, in nonsaline soils the highly adapted citrus nematode appears to suppress the fungus in fibrous roots, because the female nematodes protect their long-term feeding sites from microbial invasion. Thus, where the two pathogens are concomitant in Florida orchards, roots infected with the citrus nematode support less infection by *P. nicotianae* and smaller populations of the fungus than roots where the nematode is well controlled with nematicides (see Chapter 13).

Another pest in Florida orchards is larvae of the root weevil *Diaprepes abbreviatus,* which cause severe damage to structural roots of citrus trees. Larval feeding deeply etches the root bark, exposing the cambium to infection by *P. nicotianae* and *P. palmivora,* which results in bark rot and girdling of the structural root. Trees on rootstocks susceptible to *Phytophthora,* such as sour orange and Cleopatra mandarin, decline more rapidly when attacked by both *D. abbreviatus* and *Phytophthora* species than trees with larval damage alone, especially in heavier, wetter soils in lower areas of the orchard.

Disease Management

Management of Diseases Caused by *Phytophthora* Species

Monitoring

For proper disease management, it is desirable to know whether *Phytophthora* species are present and, if so, at what level. Fruit baiting and leaf baiting are used for detection of *Phytophthora* species. These assays are qualitative, but they are relatively simple and require minimal equipment and supplies (Box 12.1).

For quantitative measurement of propagule density in soil, culture media have been developed for the selective isolation of *Phytophthora* species (Box 12.1). Methods of sample collection and handling are standardized, so that propagule counts can be compared throughout a citrus-growing area (Box 12.2). The propagule density is highest where the fibrous root density is greatest. Thus, populations diminish with depth and with distance from the tree. For routine determinations of propagule density, samples can be collected at random in the orchard.

Small amounts of soil containing fibrous roots are collected from 20 to 40 locations within a 10-acre area; the samples are taken under the canopy, within the drip line of the tree, or near irrigation emitters where roots are most numerous. The samples are combined in a resealable plastic bag to retain soil moisture and must be kept cool but not refrigerated for transport to the laboratory. The same soil sample should be tested for populations of the citrus nematode, *Tylenchulus semipenetrans* (see Chapter 13).

Propagule densities vary seasonally and from year to year. In most Florida and California orchards the density ranges from one to 20 propagules per cubic centimeter of soil, but it occasionally reaches 100 to 200 propagules per cubic centimeter. Precise thresholds for damaging populations of the fungus are difficult to establish, but populations of less than five propagules per cubic centimeter are considered insignificant, and populations in excess of 10 to 20 propagules per cubic centimeter are considered potentially damaging.

Specific polyclonal and monoclonal antibodies have been produced against *Phytophthora* species, and diagnostic kits have been developed to perform enzyme-linked immunosorbent assay (ELISA) for the detection *Phytophthora* species in roots and soil debris (Box 12.1). ELISA is a highly sensitive method and can detect *Phytophthora* species at lower population densities than dilution plating on selective media. Theoretically, tissue baiting and soil testing by ELISA should be more sensitive than the selective medium assay, since they utilize larger volumes of soil.

Nursery Sanitation

Citrus nurseries should be kept free of *Phytophthora* species to avoid disseminating the fungus to orchards. Citrus seed is treated at 121°F for 10 minutes prior to planting, to avoid introducing these fungi. Previously, seed was commonly planted in outdoor beds, where *Phytophthora* species and other fungi often caused damping-off of seedlings. Today seed is

Box 12.1

Assay Methods for the Detection of *Phytophthora* Species

Method	Description	Comments
Leaf baiting in soil or water samples	Representative water and soil samples, thoroughly mixed, are divided into 5-gram subsamples in the compartments of an ice cube tray. The compartments are then filled with distilled water, which is mixed with the subsamples. Four leaf disks, ¼ inch in diameter, are floated on the surface. The disks are incubated at 82°F in the dark for 48 hours and then plated on PARPH selective medium and incubated for 72 hours. *Phytophthora* colonies on the medium are identified by comparison with known cultures.	Leaf and fruit baiting is the most sensitive technique for the detection of *Phytophthora* species and is especially useful for detecting contamination of irrigation water in nurseries and for checking nursery crops for infestation. The assay does not give information on the population density or the level of damage to citrus crops.
Soil dilution plating	Representative soil samples are thoroughly mixed in a resealable plastic bag. The mixed soil is poured into a 6-ounce Styrofoam cup with drainage holes in the bottom. The soil in the cups is watered until it is saturated and then is allowed to drain. After 24 to 48 hours of incubation, 10 cubic centimeters of soil is removed and mixed with 90 milliliters of 0.25% water agar. Five to 10 samples (1.0 milliliter) of the mixture are individually plated on PARPH selective medium and incubated for 48 to 72 hours at 82°F. *Phytophthora* colonies on the medium are counted, and the counts per plate are totaled and expressed as propagules per cubic centimeter of soil.	The number of propagules per cubic centimeter of soil is related to the level of root damage in nursery soils and reduction of fibrous root density in orchards. The decision to treat an orchard for root rot can be based, in part, on propagule counts. Generally, with fewer than five propagules per cubic centimeter, no need for treatment is indicated, but a count of more than 10 to 20 suggests the potential for root loss and tree response to fungicide treatment.
ELISA (enzyme-linked immunosorbent assay)	Fibrous roots are cut into 0.5-inch lengths and mixed thoroughly. A 1:5 dilution (weight/weight) of roots with extraction buffer is placed in a thick, resealable bag, and a hand-held roller is run over the roots to grind them into a pulp. The extract is dispensed into precoated immunoassay wells. The assay is performed as directed by the kit instructions. The concentration of *Phytophthora* protein in each sample is calculated from a standard curve of values produced by a standard preparation of soluble protein from *P. nicotianae.*	ELISA is useful for rapid detection of *Phytophthora* species at low levels, if the sample of roots is large enough to be representative. The preparation of root samples is labor-intensive, but the assay is very sensitive, and results are obtained the same day. Commercial kits for single or multiple assays provide an effective measure of the activity of *Phytophthora* species in roots. However, ELISA is much more expensive than other methods. A simple method for detecting *Phytophthora* species on-site, using no specialized equipment, is available.

planted in sterilized soilless media in containers in greenhouses and shadehouses, to eliminate most pathogens at the seedbed stage.

Optimally, the site selected for a field nursery or greenhouse should be well-drained land, protected from the wind, at some distance from existing plantings. If a site must be used repeatedly as a citrus nursery, it should be fumigated, especially if disease problems have been observed in previous crops. Preplant application of a mixture of 98% methyl bromide and 2% chloropicrin at 500 to 600 pounds per acre or metam-sodium (42% active ingredient) at 75 gallons per acre reduces pathogen populations substantially but does not com-

Box 12.2

Sampling for Monitoring Soil Populations of *Phytophthora* in Mature Orchards

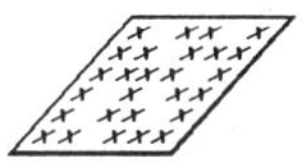

Randomly select 20 to 40 locations within 10-acre area of orchard in mild to moderate tree decline; avoid trees in severe decline.

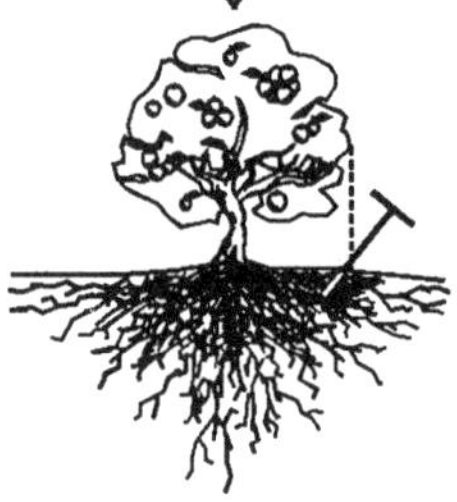

Individual samples are collected within the tree dripline or near the irrigation emitter where the roots are concentrated.

composite samples

Samples are mixed into a resealable plastic bag to retain soil moisture; do not refrigerate or overheat samples.

ship to lab within 24 to 48 hours

Mixed soil is poured into a styrofoam cup with drain-holes, water is added to saturation and allowed to drain.

24 to 48 hour incubation at 72° F

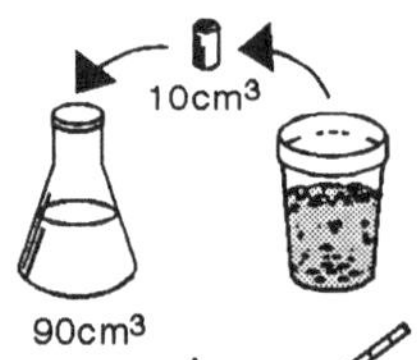

Capful of soil is removed and mixed with 0.25% water agar.

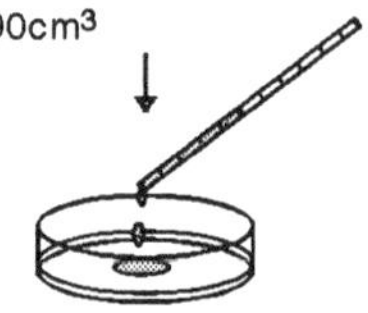

5-10 1.0mL samples are pipetted and spread on PARPH selective medium.

incubate 48 to 72 hours at 82° F

Colonies per plate are counted and totalled to express propagules/cm^3 soil.

Populations in excess of 10 to 20 propagules/cm^3 are considered potentially damaging depending on orchard disease history, soil type, topography, irrigation methodology and rootstock cultivar.

pletely eradicate *Phytophthora*. Repeated fumigation of nursery soil eliminates beneficial mycorrhizal fungi and thus impedes the uptake of phosphorus, zinc, and copper. This may result in stunting of seedlings, a condition that may be partially overcome by a proper fertilization program. Trifoliate orange and its hybrids are generally less susceptible to fumigation-induced stunting than sour orange, Cleopatra mandarin, and other rootstock species.

Field nurseries and greenhouse production facilities can be kept free of *Phytophthora* species by following certain sanitary practices (see also Management of Black Root Rot). Plant material from sources not known to be free of *Phytophthora* species should not be allowed in clean facilities. To the extent possible, all equipment should be dedicated to a particular site; equipment from one site should be disinfested before entering another site. The site should be adequately fenced to direct all traffic through entrances with copper footbaths and vehicle wheel dips (2.5 pounds of copper fungicide per 100 gallons of water). Irrigation water from ponds and canals should be monitored and treated if contaminated. Well water is free of *Phytophthora* species unless contaminated at some point by runoff from citrus orchards. The site should be gently sloping in only one or two directions to control runoff within the site. Care must be taken not to allow runoff water from contaminated areas to enter *Phytophthora*-free zones. If a small portion of a nursery becomes infested, the trees should be destroyed, and the site fumigated. If not destroyed, infected trees should be isolated from other areas of the nursery, and no traffic should be allowed between clean and infested areas.

The systemic fungicides metalaxyl and fosetyl-Al are highly effective against *Phytophthora* species and provide control of diseases caused by these fungi in the field. However, they are often used routinely by nursery owners to suppress pathogen populations and reduce root damage in nursery stock. Fungicide management of *Phytophthora* diseases in citrus nurseries

Citrus nurseries should be kept free of *Phytophthora* species to avoid disseminating the fungus to orchards.

is not recommended. In infested nurseries, fungicides may suppress populations temporarily, so that trees appear healthy and free of damage, but pathogen populations can rise sharply after the trees are transplanted, and they may be seriously damaged if planted at fumigated or clean sites conducive to root rot. The use of fungicides as a substitute for proper sanitation leads to the development of resistance to metalaxyl in *Phytophthora* species. Metalaxyl-resistant strains exported to orchards persist in field populations, and thus the fungicide is no longer effective at those locations.

Resistance

Most commonly planted scion cultivars are susceptible to bark infection, but most rootstocks are at least moderately tolerant (Table 12.1). Nevertheless, rootstocks vary in their susceptibility to diseases caused by *Phytophthora* species. Resistant rootstocks are the best solution for the control of *Phytophthora* diseases; however, some highly resistant rootstocks are susceptible to other diseases or are horticulturally unacceptable.

Virtually all rootstocks become infected with root rot following artificial inoculation and in orchard soils. Resistance, as it applies to foot rot, has not yet been defined for root rot. However, some rootstocks are considered tolerant, rather than resistant: they become infected but do not lose many roots, either because infected roots do not rot or because they generate new roots to maintain root mass density. Susceptible young roots of all rootstocks support equally high populations of *Phytophthora* species. As the roots of tolerant rootstocks age, the pathogen population declines in rhizosphere soil.

Table 12.1. Susceptibility to foot rot in common scion and rootstock cultivars of citrus

	Susceptibility
Scions	
Grapefruit	Very susceptible
Lemons	Very susceptible
Limes	Very susceptible
Mandarins, tangerines, and their hybrids	Susceptible
Sweet oranges	Very susceptible
Rootstocks	
Alemow	Moderately resistant
Carrizo citrange	Tolerant
Cleopatra mandarin	Tolerant
Rangpur	Tolerant
Rough lemon	Tolerant[a]
Sour orange	Moderately resistant
Sweet orange	Susceptible
Swingle citrumelo	Resistant
Trifoliate orange	Resistant
Volkamer lemon	Tolerant

[a] Most selections are tolerant, but some are susceptible.

Most commercial rootstocks are resistant to bark infection by *Phytophthora* species and foot rot (Table 12.1). Carrizo citrange, sour orange, and Cleopatra mandarin are intolerant of root rot. Trifoliate orange and its hybrid Swingle citrumelo are considered highly resistant to bark infection and are tolerant of root rot. Volkamer lemon is intermediate in resistance to foot rot and tolerance for root rot. Root rot tolerance in trifoliate orange and Swingle citrumelo is related to their ability to regenerate roots in *Phytophthora*-infested soil better than susceptible rootstocks.

Damaging populations of *P. nicotianae* (more than 10–20 propagules per cubic centimeter of soil) have been detected on rootstocks intolerant of root rot in citrus groves, but the propagule density on tolerant rootstocks is rarely that high. Rootstocks that do not support high populations on their roots are resistant to infection of the rootstock bark, and susceptible scions grown on them are less likely to experience foot rot. The resistance or tolerance of rootstocks to *Phytophthora* diseases appears to be controlled by several plant genes. Multiple-gene resistance of rootstocks to *Phytophthora* diseases is a durable and long-lasting form of biological control.

Management of Foot Rot and Gummosis in Young Orchards

In orchards with a history of root rot and foot rot, resistant rootstocks are recommended for replanting. Cleopatra mandarin and sweet orange are avoided, because they are highly susceptible to root rot and prone to foot rot.

The taproot and fibrous roots of planting stock should be tested and inspected for root rot in the nursery or orchard before planting, to determine the potential for foot rot in the young trees. In existing orchards, replant sites can be sampled for pathogen populations to determine disease potential. Where foot rot epidemics occur, soil samples should not be collected beneath severely declining trees, because fibrous roots supporting *Phytophthora* species will have died, and the pathogen population may be small or undetectable.

Except in sweet orange and Cleopatra mandarin (the most susceptible rootstocks), in most cases foot rot can be prevented by cultural practices. Budding rootstock seedlings well above the soil line will help reduce foot rot incidence. In areas with ample rainfall and soils conducive to high fungal populations, trees are often budded 20 inches or more above the soil line. In locations where frequent freezes occur, trees must be budded lower, so that their trunks can be wrapped to protect them from freeze damage.

Trees should be planted with the bud union well above the soil line to avoid contact between the susceptible scion bark and infested soil. Adequate soil drainage is essential; poor drainage and overwatering promote the buildup of pathogen populations in the soil, increasing the risk of foot rot infection. The soil surface under the tree must be kept clear and dry. Injuries to the trunk bark must be avoided, since they provide entry points for pathogens. The use of herbicides in citrus orchards has helped prevent foot rot damage by lowering the humidity under the trees and by eliminating wounds produced by hoes, mowers, and other implements. However, the use of trunk wraps for freeze protection and trunk sprout control creates conditions highly favorable for foot rot development. In Florida, control of imported red fire ants and other insects is essential to deter nesting under wraps and feeding damage to moist bark. Prolonged wetting of the trunk can be avoided by irrigating with sprinklers in the morning so that bark has time to dry during the day.

Chemical control may be warranted at either the preplant or the postplant phase of orchard establishment if cultural controls are inadequate. Preplant fumigants may be used when the planting site has a history of problems with *Phytophthora* species, citrus nematode, root sprouts or weeds. Fallowing a planting site for six to 12 months effectively reduces the populations of *Phytophthora* species and citrus nematode as citrus roots die and decompose. Soil solarization with clear plastic may aid in accelerating this process.

The use of postplant fungicides, metalaxyl and fosetyl-Al, in young orchards should be determined by rootstock susceptibility, the likelihood of infestation in the nursery, and the history of *Phytophthora* diseases at the site. Young trees on susceptible rootstocks (Cleopatra mandarin and sweet orange) may be treated with fungicides on a preventive basis. For other rootstocks, fungicide treatments should commence after foot rot lesions develop. Fungicides applied as trunk paints or sprays are the most effective; applications to foliage or the soil surface are less effective against foot rot. Along with a fungicide program, cultural practices that tend to minimize pathogen damage should be implemented. The fungicide program should be conducted for at least one growing season for tolerant rootstocks and may continue beyond the first season for susceptible rootstocks. Both fosetyl-Al and metalaxyl are effective when applied at recommended rates, so alternating use of both materials should be considered to minimize the risk that fungi will develop resistance to them.

Management of Root Rot in Mature Orchards

Once an orchard matures and begins to bear regular crops of fruit, foot rot is usually no longer a serious problem. Loss of fibrous roots due to infection by *Phytophthora* species may still produce mild tree decline and reduced yields. Most commonly used rootstocks are tolerant of foot rot and nearly immune to bark infection, but some root infection occurs. However, populations usually remain low in orchards on trifoliate orange and Swingle citrumelo, and probably little loss is incurred unless orchards are established on soils ill-suited for these rootstocks (e.g., soils with high pH or high calcium and clay content). Some rootstocks, such as sour orange and Cleopatra mandarin, are tolerant of bark rot but nearly as susceptible to fibrous root rot as the highly sensitive sweet orange.

Damage to fibrous roots in orchards is difficult to assess directly. Where the rooting depth of citrus is limited by a high water table and restrictive soil layers, root damage occurs in saturated soil. Under conditions favoring the pathogen, *Phytophthora* species infect fibrous roots within hours and completely destroy roots within four to six weeks. Proper drainage and irrigation management are essential for regeneration of replacement roots. If populations of *Phytophthora* species are at damaging levels (more than 10 to 20 propagules per cubic centimeter soil) before water damage to roots occurs, they will increase to even higher levels during wet periods, as a result of increased root infection.

In subtropical areas with a high water table, the first step toward controlling the pathogen is installation of drainage tile to provide additional internal drainage in the soil profile and proper maintenance of drainage ditches. If fibrous root losses are not attributable to direct damage due to wet conditions, then soil populations should be monitored prior to and after treatment with fungicides. Fungicide applications are based on periodic soil sampling to indicate whether damaging populations of *Phytophthora* species are present in successive growing seasons. A fungicide program should be based on seasonal applications during susceptible root flushes in the spring (after the spring leaf flush) and fall. Application of metalaxyl or fosetyl-Al to the soil should be directed at areas of highest fibrous root density. Both fungicides are effective for root rot control when applied at recommended rates, so alternation of these materials should be considered to minimize the risk of fungicide resistance.

In arid climates, proper irrigation management is essential for the control of *Phytophthora* disease. Generally, flood and furrow irrigation are difficult to manage, because soils are usually too wet initially and too dry later. Citrus should be irrigated optimally to meet the evapotranspiration demands of the tree. Methods are available for precisely irrigating trees with microsprinklers or drippers using tensiometers or evapotranspiration data to determine the timing and amount of irrigation. In fine-textured soil, irrigation water should be applied to wet the top 24 to 36 inches of the soil profile, and the top 12 inches should be allowed to dry to –600 to –700 kilopascals before rewatering. This wetting and drying cycle can damage sporangia and zoospores of *Phytophthora* species in the upper levels of the soil profile, where they are most prevalent. This irrigation cycle also allows the rapid regeneration of new roots to replace those damaged by *Phytophthora* species. If the pathogen is damaging fibrous roots, the time between irrigations should be extended, with more water applied at each irrigation, in compensation. When the upper level of the soil profile dries, deeper roots with less pathogen damage extract water from greater depths in the profile. Trees should not be underirrigated in an attempt to control *Phytophthora* species. Underwatering stresses the trees, reduces yields, and increases the susceptibility of roots to root rot damage when the trees are subsequently irrigated. Salt accumulations aggravate Phytophthora root rot, and occasional heavy irrigation may be necessary to prevent salt buildup in the root zone.

In orchards where damaging populations of both the citrus nematode and *Phytophthora* species are present, management of both pathogens does not substantially increase yields over those obtained with management of only one of the pathogens. The lack of additional response to treatment with nematicide plus fungicide is apparently due to the release of fungal activity from interference by the citrus nematode. At present, given their costs and efficacy, the use of registered fungicides is favored over nematicide treatment.

Management of Brown Rot and Canopy Blight

Where *P. nicotianae* is the primary species present, brown rot and canopy blight are seldom of much importance. Only occasional losses of fruit near the soil are incurred. However, where *P. citrophthora* and *P. palmivora* occur in rainy areas, these diseases may be severe. Usually, the primary problem is loss of fruit to brown rot, which may continue to develop and spread in packing boxes during storage or transport. Leaf and twig blight are seldom serious, but if the fungus infects large branches, cankers may be formed, and entire branches may be girdled, especially in highly susceptible species, such as lemons.

Rain providing a long period of leaf and fruit wetness is needed for infection and multiplication and spread of the pathogen on fruit, which becomes susceptible as it reaches maturity. Copper fungicides applied prior to the beginning of the rains are usually effective. The systemic fungicide fosetyl-Al provides effective control of brown rot when applied to the canopy up to 12 weeks prior to infection. Because of the prolonged residual activity of fosetyl-Al, a single application controls *Phytophthora* species for the entire period during which the fruit is susceptible. Postharvest fungicides used in packinghouses are ineffective against brown rot, but preharvest applications of fosetyl-Al control the disease (see Chapter 16).

Management of Black Root Rot

Commercial sources of potting media containing peat are usually free of *Phytophthora* species but may contain *T. basicola*. Batches of potting media should be checked by soil sampling and testing on culture media selective for *T. basicola*. A potting medium should contain milled bark or an equivalent material to promote the stability of its physical properties while maintaining air-filled porosity. It should have adequate water-holding capacity but should also promote drainage, so that roots receive proper aeration and conditions favoring root rots do not develop.

Reused containers, bench tops, and greenhouse walls should be treated with prescribed disinfectants between crops. Styrofoam containers should be avoided, because infected roots can grow into their walls, beyond the reach of disinfectants. At transplant, root systems should be inspected for symptoms of root rots (water-soaked, threadlike root tips). Budding of seedlings on weak root systems can result in bud collapse or seedling death. The pH and electrical conductivity of the medium should be monitored to avoid high pH and excessive salts, which increase the severity of root rot.

The use of potting media with proper air-filled porosity, containers free of contamination, and judicious watering prevent the establishment of black root rot and damage from the disease, despite the presence of airborne spores of *T. basicola*. Under these conditions, fungicides are not necessary. Under conditions conducive to disease, damage varies considerably among seedlings; some cells contain large seedlings that dry out next to heavily damaged, stunted seedlings that are overly wet. Thus, even careful water management does not compensate for the loss of roots due to disease and poorly drained media.

All rootstocks are susceptible to damage from black root rot under conditions highly conducive to the disease. Cleopatra mandarin appears to be the most susceptible, particularly in the production of winter crops for spring budding. If winter crops of seedlings are to be produced, greenhouses should be well illuminated and temperatures maintained to encourage vigorous root and shoot development. Root rot tolerance depends on the regeneration of replacement roots. Fungicides are ineffective as curatives when symptoms are well established.

Management of Mushroom Root Rot

Survival of the mushroom root rot pathogens depends on how long it takes food reserves for the fungus to be depleted in infected roots and stumps after the vegetation has been cleared. To minimize the risk of infection of newly established citrus trees, land should be cleared well in advance of orchard planting.

Once mushroom root rot is established, the spread of the disease is difficult to contain. Isolating infected trees by trenching to restrict disease spread is seldom effective, because the detection of symptomless trees outside an obviously diseased area requires extensive excavation. In practice, declining trees are removed with as much of the root system recovered as possible. Replant sites are left fallow as long as possible (more than one year). The soil may be injected with methyl bromide to hasten the eradication of any remaining fungus at the replant site, but this is expensive and not always completely effective.

Where there is a high incidence of tree loss due to mushroom root rot, the best long-term solution is to remove the entire orchard, let the site lie fallow for a year or more, and then replant the area. A short-term solution to extend the life of an infected tree is to remove the soil around the crown of the tree to allow the root bark to dry out and prevent the fungus from advancing and further girdling the roots or trunk.

Management of Dry Root Rot

Secondary infection of root and trunk tissues by *F. solani* is always associated with primary stress factors. Prevention of root injuries and elimination of chronic stress due to, for example, waterlogged soil or poor soil aeration allow normal resistance mechanisms to operate to reduce infection by *F. solani*. In waterlogged soil where *Phytophthora* species and poor aeration predispose roots to dry root rot, removal of the soil from around the root crown slows disease progress by reduction of foot rot damage and promotion of wound healing.

Larry W. Duncan
University of Florida
Citrus Research and Education Center, Lake Alfred

CHAPTER 13

Nematode Diseases of Citrus

Plant-parasitic nematodes are microscopic roundworms that infect roots and occasionally aboveground parts of plants. They belong to a phylum which also contains an amazing number and variety of animal parasites as well as species that live free in soil and water, feeding on other forms of life. As such, nematodes are an important component of food webs in all habitats on earth.

Like all nematodes, the plant parasites are essentially aquatic organisms, existing in soil water when they are not inhabiting plant tissues. However, many plant-parasitic nematodes also have the unusual ability to survive in the absence of water, in a dormant state known as anhydrobiosis. Survival mechanisms such as anhydrobiosis are common among these worms and vary according to the host plant habitat with which they evolved.

Probably no crop plant species is free of nematode parasites capable of causing substantial economic loss under the right conditions. Eight genera of plant-parasitic nematodes contain species known to be pathogenic to citrus (Box 13.1). Fortunately, only one species is ubiquitous in the citrus industry worldwide, and only a few of the others are economically important in isolated regions.

For various reasons related to the ecology of its nematode parasites, citrus is ideally suited to the practice of integrated pest management (IPM). For example, citriculture is one of the few agricultural endeavors that permits effective exclusion of plant-parasitic nematodes from newly planted land. Growers who can avail themselves of this opportunity save a great deal of time and expense, eliminating the need to manage these pests in the future. In infested orchards, growers often have the options of planting resistant varieties of rootstocks and implementing cultural practices to mitigate problems before considering more expensive methods of management with chemicals.

The key to a successful program for nematode control is knowledge of the biology and ecology of nematode parasites that threaten orchards in a particular region. In the remainder of this chapter, we shall consider some of those aspects of the most common species of nematode pathogens of citrus. Other, less common species are described in less detail. Finally, we shall describe how this information is utilized in IPM programs to avoid problems due to nematodes, or to reduce their impact on the growth, yield, and fruit quality of citrus.

Important Diseases Caused by Nematodes

The Citrus Nematode, *Tylenchulus semipenetrans*

The citrus nematode, *Tylenchulus semipenetrans,* is the only nematode that is ubiquitous in commercial citrus-growing regions worldwide. Unlike other nematode pathogens of citrus, it has a narrow host range and a highly refined host-parasite relationship, which results in relatively mild symptoms in the plant. These characteristics suggest that citrus and the citrus nematode have coevolved to a significant degree.

Citrus nematode populations build slowly when infected trees are planted in virgin soil. Symptoms often do not become evident for eight to 10 years after planting. The nematode also moves very slowly from tree to tree. This low rate of population growth in newly infested orchards and the slow development of symptoms in the host are the basis of the name of the disease caused by these nematodes: citrus slow decline. If an old, heavily infested orchard is replanted before the citrus nematode population declines, the young trees may be heavily attacked, with correspondingly more severe symptoms. This situation is sometimes referred to as a "replant problem." Replant problems may also involve residual populations of *Phytophthora* species.

Symptoms of slow decline are similar to those caused by other biotic and abiotic agents that impair root function. The tree canopy appears thin, with bare ends of branches near the top of the tree. Leaves may become slightly chlorotic and somewhat smaller than normal. Fruit size is reduced. Leaf analysis may reveal potassium deficiency and high concentrations of chloride. In orchards with root disorder due to other causes, patches of nematode-infected trees may appear no different from uninfected trees; nevertheless, the root density and fruit yield of infected trees are often reduced. Usually, there is a direct relationship between nematode population density and tree health. Large, healthy-appearing trees with abundant roots may support large numbers of citrus nematodes, until they eventually reduce the root system and the nematode population declines as a result. Soil adheres to the nematode's gelatinous egg masses attached to the feeder roots, often giving heavily infected roots a dirty appearance (Fig. 13.1B).

The symptoms of citrus slow decline can be mild compared to those caused by other nematode species. They are most noticeable on well-tended trees, following years of heavy nematode infection. Periodic soil sampling is the best way to detect *T. semipenetrans* before its population reaches a high level and is the only method of diagnosing its presence with certainty.

Most studies estimate yield losses due to the citrus nematode to be in the range of 10–30%, depending on the level of infection. Populations of the nematode appear to reach their highest and most damaging levels in dryland and Mediterranean growing regions, rather than in the humid subtropics. Economic losses due to citrus nematode damage can be substantial in citrus marketed as fresh fruit rather than for juice processing; most of the loss is the result of a reduction in fruit size, rather than reduced yield.

Burrowing Nematodes, *Radopholus* Species

Spreading decline of citrus, caused by the citrus race of the burrowing nematode *Radopholus similis* (previously known as *R. citrophilus*), was once considered the most serious disease of citrus growing on the deep sandy soils of the Central Ridge region of Florida. Over a period of several years, trees become increasingly debilitated during the dry winter months and recover briefly during the summer rainy season. The citrus race of *R. similis* is known to exist only in Florida. A recently described species, *R. citri,* appears to cause very similar symptoms in citrus growing in light-textured soils in Indonesia.

Symptoms of spreading decline are similar to but more pronounced than those of slow decline. The tree canopy becomes very thin, with a large number of bare branch ends. Eventually, entire limbs may die. During prolonged drought, infected trees wilt prematurely. The fruit are small, and fewer mature fruit remain on the tree. The population density of *R. similis* is highest in roots below a depth of 12 to 24 inches. Therefore, the root density of an infected tree may be normal in the upper 12 inches of soil, whereas below that depth it may be reduced to less than 10% of that of an uninfected tree. Symptoms generally do not occur until trees are more than five years old, when they begin to need a deeper root system. Symptoms of spreading decline have been mitigated somewhat in recent years as a result of changing cultural practices, described below (see Epidemiology). However, the disease can reduce fruit yield more than 60%, making production unprofitable.

Spreading decline symptoms are sometimes confused with those of citrus blight. However, unlike trees with spreading decline, trees with citrus blight exhibit symptoms in sections of the canopy and have characteristic upward-pointed, small leaves. Furthermore, the flow of water in the trunk xylem, as determined by water injection tests, is restricted in trees with citrus blight, but not in trees with spreading decline.

Burrowing nematodes move through a grove at a rate of up to 60 feet (18 meters) per year, rapidly spreading the disease. Once they have been introduced into an orchard, their progression down and across tree rows and even across roads to adjacent orchards is difficult to halt.

Lesion Nematodes, *Pratylenchus* Species

Several species of lesion nematodes (*Pratylenchus* species) cause symptoms collectively known as citrus slump. When *P. coffeae* is involved, the disease can be as severe as spreading decline. Aboveground symptoms in mature trees are indistin-

Box 13.1

The Nematode Pests of Citrus

Listed below are the species of plant-parasitic nematodes known to reproduce on *Citrus* species and damage citrus trees.

Major pests are those known to cause significant economic losses in citrus in more than a single, isolated region of the world. Minor pests are those which do not cause significant economic losses or are restricted to relatively small geographic areas.

Nematode species	Common name	Areas reporting damage to citrus
Major pests		
Belonolaimus longicaudatus	Sting nematode	Florida
Meloidogyne species	Root-knot nematodes	China, Taiwan, India
Pratylenchus coffeae	Coffee lesion nematode	Japan, India, Taiwan, Brazil, Florida
Radopholus similis	Burrowing nematode	Florida
Tylenchulus semipenetrans	Citrus nematode	Worldwide
Minor pests		
Hemicycliophora arenaria	Sheath nematode	California
H. nudata	Sheath nematode	Australia
Paratrichodorus lobatus	Stubby-root nematode	Australia
P. minor	Stubby-root nematode	California
Pratylenchus brachyurus	Meadow lesion nematode	Florida
P. vulnus	Lesion nematode	Italy, California
Radopholus citri	Burrowing nematode	Indonesia
Xiphinema brevicolle	Dagger nematode	Israel
X. index	Dagger nematode	Israel

guishable from those of spreading decline, except that citrus slump does not spread as rapidly. Unlike burrowing nematodes, *P. coffeae* freely parasitizes roots throughout the soil profile. Thus, both young and mature trees are severely debilitated.

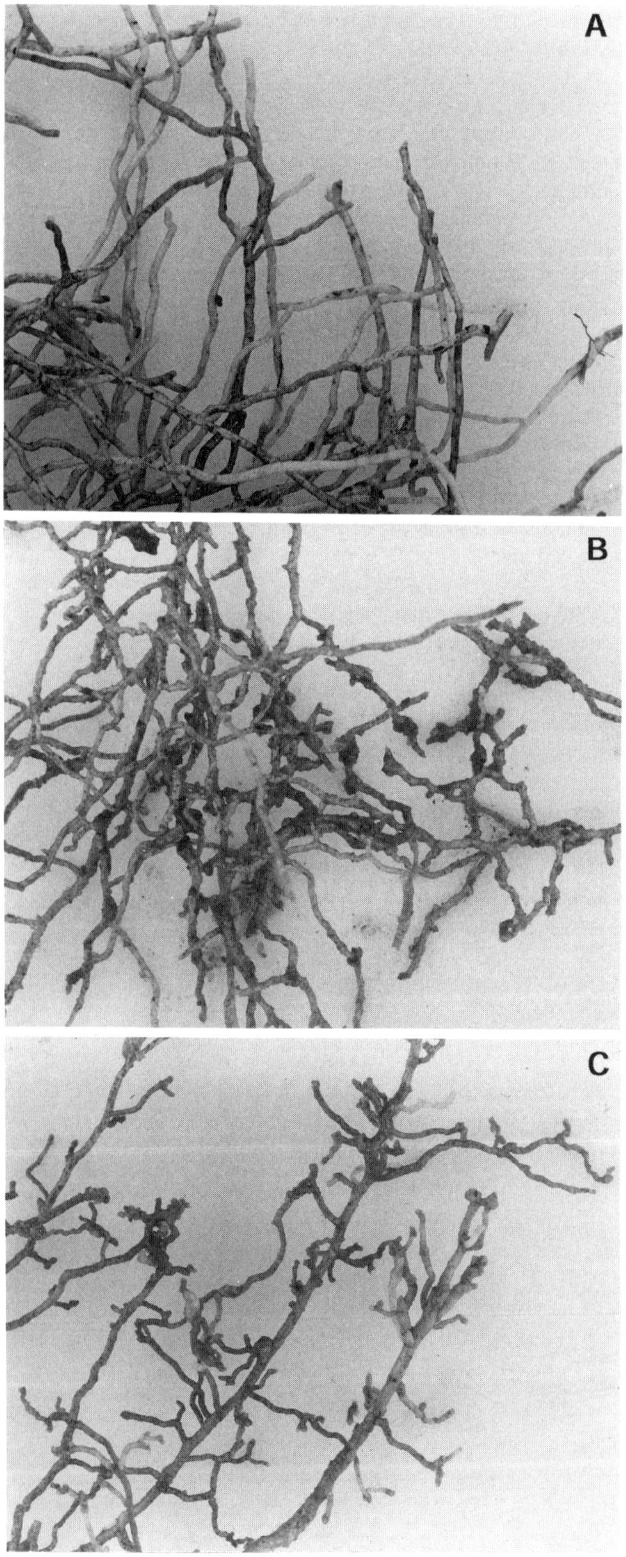

Fig. 13.1. Fibrous root of citrus. **A,** Uninfested roots. **B,** Soil adhering to gelatinous egg masses of the citrus nematode, *Tylenchulus semipenetrans,* causing a dirty appearance. **C,** Stubby roots and tip swelling caused by the sting nematode *Belonolaimus longicaudatus.*

P. coffeae attacks a large number of plant species worldwide, but it has been reported to cause damage to citrus only in Brazil, Japan, India, Taiwan, and Florida. Relatively few orchards are known to be infested with *P. coffeae* in Florida. Although the nematode occurs in some imported ornamental plants in the state, it has not been detected on native plants outside of infested citrus orchards. In Florida orchards, *P. coffeae* occurs primarily in the light, sandy soils of the Central Ridge.

P. brachyurus and *P. vulnus* appear to be pathogenic primarily in seedlings and young trees. *P. brachyurus* is widespread in Florida orchards, where its population density tends to reach about 10% of that of *P. coffeae.* Nevertheless, the growth of young trees can be temporarily slowed by *P. brachyurus,* and older trees occasionally support this species in large numbers. *P. vulnus* is highly pathogenic in nursery stock in Italy, where it is a serious pest. It appears to cause damage only to sour orange rootstock and does not affect mature trees.

Sting Nematodes, *Belonolaimus* Species

Several species of sting nematodes have been found associated with citrus, the most prevalent being *Belonolaimus longicaudatus.* It exists only in the eastern and central part of the United States and thus creates problems in citrus only in Florida. *B. longicaudatus* is most damaging to plants growing in light, sandy soil and is present in 60–80% of orchards in the Central Ridge in Florida. Previous and ongoing research findings suggest that races of sting nematodes exist, some of which are more pathogenic to citrus than others.

Sting nematodes are ectoparasites and can severely stunt the growth of young trees in nurseries and orchards. During the first few years of growth, young trees growing in heavily infested soil may be less than half the size of uninfested trees, and fruit yield can be depressed by more than 60%. Chlorosis is more apparent in infested trees growing on rootstocks such as Carrizo citrange, which exhibit chlorosis in winter.

Sting nematodes are the only nematode species whose damage to citrus roots is a reasonably certain diagnostic feature. Lateral fibrous roots damaged by sting nematode feeding are short and stubby (Fig. 13.1C). The nematodes feed at young root tips, causing a cessation of growth and sometimes slight swelling. Aboveground symptoms are less evident in mature trees, but their root symptoms do not differ from those of young trees. No studies of crop losses due to sting nematodes in mature orchards have been published.

Root-Knot Nematodes, *Meloidogyne* Species

Root-knot nematodes (*Meloidogyne* species) cause more crop loss worldwide than nematodes in any other genus. They have a vast host range and are common pests in all parts of the world, particularly in warm climates. Therefore, it is noteworthy that they have been reported as pests of citrus only in parts of China, Taiwan, and India. Species described in China include *M. fujianensis* and *M. oteifae.* A species found in Taiwan has been given the common name "Asiatic pyroid citrus nematode."

The disease caused by these nematodes is called root knot because of the galls that form on infected roots. It is a serious disease, and restricting the spread of root-knot nematodes in citrus should be a major goal of international quarantine programs.

Up to 60% of citrus surveyed in one region of China was infested with *M. fujianensis.* Root knot is most common in sandy soils. The disease causes tree stunting and corresponding yield loss. Severe vascular dysfunction occurs as a result of disruption of the stele of the fibrous root by the feeding cells of female nematodes. Leaves become small and chlorotic, and the tree wilts quickly. Galls are numerous and readily apparent on roots, which may also have swollen or abnormally branched tips.

Even where citrus is not a host of root-knot nematodes, they occasionally cause problems. When a dense population of root-knot nematodes is present on adjacent host plants, they can penetrate citrus roots and impair root health. However, they fail to develop and do not reproduce in citrus. The problem can be eliminated by removal of the host plants. Nevertheless, if problems of this sort are noted, it is prudent to check the citrus root systems for egg masses, to guarantee that nematodes capable of reproducing on citrus have not developed or been introduced. At least one such population was detected in California and was subsequently eradicated.

Other Nematode Parasites of Citrus

A sheath nematode, *Hemicycliophora arenaria,* is confined to the desert valleys of southern California, where it is common on native plants. Its host range is variable; it can damage lemon, lime, and mandarin nursery stock, but trifoliate orange, sour orange, grapefruit, and sweet orange are resistant. Large numbers of sheath nematodes aggregate at root tips, where their feeding causes hyperplasia and distinctive round galls on the root terminals. Similar symptoms and damage caused by *H. nudata* have been noted in Australia.

Dagger nematodes, *Xiphinema* species, are some of the most common plant-parasitic nematodes in the citrus rhizosphere, particularly in light-textured soils. Tree and root damage associated with these ectoparasitic nematodes has been reported, but few studies of their pathogenicity have been conducted. In a study in Israel, *X. brevicolle* and *X. index* reduced seedling growth in pots by nearly half; they caused necrosis of the epidermis and cortex of feeder roots and markedly reduced root abundance. No efforts to manage these nematodes have been reported.

Paratrichodorus and *Trichodorus* species are also commonly detected at low levels in citrus orchards. In citrus growing in pots, *P. porosus, P. lobatus,* and *P. minor* cause stubby-root symptoms similar to those due to sting nematodes but without necrosis. These nematodes can attain high population densities following fumigation in nurseries. In Australia, population densities as high as 300 nematodes per 100 cubic centimeters of soil have been associated with damage to nursery stock. Rarely, however, are densities high enough to warrant management measures.

Epidemiology

Nematode Life Cycles and Feeding Habits

The generalized life cycle of a plant-parasitic nematode begins when first-stage juveniles molt within the egg. Second-stage juveniles hatch and migrate to a suitable feeding site on or in roots. As they feed, they pass through a third and a fourth juvenile stage and finally molt to reach the adult stage. Among the nematodes that parasitize citrus, individuals of the same sex are monomorphic in most stages, but several species are polymorphic, with adult forms that are very different from the juvenile forms. Some of the nematodes that parasitize citrus are amphimictic (they reproduce by mating), while others reproduce by obligate or facultative parthenogenesis (the female produces progeny without mating). The length of the life cycles of these nematodes is temperature-dependent, but generally it takes 20 to 30 days to complete a life cycle under optimum conditions.

Plant-parasitic nematodes can be classified according to their feeding habits (Fig. 13.2) as ectoparasites (which feed at the surface of roots) or endoparasites (which feed within roots) and as migratory or sedentary.

Migratory ectoparasites, such as sting nematodes, spend their lives migrating through the soil. They feed by inserting their stylets into roots, secreting enzymes into plant cells to digest nutrients extraorally, and then withdrawing the digested nutrients.

Migratory endoparasites, such as lesion nematodes and burrowing nematodes, burrow into and through the cortex of roots. These nematodes spend their entire life cycle in the roots of host plants and pass through the soil only to locate new feeding sites. They also acquire nutrients from plant cells by extraoral digestion.

Sedentary endoparasites and semiendoparasites spend their juvenile stages in the soil but eventually penetrate roots partially or completely, before they molt to the adult female stage. Adult females are sessile, requiring that the plant form highly specialized feeding cells which nourish them for the remainder of their lives. In addition to digestive enzymes, these nematodes secrete proteins that cause certain plant cells to develop into specialized nutrient sinks.

Citrus Nematode

Citrus nematode juveniles migrate and feed on root epidermal cells for several weeks after hatching. Eventually, the anterior portion of the young female penetrates to the root cortex. The female secretes enzymes that cause six to eight of the cortical cells to develop into nurse cells, specialized feeding cells from which the nematode will derive nourishment for the remainder of its life. The nematode also produces several enzymes to digest starches in the nurse cells. As the female matures, the posterior portion of the body enlarges outside the root, and the nematode begins to lay eggs, which remain attached to the body in a gelatinous matrix. Soil particles adhere to these egg masses, giving infected roots a dirty appearance even after they have been rinsed (Fig. 13.1B). A single female can produce hundreds of eggs in a lifetime. The citrus nematode is facultatively parthenogenic. The male spends its life in the soil and does not feed.

The population density of the citrus nematode varies with the season of the year in humid tropical and subtropical areas; it varies much less in drier climates. In humid regions, the population density increases in the autumn and early spring and declines during the summer. It may also decline somewhat in winter, depending on the temperature. Some of the factors that regulate citrus nematode populations are described below (see Environmental Interactions).

Citrus nematode populations can reach such high levels that the carbohydrates and perhaps other nutrients consumed by the nematode can amount to a significant portion of those produced and stored by the plant. When heavily infected trees initiate flushes of root growth more frequently than normal, they use additional carbohydrate. The nematode also creates an imbalance in the partitioning of nutrient elements throughout the plant. Most notably, potassium becomes deficient in the leaves and roots of heavily infected trees, and chloride and so-

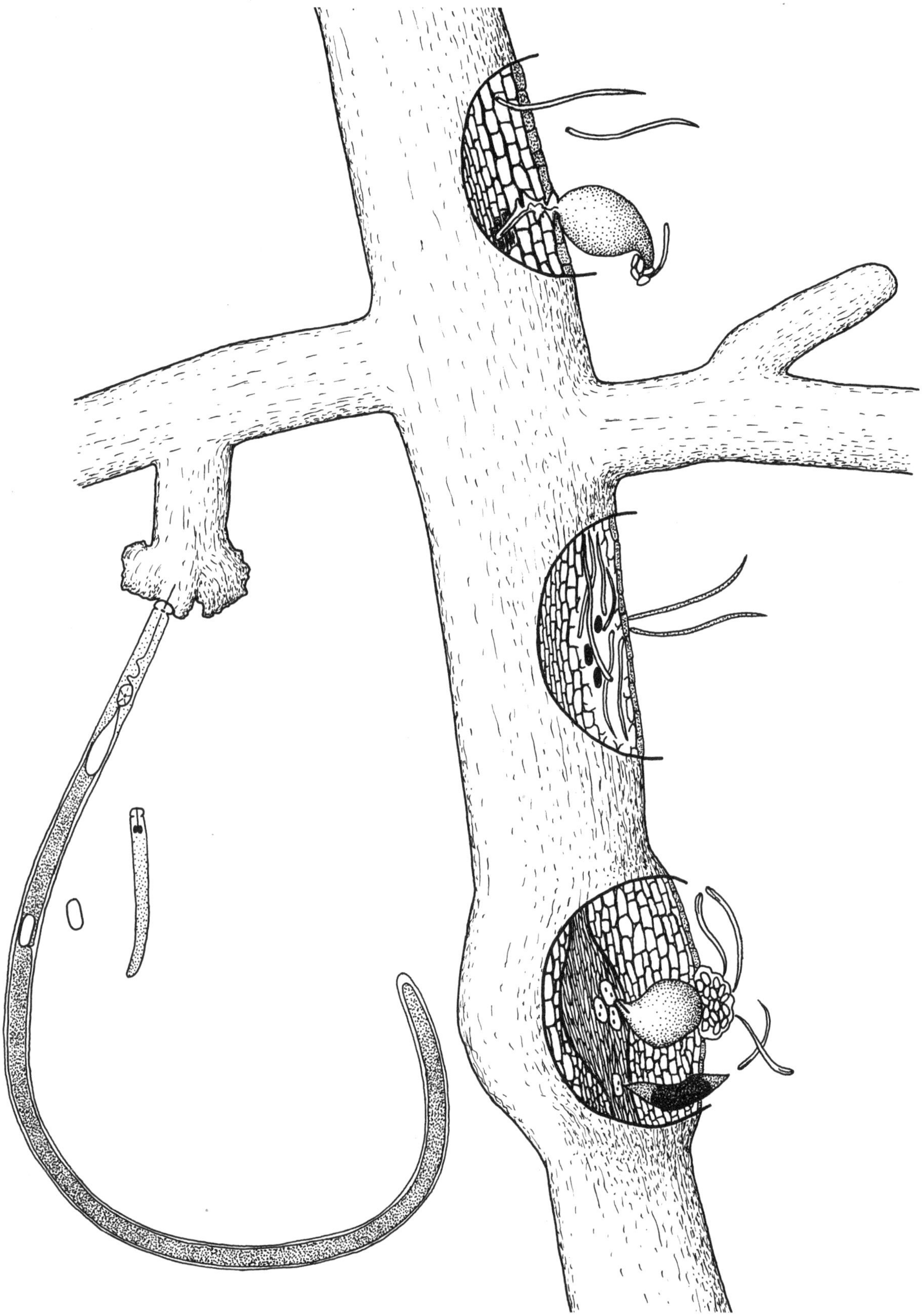

Fig. 13.2. The common nematode pathogens of citrus. The sting nematode *Belonolaimus longicaudatus* is an ectoparasite that feeds on young fibrous roots, causing swelling at the root tip (left). Root-knot nematodes (*Meloidogyne* species) are sedentary endoparasites that create specialized feeding cells (giant cells) in the vascular system of roots and cause galls due to cell proliferation in roots (lower right). Lesion nematodes (*Pratylenchus* species) and the burrowing nematode *Radopholus similis* are migratory endoparasites that tunnel through cortical cells, creating entry wounds for secondary pathogens (middle right). The citrus nematode, *Tylenchulus semipenetrans,* is a semiendoparasite that creates feeding cells (nurse cells) in the cortex of fibrous roots (upper right). (Drawing by Nancy Browning)

dium can accumulate to toxic levels in leaves. Deficiencies of micronutrients, such as zinc, copper, and manganese, often develop in the leaves of heavily infected trees.

Despite the adverse effects of parasitism by the citrus nematode, it forms a relationship with the host that in certain respects is better adapted to the commensal existence of both the parasite and host than the relationships formed by parasites with broader host ranges. Unlike other sedentary parasites of citrus, such as root-knot nematodes, the feeding site of the citrus nematode is in the cortex, where the vascular function of the root is not disrupted. Secondary pathogens do not invade tissues infected by the citrus nematode at a rate comparable to the secondary invasion of tunnels created by burrowing and lesion nematodes. The citrus nematode infects behind the zone of root elongation, so that it does not cause root growth to cease, unlike ectoparasites, such as sting nematodes. Thus, a citrus tree can support a very large number of citrus nematodes without incurring the severe damage that would result from infection by a comparable number of any other nematode parasite.

The citrus nematode, *T. semipenetrans,* can reproduce on several species of woody plants, including persimmon, grape, and olive, in addition to citrus. However, biotypes of the nematode have been described, which differ in their host ranges. The citrus biotype reproduces well on all of the above-mentioned plants, but not on trifoliate orange and some of its hybrids. The Mediterranean biotype is very similar to the citrus biotype but cannot reproduce on olive. The trifoliate orange biotype can infect *Citrus* species, trifoliate orange, and grape, but not olive. The recently described species *T. palustris* and *T. graminis* (formerly recognized as a "grass biotype" of *T. semipenetrans*) do not infect citrus.

Citrus nematodes are disseminated primarily on infected planting material. Secondary sources include water flowing between orchards and agricultural implements and vehicles. If not transported by water or tillage, citrus nematodes do not migrate rapidly between or even within orchards. Even in mature orchards, it is not uncommon to find patches of infected trees randomly mixed among uninfected trees. This tendency to remain localized provides growers with an opportunity to minimize the extent of citrus nematode infestations through proper sanitation practices, described below (see IPM for Nematodes).

Burrowing Nematodes

In most life stages, burrowing nematodes remain within the fibrous roots of the tree. Nevertheless, rapid root decay due to secondary pathogens and the wide host range of these nematodes, including many woody and herbaceous plants, promote the rapid migration for which they are known. Burrowing nematodes primarily inhabit cortical tissue, but they can also damage vascular tissues. They are vermiform in all stages of the life cycle, but the head region of the adult male is morphologically distinct from that of other life stages. The male does not feed. Burrowing nematodes are facultatively parthenogenic. The female lays several eggs per day, and the generation time at optimum temperature (75°F) is 19–20 days. Populations tend to be low in summer and highest in autumn and spring.

The propensity of burrowing nematodes to infect citrus roots deep in the soil is the reason that spreading decline occurs primarily in trees growing in deep sands. The root system of citrus can grow to depths of more than 20 feet when unimpeded by a high water table or clay layers. In deep, light-textured soils the lower portion of the root system can support the transpiration needs of the tree during the dry winter and spring seasons in Florida. Nematode-infected trees do not have this capacity, so that symptoms of spreading decline are much more apparent in the dry season and are primarily the result of water stress. A low rate of infection of roots near the soil surface permits the tree to compensate for the loss of deeper roots by slightly increasing the number of shallow roots. The widespread adoption of irrigation during the past 30 years appears to have mitigated the damage caused by burrowing nematodes, by providing year-round adequate moisture in the shallow rooting zone. Discontinuance of disking in favor of herbicides for weed management has also reduced damage to surface roots.

Although *R. similis* has a very wide host range, only populations from Florida are known to parasitize citrus. Thus, they are designated the citrus race of the burrowing nematode. For a time the citrus race of burrowing nematode was elevated the status of species and was called *R. citrophilus*. However, recent research indicates that *R. similis* citrus race is the proper taxonomic designation.

Populations of *R. similis* citrus race are also known to differ in their ability to reproduce on different citrus rootstocks. Some populations reproduce on and damage all known citrus varieties except Milam lemon, Ridge Pineapple and Albritton sweet orange, and some selections of Carrizo citrange. Other populations are pathogens of all these species.

Lesion Nematodes

The life cycle of lesion nematodes is very similar to that of burrowing nematodes. They migrate between and through cortical cells, creating cavities and lesions that result in rapid root decay. They rarely penetrate vascular tissues. Population growth and ceiling population densities in roots are very high, making it perhaps the most pathogenic nematode attacking citrus. P. coffeae appears to be obligately amphimictic. Males constitute 30–40% of the population and enter and feed in the root cortex. *P. brachyurus* is primarily parthenogenic, and males of this species are rare.

In Florida, a number of common weed species in infested groves are hosts of *P. coffeae*. Moreover, populations of *P. coffeae* have been found on several imported plant species outside citrus orchards. However, no known population of *P. coffeae* from ornamental plants in Florida is capable of infecting citrus, which suggests either the existence of biotypes of the nematode or a need to reevaluate the taxonomy of this group.

Sting Nematode

B. longicaudatus is obligately amphimictic. Males are common and are morphologically similar in all life stages. The nematode has a wide host range, including weed and crop plants, and is commonly found in sandy soil throughout the citrus-growing areas of Florida. In orchards, its population density is often higher on cover plants in row middles than in the tree row. The population density is usually highest in spring and autumn. However, in row middles seasonal patterns of population change are affected by plant succession, as more or less favorable hosts become dominant during the course of the year.

Environmental Conditions

Many biotic and abiotic factors influence the growth and pathogenicity of nematode populations. However, no factor affecting nematodes operates independently. In order for nematodes to respond fully to optimum conditions of one factor, others cannot be limiting. Therefore, identifying the nature of nematode population dynamics in the field is a challenging problem. Some of the major factors known to regulate nematode parasites of citrus are described below. The general interactions between these factors varies in different citrus-growing regions.

Climate

The ranges of temperature and precipitation in citrus-growing regions around the world probably do not preclude the existence of any of the major nematode pathogens of citrus. Therefore, climate mainly influences the damage caused by nematodes by affecting their population density. The citrus nematode is the only nematode pathogen of citrus whose distribution is wide enough to compare the effects of climate. Generally, reported population densities of citrus nematodes are higher in dryland and Mediterranean climates than in the humid tropics and subtropics. Not surprisingly, this nematode is sometimes considered to be of greater economic concern in the former regions.

Soil Texture

The distribution and population density of all plant-parasitic nematode species are profoundly affected by soil texture. Light-textured, sandy soils favor population development of *R. similis, B. longicaudatus, Meloidogyne* species, and several of the minor pathogenic species and promote the development of tree symptoms. Most of these nematodes do not cause significant problems in citrus growing in finer-textured soils. *P. coffeae* is pathogenic to citrus in light and fine-textured soils. The citrus nematode population density increases more quickly in heavier soils with moderate amounts of clay and silt (up to 15%), although it can eventually reach a high level in sandy soils as well. Soil with a high organic matter content also favors the growth of the citrus nematode population.

Soil texture has major direct effects on nematodes by determining the water-holding capacity and cation exchange capacity (and its affect on salinity) of the soil. The size of pore spaces affects the movement of nematodes. These factors probably also affect nematodes indirectly through the plant, which also has optimum soil conditions for growth. For example, less than optimum soil moisture or salinity can increase the susceptibility of roots to nematode infection (see below).

Soil Moisture

There are two main considerations regarding soil moisture in nematode management. Because soil moisture is a key element in crop production, it is critical to ensure that trees are unaffected by water problems. Nematode management is of no value if trees suffer either from drought or excessive soil moisture. A second consideration is how soil moisture affects nematode population density.

P. coffeae and *R. similis* are relatively unaffected by soil moisture, because they are located mainly within fibrous roots. If soils are fallowed to manage nematodes, moisture is more critical. Both species have the ability to survive in dry soil in a state of anhydrobiosis. However, they are active in moist soil, and in the absence of host roots the nematode population will slowly decline, as the nematodes starve.

In contrast, soil moisture has a major effect on the citrus nematode and sting nematodes in orchards. Soil moisture may be the main reason that sting nematodes are primarily a problem in young trees. They are large nematodes and require relatively high levels of soil moisture. When the soil dries beyond the permanent wilt point, the nematode either dies or moves deeper in the soil. Young trees attacked by sting nematodes remain stunted and have very low root density. The soil beneath these trees dries much more slowly than beneath larger, healthy trees, which transpire greater amounts of water. In this way, sting nematodes probably create a favorable environment in the shallow portion of the soil profile, where the fibrous root density is greatest. This favorable niche exists until the trees grow sufficiently large to cause the soil to dry between precipitation events.

In general, citrus nematode population growth is inversely related to soil moisture. Paradoxically, these nematodes have a rather poor ability to survive in dry soil. Their survival in dry soil depends on how water is distributed beneath the tree. If drought conditions prevail throughout the root zone, the tree will wilt, and the nematode population will decline rapidly. However, if there are localized areas of dry soil beneath a tree (as a result of low-volume irrigation, for example), the nematode reproduction rate is often highest in the dry zones. Unirrigated soil may contain localized dry zones during the dry season, but irrigated soil may also contain dry zones when it is not under drought conditions, because the higher root density in the irrigated zone can result in locally higher rates of transpiration. The citrus nematode flourishes in pockets of dry soil because of hydraulic lift, in which water is pulled from moist to dry soil by the action of the root xylem.

Soil moisture is probably a key factor regulating seasonal patterns of citrus nematode population density in humid production areas. For example, the summer rainy season in Florida coincides with a decline in the nematode population. In Taiwan, the spring monsoons are associated with a low population density of the nematode. Low soil moisture may also be the reason that citrus nematode populations tend to be large in relatively dry citrus-growing regions.

Soil Temperature

Since nematodes do not regulate body temperature, the ambient temperature regulates their growth and development. The cardinal temperatures for a given species are determined by the environment in which they evolved. Citrus and burrowing nematodes can reproduce at temperatures between 59 and 90°F, with 75–77°F being optimum, and little reproduction occurs at the extremes of the range. The optimum temperature for reproduction of the coffee lesion nematode, a common parasite of tropical crops, is 79–86°F.

Soil Salinity

The citrus nematode is known to cause greater damage to citrus growing under saline conditions in Texas, South Africa, and Florida. Direct exposure to salinity does not favor nematode population growth. However, if plants are stressed by salinity and then the salinity is lowered, as when salts build up in the soil during the dry season and then are washed below the root zone during the rainy season, the rate of infection by the nematode and its reproduction rate increase rapidly, and the concentration of salt ions in leaves of infected trees rises higher than that of uninfected trees. The citrus nematode affects citrus most strongly under conditions of mild to moderate salinity. At very high salinity, it is probably of minor significance in subsequent tree decline.

Soil pH

Little is known about the effect of soil pH on most nematode parasites of citrus. Generally, nematodes can be found across the range of pH that is suitable for tree growth. There is a positive relationship between the population density of the citrus nematode and the pH of the soil solution; however, no attempts have been made to manage nematodes by controlling soil pH in orchards.

Tree Phenology

The concentration of many of the nutrients available to plant-parasitic nematodes in the roots of citrus trees fluctuates during the year. Starches are abundant in roots during late autumn and winter but decline as the fruit develops in late spring and summer. Just as trees often bear more or less fruit in alternating years, so too can levels of root carbohydrates vary from year to year. The number of eggs produced by the female citrus nematode increases with increasing starch concentration in roots. Moreover, in Florida, the population density of the citrus nematode is reasonably proportional to the starch concentration in roots from season to season and year to year. Nutrient availability is greatly influenced by other factors, such as soil moisture, which probably have a greater impact on final nematode density.

The age of fibrous roots also affects nematode infection and development. As fibrous roots age, the amount of cortical tissue available for infection and feeding decreases, whereas secondary tissues and root lignification increases. Root-knot nematodes invade roots only immediately behind the growing root tip. Sting nematodes feed on actively growing root tips, and lesion, burrowing, and citrus nematodes feed only in the cortex. Therefore, nematode infection and development reach maximum levels immediately following flushes of new roots. Major root flushes alternate with shoot flushes during the growing season. There is evidence that the amount of root growth, like the carbohydrate level, alternates from one year to another, with corresponding effects on nematode density.

Interactions with Other Soil Organisms

Only a few organisms in soil have been studied in combination with plant-parasitic nematodes on citrus, and most of the information comes from experiments conducted in the greenhouse. Mycorrhizal fungi (which infect all citrus in the field) have been shown to mitigate damage to citrus by *R. similis* and to reduce the population density of the nematode. The suppression of the nematode appeared to result from the increased phosphorus provided to the plant by the fungus. When the phosphorus level in the soil is adequate, the fungus may have little affect on the nematode. The citrus nematode caused a greater reduction in the size of citrus seedlings grown in the presence of nondamaging populations of *Fusarium solani,* a nearly ubiquitous fungus. In a field study, the coffee lesion nematode and the citrus nematode appeared to compete effectively with one another, which may partly explain the low incidence of the former in Florida.

Perhaps the most common fungal pathogens found concomitantly with nematodes are *Phytophthora nicotianae* (formerly *P. parasitica*) and *P. citrophthora* (see Chapter 12). Both organisms derive nutrients from the root cortex and thus compete directly with many nematode parasites of citrus. *P. nicotianae* is likely one of the fungi responsible for secondary invasion of wounds caused by lesion and burrowing nematodes, but no study has yet shown that nematode infection predisposes roots to infection by the fungus. There is some evidence that low to moderate levels root infection by the citrus nematode may suppress the growth of the more pathogenic *P. nicotianae* and mitigate damage by the fungus. However, it has also been suggested that chemical management of *Phytophthora* species can increase nematode problems in the healthier root system, although no direct evidence of such an interaction has been published.

No biocontrol agents of nematode parasites of citrus are commercially available, but a large number of such organisms prey naturally on these nematodes. Most soils contain numerous fungi, arthropods, bacteria, protozoans, and predatory nematodes that probably reduce the severity of nematode diseases in citrus. Biotic antagonists may play a role in the seasonal regulation of these pests. The major difficulty in enhancing the level of biological control is the heterogeneity of the soil environment and the specialized niche requirements of biocontrol agents. Maintaining conditions that consistently favor another organism at the expense of parasites may not be feasible, and methods for affordable inundative treatment with biocontrol agents have not been developed.

IPM for Nematodes

Biomonitoring and Sample Interpretation

A characteristic of many nematode diseases in the field is that symptomatic plants often occur in clumps, known as hot spots (Fig. 13.3), because nematodes occur in patches across fields. However, most tree and root symptoms of nematode-induced diseases can result from other causes and are unreliable for diagnosis. Root and soil samples are essential to confirm the involvement of nematodes in tree decline. Where possible, sample collection can be scheduled at times when it will provide the maximum amount of information needed by a grower. Estimates of the population densities of nematodes and *Phytophthora* species and determinations of soil pH, salinity, and concentrations of nutrient elements in the soil can all be accomplished by analysis of a single sample.

Sampling should be the basis for all nematode management decisions, even before an orchard is planted. Preplant sampling can indicate the presence of nematodes that may injure young trees, thus providing the grower with the option of managing the population through rootstock selection and cultural or chemical means before planting. If citrus is being replanted on an old orchard site, the odds of a significant nematode problem are greater. Regular scouting of trees to detect symptoms and regular soil sampling provide some of the information needed to make timely and rational management decisions. They also provide a balanced picture of the levels of phytoparasitic nematodes at a site, because population densities fluctuate from season to season and year to year in this perennial crop.

Stratification of areas to be sampled within an orchard is useful in distinguishing sites which may respond to nematode management from those in which management is unnecessary. In infestations by the citrus nematode, uninfested areas may remain nematode-free for many years, but in infestations by more migratory species, such as the burrowing nematode, stratified samples may simply delimit rapidly changing areas of infestation. Generally, samples should encompass an area no larger than 2–5 acres. A preliminary sample from a large area can serve as the basis for future sampling of progressively smaller areas, to refine the identification of infested sites. Areas of orchards with and without symptoms of nematode infection are a good basis for sample stratification. Stratification can also be based on differences in soil texture; topography, if it influences soil moisture; rootstock resistance; or other factors that influence nematode population density.

Sampling should be the basis for all nematode management decisions, even before an orchard is planted.

Samples that are adequate for detecting most nematodes can be obtained by systematically collecting soil and roots from the upper 12 inches (30 centimeters) or so of soil (Box 13.2). Burrowing nematodes are a special case, however, because of their propensity for higher infection rates in deep roots than in shallow-growing roots. When sampling for burrowing nematodes, growers in the Central Ridge of Florida have two options. Large quantities of shallow roots beneath the tree canopy can be dug with shovels, or growers can pay the Florida Department of Agriculture and Consumer Services to collect smaller quantities of roots, by the use of special equipment, from depths of 12 to 48 inches (30 to 120 centimeters). The two sampling methods detect infestations of burrowing nematodes with comparable precision.

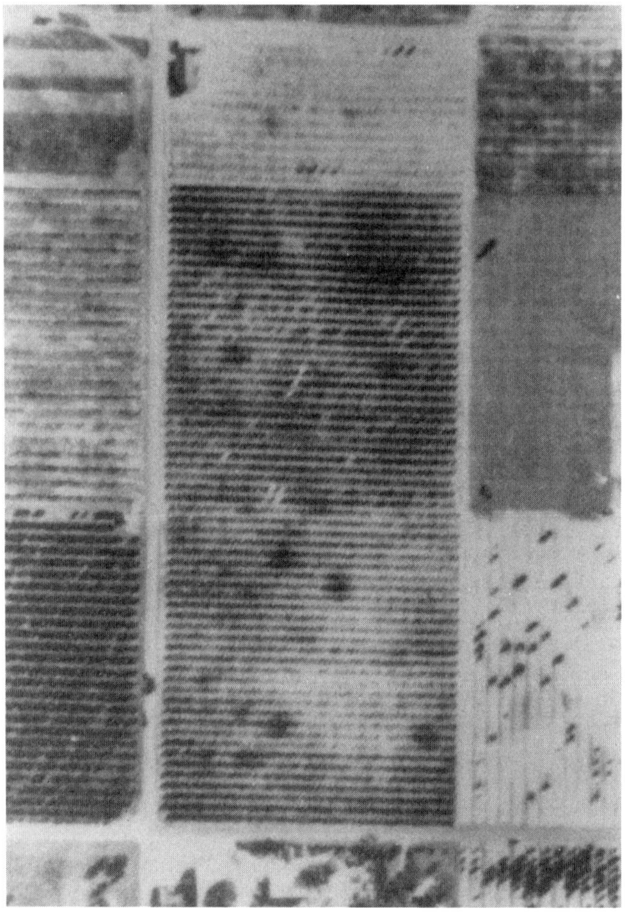

Fig. 13.3. "Hot spots" of damage to orchard trees due to the nonuniform distribution of sting nematodes in the soil. Trees in the middle of hot spots are smaller than trees in the surrounding areas, and more trees have been removed from the hot spots.

Often growers can submit soil and root samples to private companies or public agencies for analysis. Factors to consider in choosing a laboratory include the type of extraction methods available (Box 13.3), the quality of management advice offered, and the training and experience of the personnel processing the samples. The number of nematodes recovered from samples is affected by the processing method, and methods often vary between laboratories. Laboratories that separately process root and soil samples are more likely to detect burrowing and lesion nematodes than those that process only soil samples. Centrifugal flotation is more likely to recover sting nematodes than filtration in a Baermann funnel. It is important that laboratories utilize methods that permit a reasonable comparison of their results with published information regarding damaging levels of nematodes.

Detection of any burrowing or coffee lesion nematodes, without regard to their estimated population, is sufficient evidence of an economically significant problem, under the right conditions (generally, in sandy soils and, in the case of burrowing nematodes, orchards with a deep root system). The high population growth rates of these nematodes and the perennial nature of the host guarantee that, without management, their populations will reach damaging levels.

For other nematode species, knowledge of population levels is more important in management decisions. In annual crops, the relationship between the number of nematodes in soil at the time of planting and the crop yield several months later is relatively simple and consistent. However, predicting loss of citrus yield based on nematode levels is difficult, because nematode populations are measured at a single point during their naturally fluctuating population cycles. Most laboratories predict relative crop loss on the basis of published information giving ranges of nematode populations that are likely to be insignificant, probably significant, or almost certain to warrant management. The amount of data upon which to base a prediction varies with the nematode species. The citrus nematode is the most thoroughly studied in this regard, and rough threshold densities are used in many citrus industries. For example, in California the expected value of citrus nematode management is considered to be unprofitable, moderate, or high if the numbers of juveniles extracted from 100 grams of soil in early spring (February–April) are less than 400, more than 1,000, and more than 2,400, respectively. The corresponding numbers for samples taken in early summer (May–July), when populations are expected to be highest, are less than 800, more than 1,600, and more than 3,600. Similar guidelines have been developed for interpreting counts of female nematodes per gram of roots.

Sanitation

Most of the nematodes that parasitize citrus have a very limited geographic distribution. The narrow host range of the only widely distributed species, the citrus nematode, restricts it to citrus and a few other crops. Therefore, growers can often eliminate the threat of nematode problems in newly established orchards in virgin soil by following simple sanitation procedures. Once introduced, nematodes cannot be eradicated from orchards. The high, continuous cost of other forms of nematode management make sanitation one of the best investments available to growers.

The use of certified nematode-free planting stock is the simplest method of avoiding the introduction of nematodes into an orchard. Some industries require that commercial nurseries participate in a nematode certification program. The movement of sheath nematodes from California nurseries has been successfully restricted by a quarantine on infected material. In Florida, nurseries are prohibited by law from selling trees contaminated by coffee lesion, burrowing, or citrus nematodes. A cost-benefit analysis indicated that restricting the movement of burrowing nematodes during the period from 1960 to 1994 resulted in $1.4 billion of additional revenue to Florida growers, at a cost of $100 million. In the 1994–1995 season alone, it was estimated that growers received an additional $32.5 million as a result of limiting the spread of the citrus nematode, at a cost of $70,000 (see Box 4.4).

Box 13.2

Sampling to Estimate Nematode Population Density

Soil and root samples that are submitted to nematode assay laboratories for estimating population density are composed of subsamples. Small quantities of soil and roots, each obtained from many spots scattered throughout a defined portion of an orchard, are pooled together in a single sample. The contents of the sample are mixed, and only a small quantity of it will actually be processed to recover nematodes. Thus, a *subsample* is the material collected at a single spot in the orchard, and a number of subsamples are pooled together to make a single *sample* of a portion of the orchard.

A proper sampling protocol is necessary in order to obtain a reliable profile of the kinds and numbers of nematodes in an orchard. Nematodes are not distributed evenly in orchard soils. Rather, they are present in small clumps of various sizes and densities. A sample composed of too few subsamples will seldom provide a realistic estimate of the average number of nematodes at the sample site. Beyond a certain number of subsamples, the improvement in the population estimate is too small to justify the additional time and expense of collecting additional subsamples. The best sampling programs are designed to provide the minimum level of reliability needed, for the least expenditure of time and money. This is done by collecting a sufficient number of subsamples from areas most likely to contain high numbers of nematodes. The first two protocols described below can be used to sample for all of the nematode parasites of citrus except burrowing nematodes. Special sampling requirements for detecting burrowing nematode infestations are described separately.

Sampling an Orchard

Stratify the orchard according to factors likely to be related to nematode population density, including tree quality, symptoms, soil texture, groundwater level (if high), and topography. Within the selected strata, sample representative areas of approximately 2–5 acres.

Using a soil probe or a shovel, collect a small amount (a handful) of soil and feeder roots from a single spot beneath the canopy of each of 30 trees in the sample area. Take subsamples from the upper 12 inches of soil (the length of a shovel blade) at a point midway between the trunk and the drip line of the tree. The trees can be selected arbitrarily, but they should be distributed throughout the sample area.

As the 30 subsamples are collected, combine them into a single bulk sample. When the collection of subsamples is finished, each sample area will be represented by a single sample composed of soil from 30 trees.

Thoroughly mix the bulk sample, and remove about 1 quart of soil and a handful of roots. The remaining soil and roots can be discarded at the sample site.

Tightly seal the sample in a labeled plastic bag, and store it in a cooler. Refrigeration is unnecessary, but do not allow the temperature of a sample to rise above the ambient temperature. Avoid exposure of samples to direct sunlight for even a short time, and do not allow them to remain unprotected in a hot vehicle. Arrange for the samples to arrive at a laboratory as quickly as possible.

Sampling Individual Trees

To diagnose problems in a few trees, collect a sample as described above, taking a minimum of 30 subsamples of soil and roots, divided among the number of trees to be sampled. For example, 10 subsamples from each of three trees can be combined into a single sample. If a single tree is being sampled, it is sufficient to collect 16 subsamples. Take all of the subsamples at approximately the same relative distance between the tree trunk and the drip line. Randomly choosing spots to sample beneath the tree canopy is less reliable than systematic sampling based on the distance between the trunk and drip line.

Sampling to Detect Burrowing Nematodes

Growers in Florida can contract with the Florida Department of Agriculture and Consumer Services, Division of Plant Industry, to survey orchards for the presence and distribution of burrowing nematodes. Special equipment is used to sample roots as much as 4 feet deep in the soil. If a grower wishes to attempt to contain the spread of the nematode, the extent of the current infestation can be determined.

Growers can also sample for the nematode themselves at less cost, with the same level of accuracy, by obtaining large quantities of fibrous roots near the soil surface. Large samples are needed, because roots near the soil surface are less heavily infected than those deeper in the soil.

To detect burrowing nematodes, select trees with symptoms of spreading decline. Use a shovel to dig handfuls of roots, to a depth of 10 inches, beneath the tree canopy. If a single tree is being sampled, collect roots from 10 spots beneath the canopy. If a group of trees is being sampled, collect 15 subsamples divided among the trees. Combine all of the roots in a single sample. Enough roots should be collected to fill a 1-liter freezer bag.

If surface roots are being sampled, make arrangements with the assay laboratory to process the entire quantity of roots in each sample, by the root incubation method. Sampling surface roots to detect the presence of burrowing nematodes is reliable only if large quantities of roots are processed.

Rootstock certification can be profitable even in heavily infested areas, if major shifts occur in citrus-growing regions. Cold weather has forced a gradual southward shift of citrus production in Florida, and the newly planted southern part of the citrus-growing region is nearly nematode-free. A similar shift often occurs in dryland regions as irrigation is used to reclaim desert areas.

Nematodes can also enter orchards on infested implements and vehicles and occasionally in irrigation water, which may flow in canals or rivers from infested orchards to others. All such avenues of entry can be closed by careful cleaning and disinfestation and by the use of groundwater for irrigation when possible. Guidelines for nursery certification include proper sanitation methods for implements and vehicles, setback requirements to isolate nurseries from orchards and other possible sources of contamination, and regular site-sampling protocols to detect accidental introduction of nematodes before trees are sold.

The movement of sting nematodes is not restricted by the certification program in Florida, because of their widespread occurrence on many hosts throughout the state. Nevertheless, infested nursery stock should be avoided, because of the serious damage caused by large populations of these nematodes in young trees. Nurseries infested with sting nematodes can be identified by examining the official certification form, which lists all plant parasites detected. The obvious root symptoms caused by the nematode can also help growers recognize potential problems when trees are received from nurseries. Sting nematodes can exist on bare-root planting stock, and the only way to disinfest roots is to subject them to heat (see below).

Box 13.3

Nematode Extraction Methods

Some common methods used by diagnostic laboratories to extract nematodes from soil or roots are described below. There are many variations on each method. All of these techniques can potentially recover any of the plant-parasitic nematodes.

An extraction method must be adequate to detect nematodes at low population densities. Thus, depending on the biology of the nematode species, either roots or soil may contain the largest numbers of nematodes. Sieving a soil suspension before final processing permits the extraction of a larger amount of soil than direct use of a Baermann funnel, and it is necessary in detecting species whose density is typically low. The choice of centrifugation or filtering after sieving depends primarily on whether the species can easily pass through the filter. Root samples can be processed by incubation or maceration; the main difference between these techniques is that incubation, while less expensive, typically requires several days, and maceration takes only several minutes.

Method	Description	Comments	Nematodes for which the method is preferred	Nematodes for which the method is not preferred
Baermann funnel	Used to process soil samples. Nematodes migrate through a filter, which supports the soil sample, and are collected in a container of water below.	Recovers only live, active nematodes in motile stages of development. Works well for many species but is not preferred for migratory endoparasites, species that cannot easily pass through the filter, or species at low population density in the soil.	Citrus Root-knot	Burrowing Dagger Lesion Sheath Sting Stubby-root
Sieving and sucrose centrifugation Sieving and Baermann funnel	Used to process soil samples. A suspension of soil in water is passed through a fine-mesh sieve, and nematodes are separated from residual soil and debris by centrifugation or filtration in a Baermann funnel.	Centrifugation recovers live and dead nematodes, whether or not they are motile. The Baermann funnel recovers only live, motile nematodes. These methods work well for many species but are not preferred for migratory endoparasites.	Citrus Dagger Root-knot Sheath Sting Stubby-root	Burrowing Lesion
Root incubation	Nematodes are permitted to migrate out of roots and then are recovered by various methods.	Recovers primarily live nematodes in motile stages of development in roots or in egg masses on the root surface.	Burrowing Citrus Lesion Root-knot	Dagger Sheath Sting Stubby-root
Root maceration	Nematodes are separated from root tissue by macerating roots in a mechanical blender or other device.	Recovers nematodes at all life stages in and on roots.	Burrowing Citrus Lesion Root-knot	Dagger Sheath Sting Stubby-root

Resistance

Rootstocks resistant to citrus and burrowing nematodes are commercially available. Trifoliate orange and some of its hybrids are highly resistant to most populations of the citrus nematode. Swingle citrumelo, a trifoliate hybrid, is increasingly used because of its resistance to the citrus nematode and (in various degrees) tolerance of citrus tristeza virus, the exocortis and xyloporosis viroids, *Phytophthora nicotianae,* citrus blight, and low temperature. However, nematode resistance is primarily due to a single gene inherited from trifoliate orange. As noted previously, resistance-breaking biotypes of citrus nematodes exist and can develop in response to widespread planting of these rootstocks. Additional sources of resistance in a single rootstock would remedy that situation. Edaphic factors can also limit the usefulness of rootstocks derived from trifoliate orange, which grow poorly in calcareous soils and in soils with high levels of clay.

Milam lemon and Ridge Pineapple sweet orange rootstocks are resistant to burrowing nematodes. Some sources of Carrizo citrange are also resistant, and others tolerate infection by burrowing nematodes with little damage. Resistance-breaking biotypes can develop in response to resistant rootstocks. Selection of these rootstocks often depends on their susceptibility to other soilborne pathogens. Milam lemon and Carrizo citrange are very susceptible to citrus blight, which can be widespread in areas where spreading decline is prevalent. Ridge Pineapple is very susceptible to foot rot; however, in the sandy soils where spreading decline occurs, water management can help overcome this problem.

No source of resistance to sting nematodes or the coffee lesion nematode is commercially available. Cleopatra mandarin appears to be useful in managing the coffee lesion nematode, however. The nematode develops more slowly on this rootstock, and in some cases young trees on Cleopatra mandarin appear to tolerate the nematode reasonably well.

Cultural Practices

In experimental work, yields frequently increase in response to nematode management when orchards are otherwise healthy and free of other problems. When orchards suffer from salinity, other pests, or inadequate or excessive soil moisture, it is difficult to detect any benefit from nematode management. In essence, trees cannot produce more fruit if any factor is limiting. The obvious conclusion from these experiences is that nematode management should often be the final consideration in orchard management.

Certain cultural practices have a direct effect on nematode damage to trees. Avoiding practices that increase salinity can reduce damage by the citrus nematode and possibly other species. Attention to potassium and micronutrient levels in leaves can sometimes mitigate nematode-induced stress. The symptoms of spreading decline, in particular, have been significantly affected by changing cultural practices; where disking is avoided, to protect shallow feeder roots, and frequent irrigation and fertilization are provided for these roots, trees that have lost deeper roots to burrowing nematodes can still grow normally.

To slow the movement of nematodes, some growers maintain buffer zones between nematode-infested and uninfested portions of orchards. Previously, growers sampled groves intensively, removed rows of trees between infested and uninfested areas, and then treated the buffer zones with ethylene dibromide to kill weeds and citrus roots that could serve as a bridge for nematode movement. Ethylene dibromide was prohibited when it was discovered in groundwater beneath buffer zones. Herbicides can still be used to prevent weed growth in buffer zones, but citrus roots grow so deep in the soil that chemicals are unlikely to kill all of them without contaminating groundwater. Mechanical pruning of roots can prevent most lateral root growth if trenches are dug close enough to the trunk to cut the roots before they grow too deep. However, limb pruning and deep trenching (to a depth of 5 feet) are expensive, and the efficacy of the practice is not well documented.

Fallow may become increasingly important in the management of nematode parasites, since most soil fumigants are no longer available. Many root-inhabiting species cannot be eradicated from old orchard sites without many years of fallow, because remaining roots continue to sprout and harbor the pests. However, their population densities decline rapidly after just several months of fallow. In orchards infested with species such as the citrus nematode and sting nematodes, the result can be faster growth of young trees. The host range of the species should determine whether a fallow should be completely bare or simply without citrus trees. Control of sting nematodes requires bare fallow. Citrus nematode populations will decline rapidly once citrus trees are removed. There is little value of fallow in virgin soils, unless they are sandy and sting nematodes are present.

When orchards suffer from salinity, other pests, or inadequate or excessive soil moisture, it is difficult to detect any benefit from nematode management.

Heat Treatment of Roots

The root systems of infected nursery stock can be immersed in hot water to eradicate nematodes. Sting and coffee lesion nematodes have been eradicated by treatment at 120°F for 5 and 10 minutes, respectively, without affecting subsequent tree growth. Burrowing, sheath, and citrus nematodes can be eradicated by treatment at 122°F for 10 minutes. Trees should not be treated in winter, and treated trees initially require more irrigation than untreated trees.

Chemical Control

Nematicidal chemicals are used to manage populations of plant-parasitic nematodes before planting and in existing orchards. Preplant chemicals registered for use in various industries are generally fumigant compounds, such as 1,3-dichloropropene and metam sodium. Chemicals used in existing orchards are mostly organophosphate and carbamate compounds that inhibit acetylcholinesterase in the nematode.

Preplant fumigants can be beneficial when infested orchard sites are replanted. Numerous studies have demonstrated increased growth and fruit production of young trees in fumigated soil, compared to trees in unfumigated soil heavily infested with sting, citrus, or burrowing nematodes. Fumigation followed by six months to two years of bare fallow has been widely used in an effort to eradicate burrowing nematodes. However, no study has ever compared the effects of fumigation with that of bare fallow on subsequent tree growth in orchards infested with any of these species.

Soil preparation is critical to achieving good results from fumigation. Proper tilth, moisture, temperature, and surface sealing are needed to maximize the distribution and retention of fumigant in the soil and to avoid phytotoxicity when trees are planted. These conditions vary for different products, so the directions for any particular product should be carefully followed.

After planting, the nematode population density can be reduced to nondamaging levels with various organophosphate and carbamate nematicides. When treatment is effective, yield often increases by the second year following treatment. Many nematicides are formulated for application by low-volume irrigation systems, herbicide booms, or granular application equipment. When applied, these nematicides must be completely incorporated into the soil mechanically or by means of irrigation water, for the protection of people and wildlife.

A number of factors affect the efficacy of postplant nematicides. Proper placement and coverage is necessary to achieve the needed effect. Application equipment must be designed to apply and incorporate the material beneath the canopy of the tree, where the root density and the nematode population density are greatest. Most nematicides are moderately to highly water-soluble. When possible, the movement of nematicides in soil water should be managed by irrigating for short periods. Applications should be made when nematodes are active or at the beginning of a period when their population density is expected to increase. For some products, splitting the annual permitted dosage into several applications, spaced at one- to two-month intervals, is much more effective than a single application. This approach shows great promise, because populations of nematodes such as the citrus nematode can be reduced by some products to such low density that further treatment is unnecessary for several years. Increasing the interval between treatments is important, not only to reduce expenses and environmental risk, but also because accelerated microbial degradation can greatly reduce the efficacy of some nematicides over time. Populations of microorganisms capable of metabolizing nematicides increase with repeated use, and eventually the microorganisms degrade these the compounds before they can affect the nematodes.

Regular soil sampling is important in managing nematodes with nematicides. Treatment should be based on a demonstrated need to reduce the nematode population, and follow-up sampling is useful to determine the efficacy of treatment and to monitor subsequent management needs. Sometimes it is useful to leave a few untreated areas in an orchard, for comparison with treated areas. Orchard and soil conditions are so variable that careful monitoring of the responses of the nematodes and the trees to treatment is the only way to determine if a practice is profitable.

Clayton W. McCoy
University of Florida
Citrus Research and Education Center, Lake Alfred

CHAPTER 14

Arthropod Pests of Citrus Roots

The major arthropods that are injurious to plant roots are members of the classes Insecta and Acari (mites). Two-thirds of these pests are members of the order Coleoptera (beetles), which as larvae cause serious economic loss in a wide range of plant hosts. Generally, the larvae hatch from eggs laid by adults on plants or in the soil and complete part of their life cycle chewing on plant roots, and in many cases as adults they feed on the foliage of the same or other host plants. A number of arthropods inhabit the rhizosphere of citrus trees, some as unique symbionts, but few are injurious to the roots. Only citrus root weevils, termites, and ants, in descending order of importance, are root or stem herbivores. These insects are found in both arid and humid citrus-growing regions of the world. Because of their sporadic distribution among groves in a given area, they are categorized as pests of minor economic importance to citrus. However, in an infested grove, they can be devastating to the vitality and productivity and even the survival of trees of all ages. In addition, these soil insects are often associated with root diseases.

Citrus Root Weevils

Adult Identification

Citrus root weevils are beetles with specialized prognathous (snoutlike) mouthparts, in the family Curculionidae (Fig. 14.1). The snout extends forward of the eyes and bears a pair of clubbed antennae toward the end. The front wings are modified to form heavily sclerotized elytra (wing covers), which protect the hind wings at rest. Most citrus root weevils are polymorphic as adults; that is, members of the same species may be distinctly different in appearance. For example, many of the brilliantly colored tropical species exhibit wide variability in the number and color of stripes or spots on the elytra. The genetic plasticity of both temperate and tropical species of citrus root weevils has caused many unresolved and poorly understood taxonomic problems.

Geographical Distribution

Citrus root weevils are predominantly tropical; however, a few temperate species are important pests in the United States, Chile, Argentina, Australia, and New Zealand (Table 14.1). The northern blue-green citrus root weevil, *Pachnaeus opalus;* the Fuller rose beetle, *Asynonychus godmani;* and related species in the genus *Pantomorus* are found in temperate areas. Approximately 150 species have been recorded in the Caribbean region, including Florida, Central America, and South America, feeding as larvae on the roots of all species of the genus *Citrus.* No reports of root weevils on citrus in Europe, Africa, and southeast Asia have been published, although they are likely present in those regions. Currently, species in the genera *Diaprepes* and *Pachnaeus* pose the greatest threat to the citrus industry of Florida. *Exophthalmus* species pose a greater threat if introduced from the Caribbean. Other citrus root weevils of lesser importance to Florida are the little leafnotcher, *Artipus floridanus; Tanymecus lacaena;* and *Epicaerus fermidolosus.*

Generalized Life History and Injury

Most citrus root weevils are univoltine species; that is, they produce one generation per year. Soil conditions greatly influence the amount of time required to complete a life cycle. Both adults and larvae are polyphagous; that is, they feed on a wide range of host plants, in many cases totally unrelated to citrus. All citrus root weevils have a similar life cycle (Fig. 14.1). They have three immature stages: egg, larva, and pupa. Adult weevils emerge from the soil and lay eggs on host plants aboveground; the larvae enter the soil to feed on roots, and the pupa and teneral adult stages are spent belowground. Adults emerge beneath citrus trees or alternate host plants throughout the year. Peak adult emergence varies within and among species and by region but normally occurs in spring and late summer. Adults are attracted to the nonreflective silhouette of the citrus trunk and crawl up the trunk or fly short distances to the tree canopy. They prefer to rest in the shaded interior foliage of the canopy when the sun is shining, but they aggregate on new leaf flushes in subdued light or darkness to feed, mate, and oviposit. Most weevils are bisexual, with females constituting about 60% of a local population. The Fuller rose beetle and a few less important species do not produce males and reproduce by parthenogenesis. When disturbed, adults of all species drop from their resting or feeding positions to the ground.

The most apparent visible injury caused by adult weevils is notching of the leaf margins of young, tender shoots (Plate 59E). Severe leaf injury can decrease water use efficiency and photosynthesis. However, leaf injury appears to be only a minor problem associated with nonbearing trees and has no apparent effect on the yield of mature trees. Adult weevils feed less on citrus leaves as the leaves mature, and some of them migrate to alternate host plants. They live for several months and never return to the soil.

Except for the Fuller rose beetle, which prefers to lay eggs in masses beneath the fruit calyx or, less commonly, in cracks and crevices in the bark or foliage, citrus root weevils normally lay eggs in clusters between two leaves, which are secured by an adhesive substance secreted by the female during oviposition. The egg masses vary in number, shape, and color and are distributed throughout the tree canopy.

Upon hatching, first-instar larvae fall from the leaf to the soil beneath the tree and enter the soil. The behavior of the larvae in the soil is poorly understood, but their size appears to dictate the site of root feeding. Tiny hatchlings (0.5 inch long) feed on feeder roots, whereas later-instar larvae feed on the larger lateral roots, forming deep grooves as they consume the outer bark, including the cambium layer (Plate 59F). The larvae remain on the roots for eight to 15 months, depending on the species, reaching a length of 0.5–1.0 inch. Larger larvae can remove the outer bark from the crown area of the root system or girdle the trunk, thereby killing the tree. The aboveground symptoms associated with extensive damage to roots are similar to other tree declines caused by root maladies, including leaf yellowing or leaf drop, twig dieback, off-blooming, setting a very heavy crop of fruit, and (in the later stages of moribundity) fruit drop. The protracted period in which larvae feed in the soil is followed by an inactive pupal stage of two to four weeks. Following pupation, the adults either remain in the soil for a short period or ascend to the soil surface.

Species of Major Economic Importance

Sugarcane Rootstalk Borer Weevil

The sugarcane rootstalk borer weevil, *Diaprepes abbreviatus,* one of 17 species native to the Lesser Antilles, was introduced into Florida in 1964. It has been recognized as a major pest of citrus in the Caribbean region for over a century. Currently, Florida is the only state in the United States where this weevil is present. It is found in 19 counties, including all those growing citrus commercially.

The adult weevils, commonly referred to as *Diaprepes,* are the largest of the Florida weevils, with a length of 0.37 to 0.75 inch. The females are generally larger than the males. Both males and females are highly variable in color, having black wing covers with stripes that vary from white to gray to yellow to orange (Plate 59A).

Adults are slow-moving and are capable of flying short distances, usually displaying nocturnal or crepuscular activity. Males are thought to emit an aggregation pheromone in frass

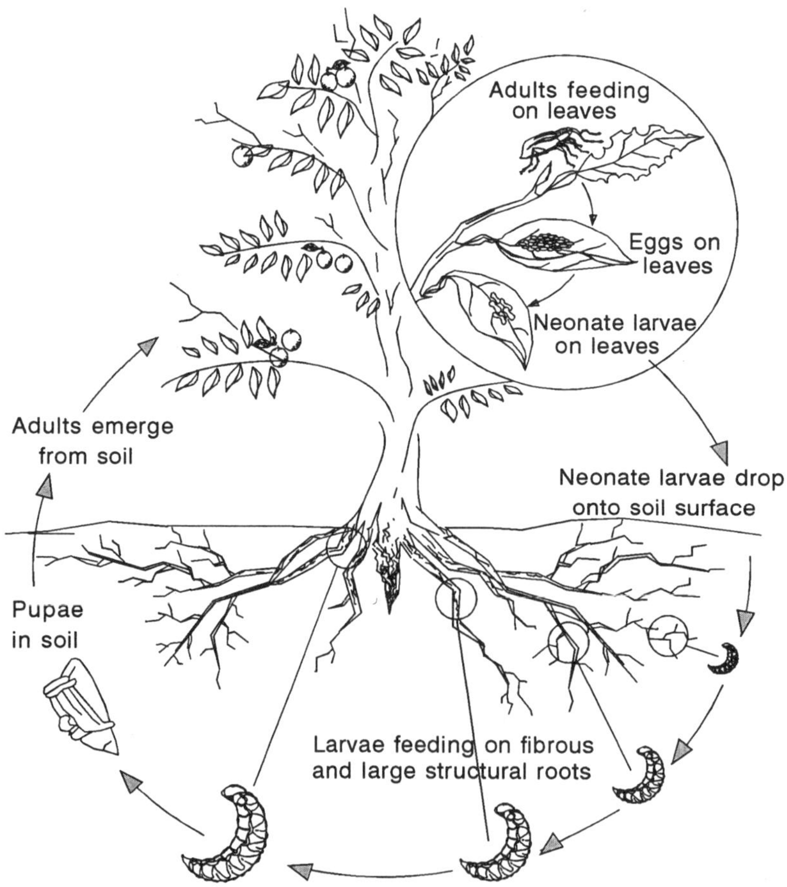

Fig. 14.1. Life cycle of *Diaprepes abbreviatus,* the sugarcane rootstalk borer weevil, on citrus. The life cycle is typical of citrus root weevils. (Drawing by Diana Drouillard)

deposited on the host plant. The pheromone attracts both sexes, resulting in a congregation of weevils on newly formed leaf flushes. This phenomenon might partially explain the aggregation of adults frequently observed on particular trees in a citrus grove.

Neither larvae nor adults are specific to citrus; they feed on a wide range of native and introduced host plants in Florida and the Caribbean region. Of the 270 plant associations recorded, many involve agricultural plants, including citrus, sugarcane, pineapple, corn, peanut, potato, and yucca, as larval hosts. In addition, many woody plants are hosts, including mahogany, live oak, dahoon holly, silver buttonwood, Brazilian pepper, pigeon pea, and *Gliricidia sepium* (quick stick, commonly grown in the Caribbean as a living fence or windbreak). Quick stick is often the preferred host, and it serves to attract weevils to the vicinity of other commercial crops. No citrus rootstock is known to be resistant to larval feeding, and rootstock research showed no inhibition of feeding on rough lemon, sour orange, Carrizo citrange, Milam lemon, Cleopatra mandarin, and many monogeneric hybrids.

Adult emergence is most frequent during the spring or late summer and lasts 10–12 weeks. The beginning of adult emergence appears to be associated with periods of frequent rainfall and peak leaf flushing in some areas, although little empirical information is available to support these observations. The female weevil can live approximately four months, slightly longer than the male, and will lay up to 5,000 eggs in masses of 30 to 250 during a lifetime. Oviposition begins three to seven days after newly emerged females feed. The eggs are smooth, white, oblong-oval, and about 0.5 inch long by 0.016 inch in diameter after deposition. They become tan with age and hatch in seven to 10 days at 78°F.

Table 14.1. Root weevils known to reproduce on and damage citrus[a]

Genus	Common name	Geographical location
Major pests		
Asynonychus	Fuller rose beetle	United States, Australia, New Zealand, Chile, Argentina
Diaprepes	West Indian sugarcane rootstalk borer	United States, West Indies
Exophthalmus	Fiddler beetle	West Indies
Naupactus	. . .	South America
Pachnaeus	Blue-green citrus weevil	United States, West Indies
Pantomorus	. . .	Central America, South America
Minor pests		
Artipus	Little leafnotcher	United States, West Indies
Campsus	. . .	Central America, West Indies, Texas
Cleistolophus	. . .	West Indies
Epicaerus	. . .	Mexico, Central America, Florida, Texas
Eutinophaea	Citrus leaf-eating weevil	Australia
Lachnopus	. . .	West Indies
Litostylus	. . .	West Indies
Maleuterpes	Spine-legged citrus weevil	Australia
Orthorhinus	Elephant weevil	Australia
Tanymecus	. . .	North America

[a] Major pests are those known to cause significant economic losses to citrus in more than a single, isolated region of the world.

The larvae are white and legless and have highly sclerotized chewing mouthparts. The larval stage is variable in length, lasting from eight to 15 months. First-instar larvae emit a defense pheromone that repels ant predators while the larvae are at the soil surface. The larvae cannot penetrate dry soil. The optimum soil moisture for penetration of sandy soils is 9% of its moisture-holding capacity. As they feed and grow, larvae molt at least 11 times, attaining a length of 0.8–1.0 inch before pupation. They move readily in sandy soils, having been found at depths of 8–10 feet. Larger larvae are particularly destructive to the crown region of the root system, frequently killing trees. This feeding behavior by late-instar larvae distinguishes *Diaprepes* from other weevils (except *Exophthalmus*) as a serious economic pest. Larvae in the last instar stop feeding and enter a variable period of quiescence; the larva then forms a naked pupa in a chamber in the soil. The pupal period lasts 15–30 days. A single generation from egg to adult in the field requires about 1 to 1.5 years, depending on soil conditions and the availability of host material on which the larvae can feed.

Fiddler Beetles

The genus *Exophthalmus* comprises 76 species of fiddler beetles, found in many countries in the Caribbean and Central America (Plate 59D). None is known in the United States. However, fiddler beetles are present in Puerto Rico, Cuba, and Hispaniola and thus pose an immediate threat to the citrus industry in Florida. In Jamaica, six species of *Exophthalmus* feed on citrus, causing severe root injury. *E. vittatus* is the most widespread on the island and is considered a major problem in commercially grown citrus. Biologically, fiddler beetles are closely related to *Diaprepes* species, feeding on similar host plants and having similar feeding behaviors and natural enemies. Fiddler beetles also prefer peanut, jackfruit, star apple, yam, mango, avocado, sugarcane, and cocoa as host plants, in addition to citrus.

Adult fiddler beetles are generally larger than *Diaprepes,* about 0.5 to 1.0 inch long, with some variation in size among species. They exhibit strong color polymorphism; the elytra (wing covers), in particular, are marked with different patterns of stripes and spots, varying from brown to blue, green, yellow, rosy, tawny, or white on a black body. The shape of the body and the genitalia vary among species.

Unfortunately, detailed studies of aspects of the biology of *Exophthalmus* have not been conducted. In Jamaica, two distinct emergence cycles have been identified, concurrent with the rainy seasons (May and October). Adult emergence starts a few days after the rains begin, when adequate moisture has percolated down to infested roots. The last-instar larvae appear to be triggered by the changing environmental conditions, and they move toward the soil surface to pupate. After about 30 days, adults begin to emerge from the soil and move to the tree canopy.

Adult fiddler beetles mate in the tree canopy shortly after emergence. After a brief preoviposition period, egg laying begins. The eggs are creamy white or grayish and oval with rounded ends. They are laid in masses of about 60, in a manner similar to that of *Diaprepes,* but in fewer numbers. One female will lay an average of 2,000 eggs over her life. Embryogenesis in *E. vittatus* takes seven to eight days. Adult females live about 110 days and males 103 days in the field.

The neonate larvae spend a brief period on the leaves before dropping to the soil. Larval behavior in the soil is similar to that of *Diaprepes;* larvae in the first two instars feed on adventitious roots and then move to lateral roots. *E. vittatus* has 10 larval instars, which complete their development on roots after

30 weeks. The last larval instar moves to the crown of the root system, where it actively feeds for about four weeks before forming a pupal chamber away from the roots, near the soil surface. The pupal period lasts three to four weeks, and the total life cycle requires about 9.5 months.

Blue-Green Citrus Root Weevils

The genus *Pachnaeus* comprises seven species of blue-green citrus root weevils, which are found on citrus in the Caribbean region and the southeastern United States. Two species, the southern citrus root weevil, *P. litus* (Plate 59B), and the northern citrus root weevil, *P. opalus,* are present in Florida, and the former also in Cuba. Adults are large, 0.30 to 0.55 inch in length, but smaller than *Diaprepes.* The coloration of the adult ranges from gray to gray-green to bright aqua. These two species are indistinguishable to the average person, because of their similarity in color and shape, but they can be readily distinguished by the degree of sinuosity of the suture between the pronotum and the elytra. *P. litus* has a visible notch on the pronotum, and *P. opalus* has a smooth pronotal edge. *P. litus* is confined mainly to coastal and interior plantings in south Florida. *P. opalus* is a more temperate species, found throughout central and north Florida and as far north as New Jersey. Where distribution overlaps, both can be found in the same citrus grove. A third species, *P. citri,* is an important citrus pest in Jamaica.

Like other citrus root weevils, the *Pachnaeus* species are polyphagous. Adult *P. opalus* has 27 known host plants, and *P. litus* has 70. Both prefer citrus but are also found on oaks, Australian pine, and woody ornamentals, such as roses. *P. litus* and *P. citri* have been observed feeding on citrus fruit in Cuba. Host plants of *P. citri* include mango, avocado, pimento, cacao, peanut, and star apple.

No detailed studies of the life cycles of these three species of *Pachnaeus* have been conducted, but from what is known they are quite similar. Both *P. litus* and *P. opalus* adults emerge from the soil continuously; peak emergence is generally from mid-May to mid-July in Florida. The emergence of *P. citri* adults is triggered by the rainy season. Adult feeding and mating behaviors are similar to those of other citrus root weevils. Adults live for up to 120 days. The female lays 30–75 whitish, elongate cylindrical eggs in a mass and can produce as many as 4,000 eggs during a life span. Hatching of *P. litus* and *P. opalus* begins after seven to 10 days and is affected by moisture. The larvae are white and legless and have well-developed, brownish black chewing mouthparts. Larvae of all species exhibit similar feeding behavior. They develop over a period of eight to 10 months. The neonate larvae of *P. litus* are highly sensitive to low soil temperatures, in the range of 60–70°F, with a survival rate up to 25% lower than that of other weevils in Florida. Root injury caused by *Pachnaeus* species is typical of that due to other weevils described previously. Injury to the crown area of the root system is generally less serious than that caused by *Diaprepes.*

Fuller Rose Beetle

The Fuller rose beetle, *Asynonychus godmani* (Plate 59C), is a ubiquitous species, native to South America. It has been reported on citrus in the United States, the Mediterranean region, South Africa, Australia, and New Zealand and on other deciduous fruit trees, numerous woody ornamentals, and other plants, such as roses and grasses. It is found on a wide range of plants in both coastal and interior areas of Florida, where it was first reported in 1916. It is very localized in citrus-growing areas of the state and frequently coexists with other weevils in infested groves. Adult feeding damage to citrus foliage and larval injury to roots generally cause little tree decline.

The Fuller rose beetle has a propensity for laying eggs under the calyx of the fruit. This habit has increased its importance as a postharvest pest, because fruit harboring eggs is considered a quarantine problem in some countries, notably Japan. In California, where the Fuller rose beetle is widespread on citrus, fruit destined for export to Japan is regularly fumigated with methyl bromide. Fumigation is expensive and is especially damaging to lemons.

The Fuller rose beetle is parthenogenic. Adult females are about 0.33 inch in length, smaller than the species mentioned above. They are brownish to gray and have a faint crescent-shaped mark on each side of the wing covers. The wing covers are fused, and the insect is flightless.

In Florida and California, the beetle produces one generation per year. Adults emerge from the soil throughout the year, but peak emergence occurs from July until the end of November. Females oviposit for about four months, depositing yellow, flattened eggs in masses of 70–100. Eggs hatch in 15–17 days, with maximum hatch occurring at high ambient relative humidity. The hatchling larvae are cream-colored except for the yellowish head capsule. They can survive soil temperatures as low as 21°F. In feeding behavior and development, the Fuller rose beetle is similar to other root weevil species. Mature larvae are about 0.4 to 0.5 inch long at pupation. The pupal period lasts about 15 days.

Economic Importance

Generally, leaf feeding by adult root weevils (Plate 59E) is of no economic importance to either young or mature trees; however, they occasionally defoliate young trees, particularly replant trees in mature groves, causing retardation of tree growth. The larvae of various key species of citrus root weevils, discussed previously, cause an estimated annual loss of $75–100 million in citrus production as a result of tree mortality and yield reduction in commercial citrus groves in the major growing regions of Florida, Jamaica, Puerto Rico, the Dominican Republic, Dominica, Martinique, and Cuba. In addition, an estimated $3 million is spent annually to fumigate fruit contaminated with eggs of the Fuller rose beetle and bound for export to Japan. Economic estimates are not available for other countries.

The larvae of various key species of citrus root weevils cause an estimated annual loss of $75–100 million in citrus production as a result of tree mortality and yield reduction.

Management

As will be pointed out in the following sections, a repertoire of historical and contemporary methods for managing citrus root weevils has been developed. They include numerous cultural, mechanical, natural, sanitary, biological, and chemical methods for the suppression of citrus root weevils at all developmental stages. No single method of control has been totally effective. The multiphasic approach of integrated pest management (IPM), utilizing both chemical and nonchemical methods, will be necessary to achieve acceptable control and, at the same time, satisfy environmental concerns into the next century.

One major deficiency in the aggressive use of chemical and nonchemical control methods is our lack of injury thresholds and economic thresholds for root weevil larvae in relation to the age and health of trees in various regions of the world. Techniques for monitoring adults of some species are in development, but none has been used to generate ecological data vital in determining injury thresholds. Preliminary research suggests that ground traps have potential in monitoring seasonal adult emergence. Such data should be helpful in timing foliar treatments for adults and soil treatments for neonate larvae invading the soil.

In California, the following monitoring program is recommended when groves are infested with adult Fuller rose beetles. Growers should sample for adult beetles from July to November by shaking or beating branches to induce beetles to drop onto an apron or tray. If beetles are found, fruit should be sampled for egg masses, especially in areas where adult feeding damage is evident. A minimum of 500 fruit in a 10-acre block, five fruit per tree from 10 trees per acre, should be sampled. Fruit at chest height is selected from different quadrants of the canopy. The stem is clipped 2 inches from the fruit; then the stem is held and twisted off the calyx. The underside of the button end and the part of the fruit that was covered by the button should be examined for the presence of egg masses. If the fruit is to be exported, the infestation level should be less than one fruit infested with a viable, unhatched egg mass per 1,000 fruit sampled at harvest.

Nursery Sanitation and Control

With their broad host range and limited mobility as adults, citrus root weevils have historically been disseminated between countries as eggs or adults on plant foliage or as larvae on the roots of container-grown nursery plants. In Florida, spread within the state has most frequently occurred by the movement of infested budded trees from the nursery to the grove. Citrus root weevil management in nurseries begins with purchasing plant material that is certified weevil-free and maintaining a weevil-free nursery. It is important to randomly inspect the canopy of all plants for leaf injury (notching), eggs, and adults. In addition, container-grown plants showing symptoms of decline should be bare-rooted and examined for larvae and feeding injury.

Sanitation is vital to any nursery. Greenhouses and shadehouses must be kept insect-proof through careful preventive maintenance. The following management program is recommended for either enclosed or outdoor nurseries subject to weevil invasion from the surrounding area. Plants should be inspected weekly for adult weevils. In outdoor nurseries, a few ground traps, such as the pyramid-shaped Tedder's trap, can be placed on the soil surface at random locations near the base of suspected host plants, to attract adults moving on the soil. Foliar chemical sprays should be applied as needed to kill adult populations on the foliage at the time of detection and thereby reduce the number of gravid females laying eggs on the plant. In nurseries with a history of weevils, soil insecticides should be applied as a liquid drench or as granules fully incorporated into the potting mix for the control of first-instar larvae invading the soil. Residual control lasts a minimum of 10 weeks but has no effect on larvae beyond the first instar. Natural enemies attack weevils at various developmental stages in nurseries, and the entomogenous nematode *Steinernema riobravis* is effective in suppressing weevil populations in containers or field soil, but eradication is required to avoid plant loss and satisfy certification requirements in Florida. In any case, pesticides should be rotated and used judiciously to prevent resistance and avoid disrupting natural enemies.

Certification programs are effective in reducing the spread of citrus root weevils from nurseries to groves. For example, the Florida Department of Agriculture and Consumer Services regulates the movement of all nursery stock originating from nurseries known to be infested with *Diaprepes*. All weevil-infested nurseries operate under a compliance agreement that allows nursery stock to be moved from the nursery only when all conditions in the agreement are met. Regulatory requirements include 1) the removal of all foliage or chemical treatment of eggs and adults and 2) larval control by approved chemical treatment of potting media in containers or by shipment of plants bare-rooted. It is imperative that nurseries do not sell liners infested with weevil larvae.

Sanitation is vital to any nursery. Greenhouses and shadehouses must be kept insect-proof through careful preventive maintenance.

Cultural Control

Improving the drainage of the soil appears to be the most fundamental strategy for minimizing root injury due to citrus root weevil larvae. Drainage ditches and raised planting beds in lowlands reduce the amount of standing water in orchards and lower the water table. As a result, the rhizosphere is larger and better-drained and favors rapid, healthy root growth free of water stress and disease. Although citrus root weevil larvae feed readily on roots growing under these improved conditions, the tree generates new roots at a more rapid rate, so that less overall tree decline occurs over time. It has been observed but not proved scientifically that improved drainage also appears to reduce the incidence of diseases caused by soilborne fungi associated with larval injury to roots. The combination of root disease and larval feeding can be devastating to a tree. Where drainage has been improved in declining groves, the regular use of a combination soil fungicide-insecticide can significantly improve canopy leaf density and productivity, particularly when declining trees are also rejuvenated by aggressive canopy pruning and regular tree care.

Bare rooting is another method of limiting crown injury caused by *Diaprepes* larvae, practiced in the Caribbean and occasionally in Florida. The upper 4–6 inches of soil is removed, by means of water under pressure, in an area extending 2 feet from the base of the trunk, to expose the crown region of the root system and thereby prevent larval feeding on this vulnerable part of the tree. This tactic has been used mainly to prevent tree mortality in declining young groves. Unless combined with effective control of adults and larvae on other parts of the tree, this labor-intensive practice will likely fail. Also, the formation of a basin over the rhizosphere can lead to overwatering and drainage problems. In Jamaica, where bare rooting has been widely practiced, nursery trees have been selected for lateral root development, and trees are planted on mounds before the crown is exposed.

Fertigation at monthly intervals has been successfully used by a few growers in Florida to promote the growth of fibrous roots near the soil surface after weevil larvae have destroyed the taproot and inner crown of the tree. The promotion of fibrous root growth can significantly improve tree health and average fruit production in declining groves. This strategy also will be effective for only a short time unless weevil adults and larvae are controlled by other means.

Skirt pruning can be effective in controlling the Fuller rose beetle and other flightless weevil species. In California, tree skirts are pruned and mechanical or chemically treated trunk barriers are used to prevent the invasion of trees by adult weevils. Weed control is also needed, to prevent movement into trees from the stems of grasses and other weeds.

Cover cropping has been practiced in a number of countries in the Caribbean since the beginning of the century; however, from the literature, it is difficult to determine its success. Early in the century, the leguminous plant *Tephrosia candida* was planted between rows or around citrus trees to repel adult *Diaprepes* species, but it is no longer used as a cover crop. Observations in Jamaica suggest that citrus root weevil adults are attracted to *Gliricidia* species (quick stick) more readily than to adjacent citrus trees, so that quick stick may have potential as a trap crop. However, other reports suggest that it attracts *Diaprepes* species to citrus, on which they lay eggs, and larvae hatching from the eggs are carried by the wind to infest citrus groves. Further research should resolve this contradiction reported in the literature. In Puerto Rico, another leguminous plant, pigeon pea, appears to be effective as a trap crop for adult *Diaprepes* species around citrus plantings.

In the past, deep tillage was practiced in the Caribbean to expose larvae to predators in the soil beneath the tree. However, mechanical root injury to the tree was always a concern. This cultural practice is rare today and is impractical in most commercial citrus-growing areas.

Mechanical Control

In the Caribbean, where agricultural labor is less expensive than in the United States, the large, brightly colored adult citrus root weevils in the genera *Diaprepes* and *Exophthalmus* are conspicuous enough in the tree canopy for hand-picking. It is common today to see teams of workers hand-collecting weevils from leaves or from white cloth sheets spread beneath the canopy, which catch weevils that fall from the tree when the branches are shaken. The weevils are usually collected in containers and then killed in a kerosene bath or burned. The economic benefit gleaned using this mechanical control to reduce adult populations is unknown.

In parts of the Caribbean and in Florida, some growers spread strips of black plastic, approximately 3 feet wide, on the soil on opposite sides of the trunk beneath the canopy to trap newly hatched citrus root weevil larvae as they drop from the leaves. During the heat of the day, larvae can be killed within minutes as they land on the hot plastic, or they are forced away from the tree to the row middle, where they are killed on the hot, dry sand or consumed by predators. When used over a large area, this mechanical barrier can aid weed control, but it causes problems in routine grove maintenance, particularly irrigation and fertilization. Wind can also be a limiting factor in maintaining an effective barrier.

In summary, a number of cultural and mechanical control methods have been developed to combat citrus root weevils, with different degrees of success. Most of these methods have been grower-driven and lack empirical data to support observations.

Natural and Biological Control

At least 50 species of parasites, predators, and pathogens are known to attack citrus root weevils at one or more developmental stages. Insect parasitoids have been observed attacking adults of some species and eggs of virtually all species of citrus root weevils.

Two parasitic wasps—a *Microctonus* species (an internal parasite) and the digger wasp, *Cerceris watlingensis*—deposit eggs on the body of adult *Artipus floridanus*. The larvae kill the host by invading the body cavity to feed on internal tissues. They subsequently emerge from the dead host as adults.

The hymenopterous egg parasitoids differ widely and are quite host-specific. Different species lay eggs externally on a host egg mass or internally, inside the eggs of a host species. The parasitoid larvae then develop as predators on or in the host eggs. Egg parasitism appears to vary greatly from species to species and for a given species within a citrus-growing area. Of the 13 species of egg parasitoids reported in the Caribbean region, *Aprostocerus haitiensis* (synonym *Tetrastichus haitiensis*) is most widely distributed on citrus. Virtually nothing is known about the population dynamics of these egg parasitoids and their potential as regulatory agents in the field; however, reports of egg parasitism as high as 90% in some locations have been published. There are no known parasitoids attacking citrus root weevil larvae in the soil.

At least 50 species of parasites, predators, and pathogens are known to attack citrus root weevils at one or more developmental stages.

A number of attempts have been made to introduce egg parasitoids into one country from another as classical biological control agents. Recent attempts in Florida have involved the introduction of *A. haitiensis* from Puerto Rico to control *D. abbreviatus* and the introduction of *Fidiobia citri* from California and *Platystasisus asinus* from Chile to control the Fuller rose beetle. Limited surveys suggest that none of these has become established in Florida.

Both invertebrate and vertebrate predators are known to feed on adult weevils during the arboreal portion of their life cycle. Birds, reptiles, spiders, ants, and mammals have been observed preying on adult weevils. Tree-inhabiting ants, particularly predatory species in the genera *Monomorium* and *Crematogaster*, and various spiders have been observed consuming eggs of citrus root weevils. The importance of egg and adult predation is unknown, and no attempts have been made to rear and release these predators as biological control agents or to augment their population in the field by environmental manipulation.

Numerous general predators, such as ants, ant lions, earwigs, and mites, forage for neonate weevil larvae on the soil surface beneath the tree canopy. Some larvae can repel ant predators by releasing defense pheromones. On small farms in many Caribbean countries, chickens and guinea hens are kept in groves to reduce larval numbers. Predation by insects, mites, and vertebrates can have a significant effect on larval populations, and it may be possible to enhance this form of biological control of weevil larvae by environmental manipulation (mulching, for example).

Citrus soils around the world contain microscopic nematodes that are obligate parasites of soil insects (Plate 59H). Five species of entomogenous nematodes, belonging to the families Steinernematidae and Heterorhabditidae, are known to attack emerging adult citrus root weevils and, more often, larvae in all stages. The nematodes carry mutualistic bacteria (*Xenorhabdus* or *Photorhabdus* species) that are toxic to the larvae after invasion by infective juveniles through external openings of the body. The bacteria kill the host within three days. The nematodes then feed on the bacteria, reproduce in the host, and return to the soil to resume parasitic activity. *Steinernema carpocapsae, Heterorhabditis bacteriophora,* and *H. indica* have been identified in citrus root weevil larvae in different Caribbean countries and Florida.

Fungal and protozoan pathogens have been found attacking both adult and larval stages of citrus root weevils. Viral and nonmutualistic bacterial infections have not been described. Unicellular gregarines and microsporidial protozoans infect the alimentary tract of adult *Artipus floridanus.* These nonspecific microbes rarely kill their host but will reduce the fecundity of the weevil. Entomopathogenic fungi are ubiquitous in nature. Citrus root weevils at all developmental stages except the pupa are susceptible to fungal conidia found on or in the vicinity of a dead host in soil or on plant surfaces. Upon contact with a host, a conidium attaches itself to the host cuticle and subsequently germinates. An infection peg (modified hypha) is formed and penetrates the host cuticle. Upon entry into the body cavity, hyphae proliferate within the body cavity, killing the host. In subtropical and tropical citrus areas, three species of fungi infect adult citrus root weevils: *Beauveria bassiana* (Plate 59G), *Metarhizium anisopliae,* and *Stilbella buquettii.* The first two of these species also infect larvae, as do a few species in the genera *Paecilomyces* and *Aspergillus.*

While entomopathogenic fungi and entomogenous nematodes can reduce weevil populations in the soil under natural conditions, their spatial distribution and persistence in the soil is variable. Soil texture, soil temperature, soil moisture, ultraviolet light, and various natural and artificial antagonists are important abiotic and biotic factors that influence populations in space and time. In Florida citrus soils, entomopathogenic fungi appear to be most prevalent before the summer rainy season. During the rainy season, nematodes are predominant.

Natural control of citrus root weevil larvae by entomogenous nematodes and entomopathogenic fungi is limited by abiotic and biotic factors in Florida, California, and some Caribbean and South American countries. Research on biological control is therefore focusing on the supplementation of naturally occurring fungi and nematodes with selected strains that are amenable to mass production and are applied as biopesticides for short-term pest control. In the United States and other developed countries, fungi and nematodes are produced and formulated as commercial products for sale to growers. In some Caribbean countries, cottage industries exist to produce nematodes and fungi for control of citrus root weevil larvae.

Numerous laboratory and field studies of different species of nematodes have been conducted in Florida to determine their potential as biological control agents of citrus root weevil larvae. From this research, *Steinernema carpocapsae* and *Heterorhabditis bacteriophora* were developed as commercial products and were sold to citrus growers for about 10 years for larval control. *S. riobravis,* an exotic species collected in south Texas, has since replaced *S. carpocapsae.* Laboratory and field studies clearly showed that *S. riobravis* is the superior nematode in sandy loam ridge soils, killing over 90% of root weevil larvae in some experimental trials. Although the overall survival rate of the nematodes is generally low, they are capable of sufficient vertical movement in the soil to reduce the larval population in the rhizosphere.

S. riobravis and *H. bacteriophora* are currently the only soil treatments used for the control of citrus root weevil larvae in Florida. Nematodes are applied either by microjet irrigation or with a herbicide applicator, at the rate of 200 million per treated acre, regardless of tree age and planting density. Pre- and post-treatment irrigation is required to minimize nematode mortality and ensure that nematodes move to the root zone. A minimum of two applications per year should be made, in early June (at the beginning of the rainy season) and in mid-September, after adult weevils have begun to emerge, to ensure a high nematode population density at the time when most larvae are invading the root zone. Nematodes should not be applied within four weeks of nematicide use.

In Puerto Rico, Jamaica, and other Caribbean countries where citrus is planted in soils that are either calcareous or high in clay, nematodes are not used regularly as biological control agents. It is not clear whether soil conditions or other factors are the reason. In California, some suppression of Fuller rose weevil larvae was obtained using *S. carpocapsae.*

In Florida, the fungal biopesticide Mycotrol WP, a conidial preparation of *Beauveria bassiana* formulated in medium oil, has been tested experimentally at a rate of 1×10^6 conidia per gram in combination with a sublethal dosage of the insecticide imidacloprid, as a soil treatment to control neonate larvae and emergent adults at the soil surface. Previous research, using virulent strains of *B. bassiana* and *M. anisopliae* formulated as conidial and mycelial preparations, respectively, determined that high dosages of fungi (1×10^8 conidia per gram) are necessary to achieve 70% control of the larvae and, therefore, are not cost-effective. In the Caribbean, mass production of *B. bassiana* conidia on rice has become a cottage industry. Rice granules harboring the fungus are applied directly to the soil beneath the tree to control citrus root weevil larvae.

Chemical Control

During the last century, chemical control has frequently been used to reduce populations of citrus root weevil larvae and adults in the field. Over 30 different chemicals have been applied as liquid sprays on foliage to control adult weevils, as soil drenches to control neonate larvae, and as granules or fumigants in soil treatments to control larvae and emergent adults. To control the Fuller rose beetle, chemicals have been applied as sprays to the trunk of the tree and in a band around the tree trunk to form a barrier against adults.

Until their cancellation, the chlorinated hydrocarbons aldrin and dieldrin were widely used as a soil treatment to kill neonate larvae falling to the soil surface. In recent years, organochlorines, carbamates, growth regulators, and on occasion even certain nematicides have been used against root weevil larvae and adults, but residual control has generally been short and unreliable.

A number of chemicals are recommended as foliar sprays to reduce adult populations in the tree during peak emergence from the soil. The purpose of adult suppression is to limit the number of gravid females and egg deposition and thereby reduce the number of larvae entering the soil. Foliar sprays should include a spreader-sticker adjuvant, such as medium oil, to improve coverage of new flushes. Even then, the residual activity of the chemicals begin to decline significantly after three to four weeks. Because of their short residual activity, two sprays may be necessary to maintain low adult populations during the peak emergence period, which can last 10 weeks.

Multiple pesticide applications, particularly during the summer, can interfere with the efficacy of natural enemies. Therefore, foliar sprays should be applied only when adults of the most important species are readily visible on the foliage or in ground traps.

Petroleum oil is widely used for both arthropod and disease control in Florida citrus. When sprayed on citrus foliage alone or in combination with another pesticide, medium petroleum oil appears to affect the bonding characteristic of the substance holding the leaves together around an egg mass. It alters the protection provided by the folded leaf and exposes the egg mass, thus increasing the natural mortality of the eggs through predation and desiccation.

Termites

Termites are small to medium-sized, white to yellow insects in the order Isoptera. They are similar to ants. They live in social groups and have a highly developed caste system. Some termites live in moist subterranean habitats, normally in wood buried beneath or in contact with the soil. A number of subterranean species are important to citrus, especially the eastern subterranean termite, *Reticulitermes flavipes,* in California and Florida. In the past, trees were routinely banked (that is, soil was piled to about 1 foot deep around the trunk) in the fall for cold protection, and any wood that was carelessly left in the soil served to attract termites. Termites often would colonize the soil around the tree and feed on the bark of the trunk, sometimes girdling the tree.

In Florida, pine-palmetto woodlands are a natural habitat of subterranean termites. When these areas were planted with citrus, the wood residue accumulated during clearing was often buried rather than burned, creating an ideal refuge for termite buildup and persistence. Under these circumstances, eastern subterranean termites have increased in numbers and can be found feeding on living bark of young trees both above and below the surface of the soil. They strip bark from the trunk in a ring between the soil line and the crown roots, frequently girdling the tree. Once the tree dies from girdling and the wood dries, termites return to consume the seasoned wood, which is a preferred food. Fresh bark and wood are hardly normal food for termites; however, the consumption of the wood source by a large population of termites may force them to eat fresh bark and wood to avoid starvation. The soilborne pathogen *Phytophthora nicotianae* often infects trees wounded by termites.

Management of termites in citrus groves should begin with the removal of wood residue, preferably before planting. Secondly, shallow planting is a desirable practice, to expose upper scaffold roots. After planting, tree wraps for cold protection should be avoided in infested areas. Infested groves should be regularly monitored for termites; wood bait blocks or wooden stakes should be used in monitoring. Soil should be removed to expose the base of injured trees. Granular chlorpyrifos applied to the surrounding soil at the rate of 2 ounces per tree will give about three months of protection. Cellulose baits treated with hexaflumuron have been used experimentally to eliminate colonies of *R. flavipes* from groves.

Ants

Ants are a large group of ubiquitous social insects in the family Formicidae. About 300 species have been collected from citrus groves throughout the world, many having a behavioral association with the soil. Ants may play an important role in the dynamics of citrus production and successful pest management in one or more of the following ways. Some species of ants are extremely disruptive to IPM. As they forage for food within the tree, they prevent parasitoids from engaging their host. They also protect honeydew producers, mainly homopterous insects, such as aphids, soft scales, mealybugs, and whiteflies, from their natural enemies. This is particularly true of the Argentine ant, *Iridomyrmex humilis,* and the native gray ant, *Formica cinerea,* in California. On the other hand, some ants are beneficial; species in some genera are strictly predaceous. Some ants are omnivorous, and their role in citriculture is often complex. Other species, such as leaf-cutting ants and fire ants, cause foliar damage or sting and bite humans so severely that they interfere with production and harvest. In Florida, the red imported fire ant, *Solenopsis invicta,* frequently builds nests in disturbed areas in proximity to citrus trees, often near surface irrigation lines or beneath wraps attached to the trunks of young trees for cold protection. This species often feeds so intensely on the bark and tender new leaf flush of young trees planted near ant mounds that the trees are killed, usually by trunk girdling. Severe ant infestations can cause significant tree loss in localized areas of a grove.

Although ants can be a localized pest by injuring trees, interfering with the natural enemies of honeydew-producing insects, or biting grove workers, their role as effective predators preying on other pests in the citrus ecosystem must not be overlooked. These factors must be carefully considered from an ecological perspective when chemical control in citrus groves is under consideration. For example, multiple applications of broad-spectrum chemicals are recommended for the control of fire ants on the soil in Florida. Such recommendations must carry a caution as to the effect of the treatment on nontarget pests and natural enemies. Growers can monitor their groves for honeydew-producing insects. Ants feeding on honeydew can be readily identified by examining the abdomen to see if it is swollen and translucent. Fire ants can also be identified by their sting.

In California, cultural controls such as skirt pruning of trees, removal of branches within 12–30 inches of the ground, and application of sticky materials to the trunk to prevent access to the tree are recommended for both the Argentine ant and the native gray ant. Sticky material can be used against the Fuller rose beetle, too. However, sticky barriers should be used with caution, because they can lead to sunburn under certain field conditions.

PART FIVE

Economics of Crop Production

Ronald P. Muraro
Stephen H. Futch
University of Florida
Citrus Research and Education Center, Lake Alfred

CHAPTER 15

Economic Decisions for Growers: How to Manage for Greater Efficiency

In citriculture, the 1990s have become known as the decade of the manager. As we enter an era of fluctuating prices, growers will have to work harder and manage more efficiently to maintain acceptable profit levels. Renewed emphasis on cost containment will be needed. Less attention will be devoted to maximizing output, and more will be focused on maximizing profits. Growers will need to approach purchasing, planting, and cultural practices with one eye on cost-efficient management decisions and the other on the bottom line of profit maximization. Therefore, the most important thing for growers is to know their groves and determine potential yield responses to changing inputs.

Changes occurring beyond the grove or orchard impact citrus investment decisions. Consumer issues are becoming more important. Concerns over food safety and the resulting "chemophobia" can suddenly shift demand for specific agricultural products. The availability of labor in the future is becoming a

As we enter an era of fluctuating prices, growers will have to work harder and manage more efficiently to maintain acceptable profit levels.

matter of concern. In addition, growers face increasing environmental constraints in the form of water use permits, water availability and quality issues, the availability of chemicals, and the impacts on production and profit of reduced chemical inputs.

This chapter will focus on management tools which can assist the citrus grower in making profitable economic decisions. The major topics will be 1) market allocation decisions (determining the production of fruit for the fresh fruit market or for the processed juice market) prior to the implementation of cultural and production practices, 2) budgets as a tool for planning grove operations and making capital investment decisions, and 3) examples of the use of budgets in production management decisions.

Market Allocation Decisions

Since every management decision impacts profits, each decision must be examined. While the citrus grower is faced with many management decisions about fruit production and marketing, a critical decision is the choice between the market for fresh fruit and the market for fruit to be processed as juice. A grower's decision to allocate fruit to the fresh or the processed market can substantially affect net returns, and therefore it is critical that the decision be both economically rational and based on objective information to the greatest extent possible.

The market allocation of citrus is determined at numerous decision times (Fig. 15.1). The first decision time is prior to the production season, when the grower must determine whether to produce the crop for the fresh market or the processed market. Expectations play an important role in this decision, with the outcome controllable to a certain extent. The final decision time is at harvest, when the fruit must be routed to either fresh or processed marketing channels, and it will largely be determined by cultural practices and environmental factors affecting the physical condition of the fruit.

When the first decision is being considered, a coordinated production and marketing plan must also be established. This is the time when the grower has the greatest flexibility, because a number of production practices can be controlled. If a grower decides to produce for the fresh market, certain inputs (for example, spraying and fertilizing) can be managed in an attempt to produce fruit with the desired characteristics. The cultural practices and production program necessary to prepare for the fresh market will cost more, but they are designed to produce fruit with the external qualities valued by fresh fruit buyers. In contrast, if a grower decides to produce fruit for processing, a less intensive cultural program is required, because exterior fruit characteristics are of little importance in the processed market.

The final decision is made at harvest, with an evaluation of alternatives based upon the outcome of the prior decisions. If the initial decision was to produce fruit for processing, in most cases the fruit of most varieties could not later be marketed as

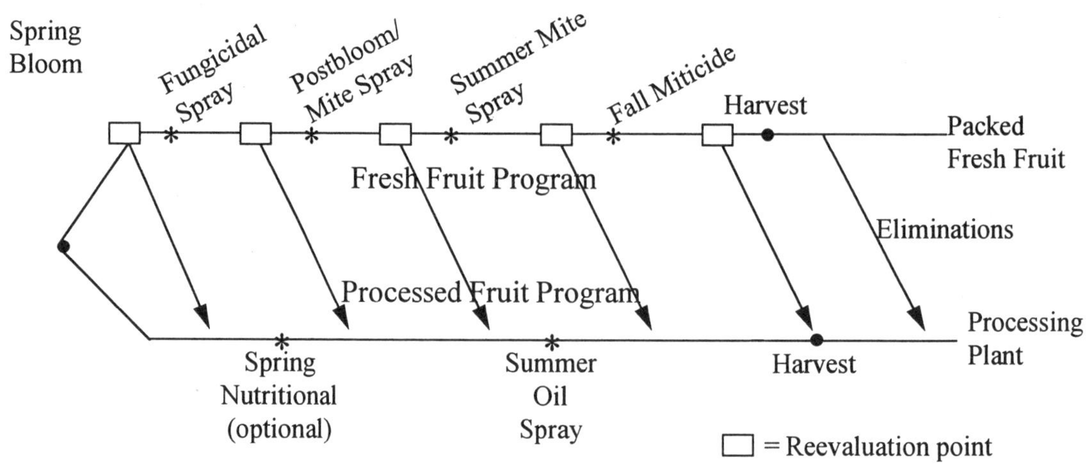

Fig. 15.1. Fruit marketing decision matrix for orange production.

fresh; it will go directly to processing. However, if the initial decision was to produce fruit for the fresh market, then a final decision of routing it to the fresh or processed marketing channels must be made. Thus, the decision at harvest is made under conditions of reduced uncertainty. At this time, the fruit characteristics, yield, prices, and costs are known, and the costs which have been incurred are not relevant to the decision. If an evaluation of the crop indicates excessive skin blemishing, which would result in a low percentage of fruit packed and sold to the fresh market, diverting the fruit to the processed market may provide a higher return to the producer. The objective now is to maximize returns, comparing fresh and processed returns.

Several types of information are needed to provide economic input into making the market allocation decision prior to the implementation of cultural practices. The following factors should be considered:

- The expected difference in the costs of cultural programs for fresh and processing fruit
- The expected price differential
- The expected yield (pounds of solids per box or per acre)
- The chances of getting fruit with the desired external characteristics (i.e., the likelihood of grade-lowering occurrences)
- The expected packout of fresh fruit, given the first and last factors listed above

The costs of a cultural program will depend on the market channel chosen, fresh or processed, and the specific needs of the citrus planting being managed as well as an evaluation of the program from fruit-set to harvest. Therefore, a grower must develop budgets, or a decision-making process, designed to maximize profit rather than maximize production.

Budgets: A Tool for Planning Grove Operations

An objective of any business should be to increase profits. In a citrus operation, higher profits can be attained through higher fruit prices, increased yields, more efficient management, reduced costs, or a combination of these factors. Fruit prices depend on supply and demand in the market, over which the grower has little or no control. The grower may have some control over fruit yield, but that is also determined by the weather, the availability of irrigation, and the age and condition of the trees. Thus, the areas over which a citrus grower has the most influence in increasing profits are improving management efficiency and decision making in an effort to minimize the cost per box produced.

As growers try to make those tough decisions to reduce costs, they need to review production practices. Through careful consideration, growers should choose production practices which will minimize or control costs without substantially reducing production potential in current and future years. Growers must remember that prices vary from year to year, not only in response to regional supply but also in response to the supply of citrus and other fruits worldwide.

The areas over which a citrus grower has the most influence in increasing profits are improving management efficiency and decision making.

Enterprise Budget

An enterprise budget, referred to here simply as a budget, is a management tool which can help a grower to improve the efficiency of the citrus operation. Used to its fullest, a budget will require the grower-manager to plan a management program thoroughly by the detailed analysis of his business operation. Developing a budget enables the grower to analyze past records (historical information and results), identify problem areas, and formulate business goals. The time frame of a budget may be monthly, annually, or longer.

Budgets are built around a list of planned production practices. Budget information helps a grower to plan fertilizer and nutritional programs, spray material purchases, and labor needs to help reduce unnecessary and costly work delays in the grove. Second, budgets project income and expenses from the operation, and these projections help the grower to plan seasonal cash flow. Third, budget preparation often uncovers forgotten costs, such as equipment depreciation, which is income tax deductible. And fourth, annual budgets can help the grower to plan credit needs. The format of a budget summary is shown in Box 15.1.

Cash Flow Budget

A cash flow budget is a basic management tool which, when used properly, is like a road map. If proper records are kept, this map will aid the grower in knowing where he has been and where he is going in the future. The budget should not be used in a fashion which gives the grower only one option, but it should get the grower to a given point in the future while at least showing other available options. This map will allow the grower to view and choose various options or routes to get to that point throughout the production year.

A beginning step in constructing a cash flow budget is to list the production practices in a month-by-month schedule. A schedule (road map) of production practices for a southwest Florida orange grove is shown in Table 15.1. There are two benefits to developing such a schedule for an individual grove. First, it shows what work is needed and when, so that operations can be planned well in advance. Second, it can be helpful if an annual cash flow analysis is developed to plan financing. The grower will benefit from developing a plan for a particular grove.

Important cultural and production practices must be carried out according to a schedule. For example, the timing of spraying directly influences success in minimizing or reducing pests and diseases that affect tree vigor, yield, and fruit quality. Building a schedule of production practices can improve the timeliness of most grove operations. Growers can categorize decisions they make by their importance. These decisions can then be organized according to when, how, and by whom they will be carried out and how much they will cost. For example, dead and diseased trees are usually removed from the grove and resets (young trees) planted and maintained between February and June, with the actual timing of removal being affected by harvesting or other factors. Equipment should be readied and labor arranged before February, so that the schedule can be met. Production practices listed on a schedule can easily be translated into budgeted costs.

Examples of estimated monthly cash flow budgets for oranges produced for the fresh fruit market and for the processed juice market are given in Tables 15.2 and 15.3, respectively. In both examples, the expected flow of revenue and anticipated monthly expenses are shown. Note that Valencia oranges are normally harvested between April and June, while Hamlin oranges are harvested between January and March. The cumulative monthly cash flows, in the bottom row of Tables 15.2 and 15.3, illustrate the usefulness of a projected cash budget for financial planning. The cash budget will indicate months in which insufficient funds will be available for paying expenditures and when funds will have to be obtained from a creditor or from savings. Also, a cash budget projection is beneficial when a grower is applying for a loan or line of credit, since it will indicate when funds will be available to make principal and interest payments.

A complete budget analysis encompasses an entire business operation. All production, income, and variable and fixed expenses associated with running a business are included in a complete budget, which would be used by a manager.

Box 15.1

Budget and Record Summary

I. Gross revenue (yield × price per box)		\$_____
II. Expense (cash and noncash costs)	–\$_____	
III. Management charge (expense) (percentage of gross sales)	–\$_____	
IV. Total specified expenses (costs) (II + III)		–\$_____
V. Return to land and trees (net profit) (I – IV)		\$_____

Partial Budget

A partial budget looks only at the expected costs of and returns from a proposed change in business practices. It is used when a proposed change will affect only part of a business. For example, Box 15.2 illustrates a partial budget to evaluate whether to install an irrigation system in a 50-acre orange grove. The budget analysis suggests that installing an irrigation system in the orange grove will be profitable on the average by \$9,230 per year—assuming the projections on added returns and costs are feasible. The investment cost of the irrigation system will be \$925 per acre, or \$46,250. However, in order to install the \$46,250 irrigation system, an additional \$30,000 in debt will be required. Will there be enough cash coming in to meet principal and interest payments (debt servicing) and also cover normal operating costs? To answer this question, an annual cash flow budget would be required.

Examples of Budgeting in Decision Making

Using a budget for a mature orange grove southwest Florida (Table 15.4), we will attempt to provide a road map which will present growers with some of the options and stimulate them to think about their travels into the future. Five major production categories will be highlighted, with some discussion of costs in each area: weed control, pest management, nutrition, pruning, and resetting. Other items may also affect production costs, but many of these may already be committed or required and are somewhat beyond manager's immediate control, such as property taxes and mortgage payments.

The examples presented in this section give estimated production costs in southwest Florida in 1996–1997. Growers should use estimated annual production costs for their own region or country to make cost comparisons.

Weed Control

In recent years, many groves have converted to low-volume under-tree irrigation systems. In a grove with one of these systems, it would not be feasible to omit a herbicide program, because of the growth of vegetation around poly tubing and emitters. The main option would be to review the cost of each weed control program and consider switching to a mix which will give acceptable weed control for a lower cost. Other op-

Table 15.1. Schedule of cultural and production practices and budget items for a mature, round orange grove in southwest Florida, 1996–1997[a,b]

	Jan.	Feb.	Mar.	Apr.	May	June	July	Aug.	Sept.	Oct.	Nov.	Dec.
Total revenue	Deposit (Hamlin)	Partial payment (Hamlin)	Deposit (Valencia)		Partial payment (Valencia)		Final payment (Hamlin)		Final payment (Valencia)			
Less: Pick and haul cost		×			×							
Grower advertising tax		×			×							
Grove expenses												
Disk												
Chop												
Mow (mechanical and chemical)		×			×	×		×			×	
Labor, general grove work, pull vines	×								×			
Herbicide (½ grove acre equivalent)[c]		×				×			×			
Spray												
Dormant												
Postbloom nutritional				×								
Supplemental miticide												
Summer oil–greasy spot							×					
Fall miticide										×		
Fertilizer												
Nitrogen (pounds per acre)	70			70			70					
Dolomite										×		
Hedging and topping			×									
Brush removal, mow brush			×									
Tree removal		×										
Young tree planting and care			×	×		×	×		×			
Microsprinkler irrigation (times per week)	1	1	2	3	3	3	2	2	2	2	1	1
Miscellaneous (clean ditches)		×										
Grove taxes, including water management											×	
Interest expense							×					
Annual principal payment on mortgage							×					

[a] Source: R. P. Muraro, F. M. Roka, and R. E. Rouse, 1997, Budgeting costs and returns: Southwest Florida citrus production, 1996–97, Univ. Fla., Inst. Food Agric. Sci., Food Res. Econ. Dep., Econ. Inf. Rep. 97-6.
[b] This is a suggested schedule of practices. Actual practices would not necessarily be done on the exact schedule shown here.
[c] The amount of herbicide materials per grove acre does not equal the amount per treated acre shown on the label. Only a strip or band is treated, and therefore it is assumed that only half of each grove acre is treated.

Table 15.2. Cash flow budget per acre for a mature Valencia orange grove in southwest Florida, producing for the fresh fruit market, 1996–1997[a]

	Jan.	Feb.	Mar.	Apr.	May	June	July	Aug.	Sept.	Oct.	Nov.	Dec.	Total
Revenue (433 boxes per acre)			$570.00		$1,420.24				$850.24				$2,840.48
Less: Harvesting costs					822.70								822.70
Grower advertising tax					125.57								125.57
Adjusted revenue			$570.00		$ 471.97				$850.24				$1,892.21
Expenses													
Mow		$ 7.37			$ 7.37	$ 11.67		$ 11.67			$ 7.37		45.45
Labor, general grove work	S 11.14								$ 11.14				22.28
Herbicide, material and application		48.45				45.86				$ 26.98			121.29
Spray													
Postbloom nutritional			$ 48.16										48.16
Supplemental postbloom miticide					63.44								63.44
Summer oil–greasy spot							$ 52.84						52.84
Fall miticide									53.27				53.27
Supplemental fall miticide											16.59		16.59
Fertilizer and lime		42.53			42.53				42.53		11.10		138.69
Hedging and topping			32.50										32.50
Brush removal, mow brush			13.68										13.68
Tree removal					18.12								18.12
Young tree planting and care			48.92	$ 6.97		6.97	6.97						69.83
Microsprinkler irrigation	5.59	5.59	11.18	16.77	16.77	16.77	11.18	11.18	11.18	11.18	5.59	$ 5.59	128.57
Miscellaneous (clean and maintain ditches, etc.)		32.61											32.61
Grove taxes											61.00		61.00
Interest on grove expenses							36.56						36.56
Annual mortgage payment (principal and interest)							530.00						530.00
Total expenditures	$ 16.73	$ 136.55	$154.44	$ 23.74	$ 148.23	$ 81.27	$ 637.55	$ 22.85	$118.12	$ 38.16	$ 101.65	$ 5.59	$1,484.88
Profit (loss)	$(16.73)	$(136.55)	$415.56	$(23.74)	$ 323.74	$(81.27)	$(637.55)	$ (22.85)	$732.12	$(38.16)	$(101.65)	$ (5.59)	$ 407.33
Cumulative cash flow	$(16.73)	$(153.28)	$262.28	$238.54	$ 562.28	$481.01	$(156.54)	$(179.39)	$552.73	$514.57	$412.92	$407.33	. . .

[a] See Table 15.1 for a schedule of cultural and production practices.

Table 15.3. Cash flow budget per acre for a mature Hamlin orange grove in southwest Florida, producing for the processed juice market, 1996–1997[a]

	Jan.	Feb.	Mar.	Apr.	May	June	July	Aug.	Sept.	Oct.	Nov.	Dec.	Total
Revenue (478 boxes per acre)	$540.00	$1,348.12					$808.03						$2,696.15
Less: Harvesting costs		906.03											906.03
Grower advertising tax		88.25											88.25
Adjusted revenue	$540.00	$ 353.84					$808.03						$1,701.87
Expenses													
Mow		$ 7.37			$ 7.37	$ 11.67		$ 11.67			$ 7.37		45.45
Labor, general grove work	$ 11.14								$ 11.14				22.28
Herbicide, material and application		48.85				45.86				$ 26.98			121.29
Spray													
Postbloom nutritional			$ 48.16										48.16
Summer oil–greasy spot							$ 52.84						52.84
Fertilizer and lime		42.53			42.53				42.53		11.10		138.69
Hedging and topping			32.50										32.50
Brush removal, mow brush			13.68										13.68
Tree removal		18.12											18.12
Young tree planting and care			48.92	$ 6.97		6.97	6.97						69.83
Microsprinkler irrigation	5.59	5.59	11.18	16.77	16.77	16.77	11.18	11.18	11.18	11.18	5.59	$ 5.59	128.57
Miscellaneous (clean and maintain ditches, etc.)		32.61											32.61
Grove taxes											61.00		61.00
Interest on grove expenses							36.56						36.56
Annual mortgage payment (principal and interest)							530.00						530.00
Total expenditures	$ 16.73	$ 154.67	$ 154.44	$ 23.74	$ 66.67	$ 81.27	$637.55	$ 22.85	$ 64.85	$ 38.16	$ 85.06	$ 5.59	$1,351.58
Profit (loss)	$523.27	$ 199.17	$(154.44)	$(23.74)	$(66.67)	$(81.27)	$170.48	$(22.85)	$(64.85)	$(38.16)	$(85.06)	$ (5.59)	$ 350.29
Cumulative cash flow	$523.27	$ 722.44	$ 568.00	$544.26	$477.59	$396.32	$566.80	$543.95	$479.10	$440.94	$355.88	$350.29	. . .

[a] See Table 15.1 for a schedule of cultural and production practices.

tions might be to reduce the width of the herbicide band, so long as this does not require additional trips through each row middle for weed control. To strive for 100% weed control is not necessary; an application may be delayed until some weeds have emerged but are not yet out of control.

Controlling weeds (mowing, applying herbicide, etc.) in a southwest Florida citrus grove could cost an average of $189.02 per acre, or 25.9% ($189.02/$731.10) of total production costs for fruit bound for the processed market (Table 15.4, column B). Some growers have considered switching to a nonresidual postemergence herbicide in an effort to reduce the cost of materials. However, this program would require more frequent application and may result in a higher annual cost of application. Growers would have to review their equipment requirements, availability or cost of contract equipment to see how this might work into their operations. The reduced cost of material may not offset the increased application cost.

Pest Management

The first step which the grower should take is to determine if a particular grove or block is to be marketed as fresh fruit or as processing fruit. The grower should choose the block or blocks which have the highest probability of making fresh fruit grade. If a block has never had a high packout or has had problems that will limit external fruit quality, then the grower should consider sending the fruit to the processing plant. By making this decision early in the production year, the grower might be able to decrease the number of spray applications by four sprays per year, thereby lowering pest management costs from $234.30 to $84.49 per acre, a reduction of $149.81 per acre (Table 15.5).

While the low-cost processing program described in Table 15.5 lists only one spray application for the production season, an additional spray could be required in some years if problems such as mite infestation or postbloom fruit drop are present at damaging levels. Also, if the grove is deficient in certain nutrients (i.e., zinc and manganese), these materials can be mixed with the summer oil spray, if necessary, and thereby the need for a postbloom nutritional spray may be eliminated in blocks where the fruit is intended for the processed market and its price is not dependent upon some external grade quality. The summer oil spray to control greasy spot and mites is the only spray required for processed juice fruit in most years. If greasy spot is left uncontrolled, the result will be fewer leaves the next season, followed by a reduction of future crops. This effect is illustrated in Figure 15.2; a grower, attempting to decrease costs, minimized the spray program in a Florida citrus grove, to the detriment of tree health, resulting in a five-year recovery period. If the grower had maintained a normal spray program in 1969 and 1970, the average yields should have remained at 600 boxes of Hamlin oranges and 400 boxes of

Box 15.2

A Partial Budget

The hypothetical budget presented below was drawn up to evaluate the costs of and returns from the proposed installation of a microsprinkler irrigation system for a 50-acre Valencia orange grove. All values are rounded.

Assumptions

- Yield will increase by 20% above the unirrigated yield of 360 boxes per acre.
- The on-tree price will be $4.37 per box.

A. Added income

1. Added returns		**2. Reduced costs**	
Revenue from irrigated orange grove (21,600 boxes × $4.37 per box)	$94,390	None	-0-
Total	$94,390	Total	-0-

Total added income (added returns + reduced costs) $94,390

B. Reduced income

1. Reduced returns		**2. Added costs**	
Revenue from unirrigated orange grove (18,000 boxes × $4.37 per box)	$78,660	Cost of irrigation ($130/acre × 50 acres)	$ 6,500
Total	$78,660	Total	$ 6,500

Total reduced income (reduced returns + added costs) $85,160

C. Estimated change in farm income per year (total added income – total reduced income) .... $ 9,230

The difference favors installing the irrigation system.

Table 15.4. Estimated production costs per acre in a mature orange grove in southwest Florida, 1996–1997[a]

	Column A: Low-cost processed fruit cultural program		**Column B: Reduced-cost fresh or processed fruit cultural program**		**Column C: Typical fresh fruit cultural program**	
Weed management and control						
Mechanical mowing of middles (three times per year)		$ 22.11		$ 22.11		$ 22.11
Chemical mowing of middles (three times per year)		23.34		23.34		23.34
General grove work (2 hours of labor per acre)		22.28		22.28		22.28
Herbicide (½ grove acre treated)[b]						
Application (three applications)	$ 24.48		$ 24.48		$ 24.48	
Material	75.34		96.81		96.81	
Total herbicide cost		99.82		121.29		121.29
Spray						
Postbloom						
Application (150 gallons per acre)			20.13		20.13	
Material			28.03		28.03	
Total postbloom cost				48.16		48.16
Summer oil						
Application (250 gallons per acre)	21.63		21.63		21.63	
Material	62.86		31.21		31.21	
Total summer oil cost		84.49		52.84		52.84
Supplemental fall miticide						
Aerial application					6.69	
Material (wettable sulfur)					9.90	
Total supplemental fall miticide cost						16.59
Fertilizer (bulk)						
Application (three applications)	20.79		20.79		20.79	
Material (15–2–15–2.4 MgO at 180, 204, and 225 pounds of N)	97.20		106.80		121.50	
Total fertilizer cost		117.99		127.59		142.29
Dolomite (1 ton applied every 3 years)						
Material and application		11.10		11.10		11.10
Pruning						
Topping ($43.75 per acre over 2.5 years)	17.50		17.50		17.50	
Remove brush from trees after topping ($21.70 per acre over 2.5 years)	8.68		8.68		8.68	
Hedging ($30.00 per acre over 2 years)	15.00		15.00		15.00	
Chop and mow brush after hedging ($10 per acre over 2 years)	5.00		5.00		5.00	
Total pruning cost		46.18		46.18		46.18
Tree replacement: three trees, 1–3 years old, per acre						
Remove four trees (pull and stack with front-end loader and burn)	18.12		18.12		18.12	
Prepare site and replant (includes three reset trees)	41.96		41.96		41.96	
Supplemental fertilizer, tree wraps maintenance, sprout, etc. (trees 1–3 years old)	34.84		34.84		34.84	
Total tree replacement cost		94.92		94.92		94.92
Irrigation						
Microsprinkler system	128.68		128.68		128.68	
Clean ditches (weed control)	11.28		11.28		11.28	
Ditch and canal maintenance	12.28		12.28		12.28	
Water control (pump water in and out of ditches and canals)	9.05		9.05		9.05	
Total irrigation cost		161.29		161.29		161.29
Irrigated processed fruit production costs		$683.52		$731.10		
Supplemental late postbloom spray						
Application (250 gallons per acre)			21.63		21.63	
Material			41.81		41.81	
Total supplemental postbloom spray				53.27		53.27
Early fall miticide spray						
Application (250 gallons per acre, aerial)			21.63		21.63	
Material			31.64		31.64	
Total fall miticide cost				53.27		53.27
Irrigated fresh fruit production costs				$847.81		$879.10

[a] Estimates are for a grove 10 years old or older.

[b] The amount of herbicide materials per grove acre does not equal the amount per treated acre shown on the label. Only a strip or band is treated, and therefore it is assumed that only half of each grove acre is treated.

Valencia oranges per acre, with a combined average increase in annual net returns of $550 per acre in 1969–1970. However, full recovery to the pre-1969 average yields did not occur until 1975. The annual potential yields during the recovery period are shown in Figure 15.2.

The remainder of this section will focus on spray options primarily available to growers who decide prior to the production season to grow fruit for the processed market. The goals of the production system must be evaluated, and the pest management program must be shaped to meet these goals. Simply stated, there are three main goals in the production of fruit for processing:

1. To protect the flushes, which are the driving force behind production
2. To maintain the yield potential of the block by protecting against fruit drop and reduction in fruit size
3. To accomplish the first two goals at minimum cost and with minimum inputs into the system

Table 15.5. Estimated costs for pest management of fresh and processed oranges, 1996–1997[a]

Spray	Fresh fruit program	Minimum processed fruit program
Postbloom (miticide + fungicide)	$ 48.16	$ 0.00
Late postbloom (miticide + fungicide)	63.44	0.00
Summer oil (greasy spot + miticide)	52.84	84.49
Early fall miticide	53.27	0.00
Supplemental fall miticide	16.59	0.00
Total	$234.30	$84.49

[a] See Figure 15.3.

In order to produce a profitable processing fruit crop, citrus growers must maintain yields and internal quality at the least possible expense. This does not necessarily mean less irrigation, nitrogen, or weed control, but it certainly may mean less control of diseases and arthropod pests. The important feature is maintaining a healthy tree which will then produce the desired yield and internal quality.

When the goals of protecting foliage and fruit yield are considered, it becomes clear that only a few pests are of recurring primary importance, namely greasy spot of summer flush, citrus rust mite, and spider mites. A high incidence of greasy spot or high mite populations can eventually reduce foliar growth and productivity and thus must be suppressed at times. While not generally as important, other pests are also managed on a case-by-case basis.

Depending on the pest or disease pressure and the crop involved, the management strategy is varied. Figure 15.3 portrays a range of treatment options, from a single summer treatment to the equivalent to a fresh fruit production program. Proceeding from left to right in this figure, the intensity of inputs and cost of the intervention increase. The fundamental distinction between this approach and a management program for fresh fruit production is that when fruit is grown for the processed market, *the grower begins with a minimal pest and disease control program, i.e., no spray, and builds the program as the need arises.* If no pest or disease problems develop, then there is no intervention, and the savings in pest management costs are considerable. Therefore, *the success of this approach is dependent on an effective pest-monitoring program.* As the season progresses, if the grower determines that a grove will produce a very low packout because of the size or shape of the fruit or rind blemishes, then the fresh fruit spray program may be discontinued, potentially eliminating several sprays and saving the associated cost of each spray.

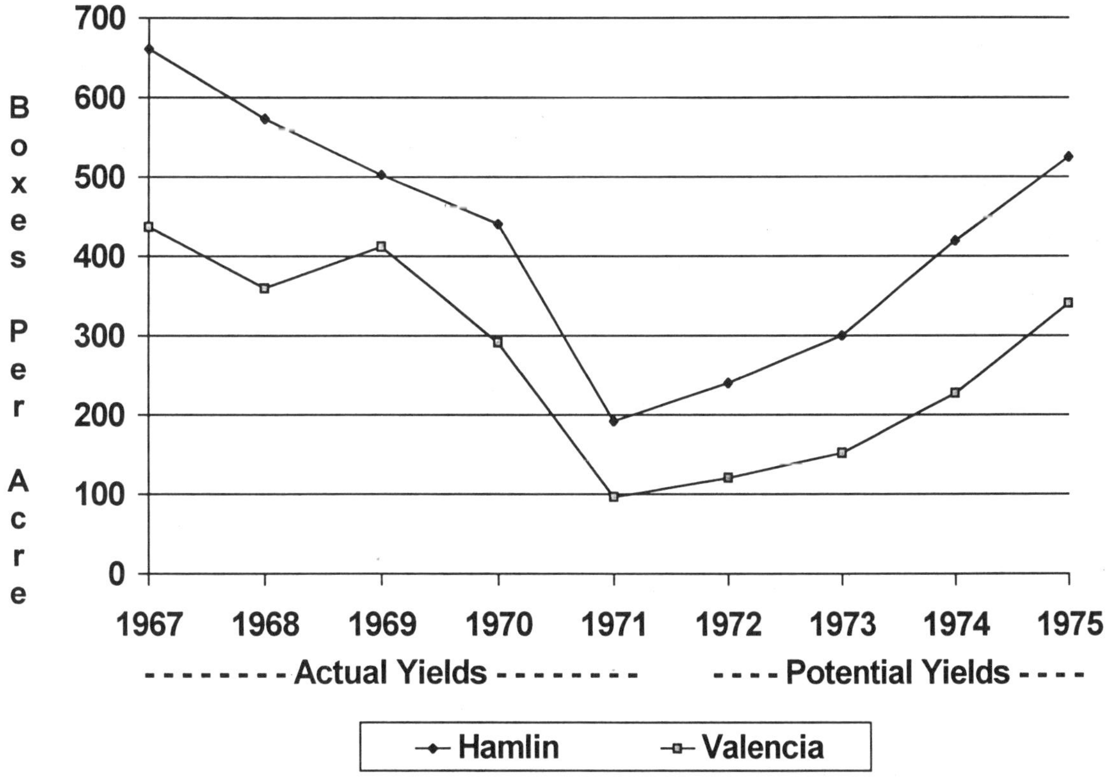

Fig. 15.2. Effects of minimizing control of greasy spot to decrease costs. Spraying to control the disease was reduced in 1969 and 1970.

Nutritional Program

Growers need to review their program to determine if the annual application of fertilizers is in the recommended range. Researchers at the Institute of Food and Agricultural Sciences of the University of Florida, have determined that a producing grove should receive fertilizer at rates based upon the grove's historical and future yield potential (Fig. 15.4). Wherever possible, fertilization rates should be based on the yield response of a citrus grove. For example, Table 15.6 summarizes the marginal yield response of a Hamlin orange planting with respect to nitrogen fertilization. In this experiment, trees were fertilized with nitrogen applied at different rates, ranging from 70 to 250 pounds per acre. In this grove, with its particular management program and its soil and site characteristics, yield (pounds of solids) increased with fertilization at rates of up to 190 pounds of nitrogen per acre. At rates between 190 and 220 pounds per acre, yields began to decline. Likewise, the net return per acre increased with fertilization at rates of up to 190 pounds of nitrogen per acre and then decreased with amounts between 190 and 200 pounds per acre. This experiment demonstrated that overfertilization with nitrogen can reduce both yields and returns.

The current thinking of many researchers is that most orange groves should be capable of producing excellent crops with not more than the 200 to 240 pounds of nitrogen per acre, and similarly for grapefruit groves with not more than 160 pounds per acre. Growers should consider all methods of application, including not only dry broadcast application but also fertigation and boom application with herbicides (see Chapter 6).

If a grower is currently applying nitrogen at a rate within the recommended range and does not exceed the upper limit, then greatly decreasing the fertilizer rate in an effort to reduce costs might result in lower yields over time. However, if yields are expected to be low in a certain year (because of postbloom fruit drop, heavy pruning, alternate bearing, frost damage, etc.), then the grower should consider reducing fertilizer in that year, but only to a level which will not affect normal vegetative growth. Also, the nutritional requirements of young, developing citrus trees differ from those of mature, productive trees (see Chapter 6), and nutritional programs should be designed accordingly.

Growers should also be aware that irrigation scheduling can affect the movement of nutrient elements through the soil. Irrigation should be scheduled in such a way as to apply only enough water to fill the root zone; excess water will move nutrients beyond the root zone.

Fertilizer should be applied in areas of greatest root concentration. Application of fertilizer in the row middles (with little fertilizer getting under the tree canopy, where the roots are in greatest concentration) may result in low utilization of the fertilizer and increased leaching.

One way to refine a nutritional program is to collect and analyze annual leaf samples which indicate the nutritional status of the grove and may suggest a modification of fertilizer inputs (see Chapter 6). For example, in 1996–1997 the typical annual cost of fertilizer in an orange grove in southwest Florida was $127.59 (Table 15.4, column B) or 17.5% ($127.59/$731.10) of the total annual production cost of processed fruit. If the grower has overfertilized, savings can result in balancing the trees' needs with the material applied.

Pruning

Hedging and topping should be done before crowding between rows or height becomes a problem, so that only small branches are cut and crop reduction is minimal. If a grower puts off pruning too long, the cost of the operation increases, as does the cost of removal of the prunings. Also, the larger the wood that is cut, the more rapid the regrowth will occur, and the regrowth tends to be vegetative. Once a hedging or topping program has been initiated, it is wise to stay on a schedule which maintains a desirable tree size. The average

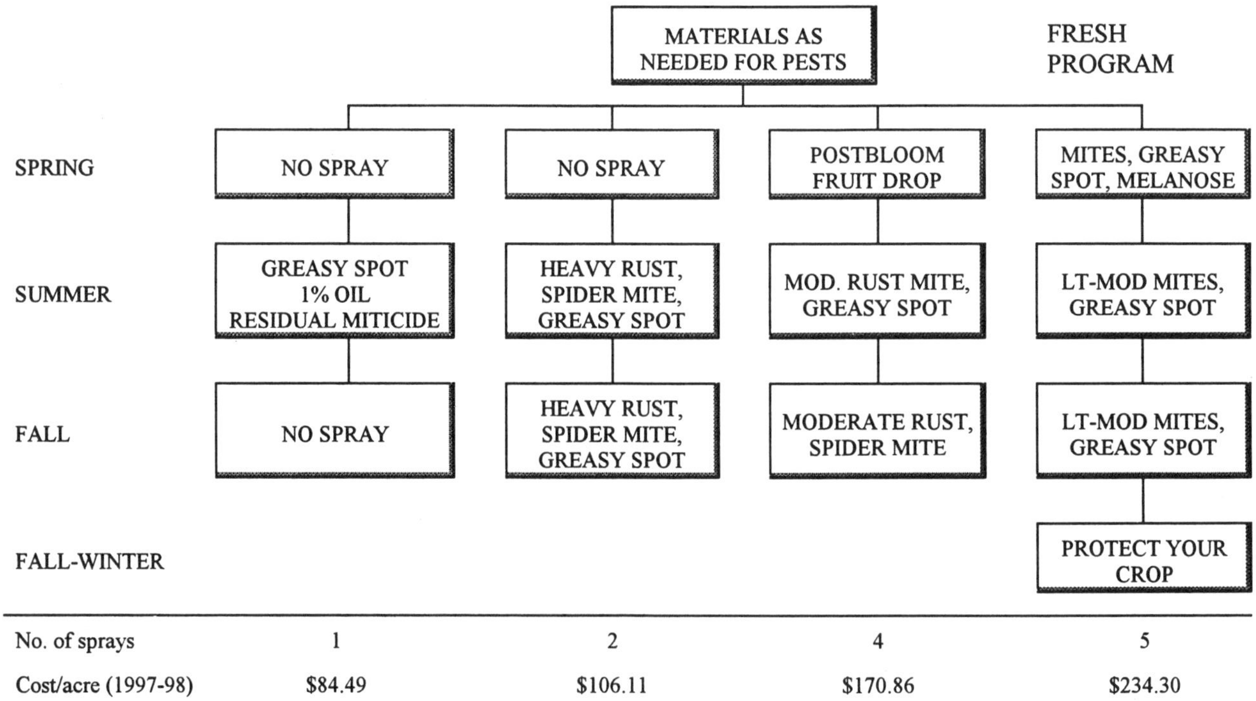

Fig. 15.3. Citrus pest management options for pesticide use. (Reprinted, by permission, from H. W. Browning and R. P. Muraro, 1993, Approaches to integrated pest management on citrus fruit destined for processing, Citrus Ind. 74(3):49–56)

annual cost of hedging, topping, and brush removal in orange groves in southwest Florida was $46.18 per acre in 1996–1997 (Table 15.4).

Resetting and Tree Replacement

Growers should maintain an active reset program to maintain long-term productivity. The longer a resetting program is omitted, the greater the impact on production. In most cases, most production practices cost the same whether a few sites are vacant or planted with resets. For example, the cost of mowing, irrigation, and herbicide application and taxes would all be about the same, regardless of the number of reset or vacant sites. A grower may look at blocks with a large number of trees to be reset and reset those blocks first. Generally, the greater the number of reset trees per acre, the lower the cost of maintaining each tree, because of economies of scale. If five trees per acre are reset, the cost per tree is lower than if only one tree per acre is reset, because of the travel time in getting to each tree (Table 15.7).

In most situations, replacing all citrus trees as they are removed will result in higher total production and returns, but a grower should evaluate total tree replacement in a high-density hedgerow planting. Depending upon the size and health of the adjacent remaining trees, the crowding effect may result in poor growth of trees. Therefore, in high-density hedgerow plantings, the impact of potential lower yields and returns should be considered with regard to the total cost of tree replacement.

Importance of Good Data on Production, Consumption, and Costs

As citrus producers increasingly have to compete in a global market, reliable cost and market information for decision making is of growing importance. To maintain a competitive advantage, citrus producers must understand the production, marketing, and distribution of products from other citrus-producing

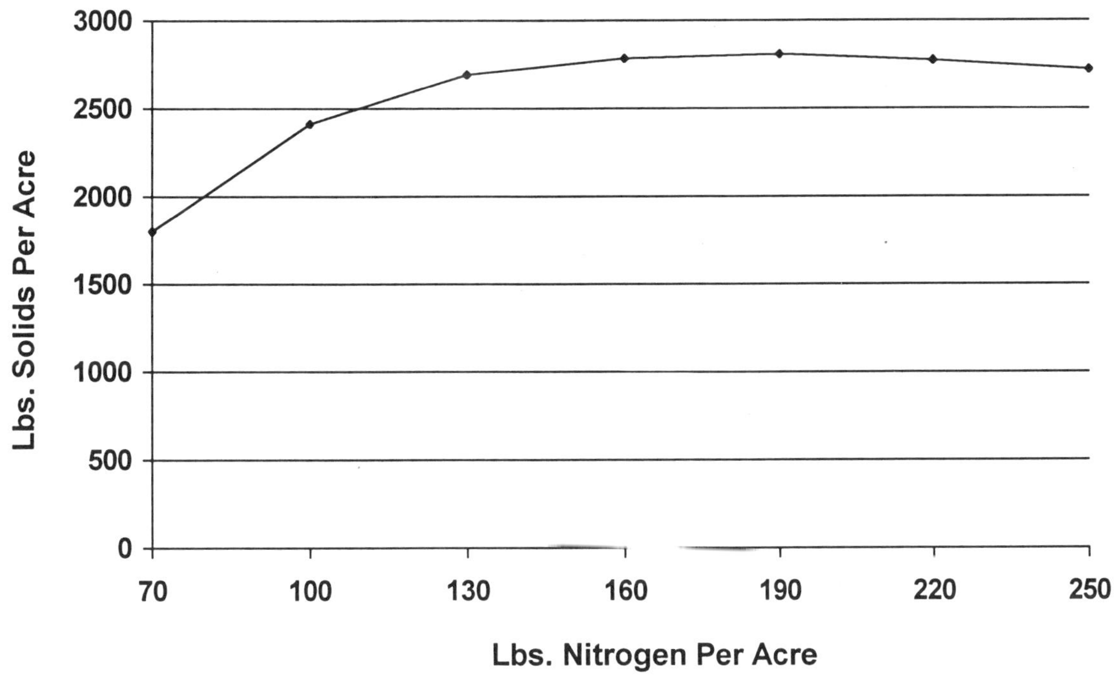

Fig. 15.4. Yield response of Hamlin orange (in pounds of solids per acre) in relation to nitrogen fertilization, 1972–1973 to 1976–1977. At the beginning of the experiment, the trees were 12 years old and planted at a density of 70 trees per acre. (Data from R. C. J. Koo, University of Florida, Citrus Research and Education Center, Lake Alfred)

Table 15.6. Marginal analysis of increasing nitrogen fertilization of Hamlin oranges[a]

Nitrogen applied (lb per acre)	Cost of additional nitrogen at $0.25/lb ($)	Total solid yield (lb per acre)	Additional yield from additional nitrogen (lb)	Value of additional yield at $0.85/lb ($)	Net return from additional nitrogen ($ per acre)
70	7.50	1,802	. . .	. . .	. . .
100[b]	7.50	2,412	610	518	510
130	7.50	2,693	281	239	232
160[b]	7.50	2,785	92	78	70
190	7.50	2,809	24	20	12
220[b]	7.50	2,775	–34	–29	–36
250	7.50	2,721	–54	–46	–54

[a] Source: J. A. Stricker and R. P. Muraro, 1981, Determining most profitable level of nitrogen fertilization for citrus, Proc. Fla. State Hortic. Soc. 94:8–11.

[b] Interpolated from original data.

Table 15.7. Estimated cost (per tree) of planting and maintaining reset citrus trees through the age of three years, 1996

	Number of resets or replacement trees per acre				
	1–2	3–5	6–10	11–25	26+
Year 1					
Tree removal (**A**)	$ 5.21	$ 4.53	$ 3.62	$ 2.93	$ 2.34
Tree cost (container tree)	4.00	4.00	3.75	3.75	3.75
Site preparation[a]	5.12	4.44	3.76	3.48	2.73
Planting tree and first watering	2.37	2.05	1.73	1.61	1.26
Total planting cost (**B**)	11.49	10.49	9.24	8.84	7.74
Supplemental fertilization (four times), application and materials	1.26	1.10	1.01	0.92	0.84
Supplemental spraying, application and materials	0.45	0.39	0.36	0.33	0.30
Spot herbicide, application and materials	0.21	0.18	0.16	0.14	0.13
Tree wrap (corrugated)	1.00	1.00	1.00	1.00	1.00
Sprouting and pruning	0.38	0.38	0.35	0.35	0.32
Miscellaneous	0.16	0.15	0.14	0.14	0.13
Supervision and overhead	0.26	0.24	0.23	0.22	0.20
Total tree care cost, year 1 (**C**)	3.72	3.44	3.25	3.10	2.92
Total cost, year 1 (**A + B + C**)	$20.42	$18.46	$16.11	$14.87	$13.00
Year 2					
Supplemental fertilization (three times), application and materials	$ 1.70	$ 1.51	$ 1.32	$ 1.12	$ 1.02
Supplemental spraying, application and materials	0.52	0.46	0.40	0.34	0.31
Spot herbicide, application and materials	0.20	0.18	0.16	0.14	0.13
Sprouting and pruning	0.45	0.45	0.38	0.38	0.34
Miscellaneous	0.14	0.13	0.11	0.10	0.09
Supervision and overhead	0.23	0.20	0.18	0.16	0.14
Total cost, year 2	$ 3.24	$ 2.93	2.55	2.24	2.03
Year 3					
Supplemental fertilization (three times), application and materials	$ 2.31	$ 2.06	$ 1.77	$ 1.51	$ 1.28
Miscellaneous	0.12	0.10	0.09	0.08	0.06
Supervision and overhead	0.18	0.16	0.14	0.12	0.10
Total cost, year 3	$ 2.61	$ 2.32	2.00	1.71	1.44
Total three-year cumulative cost	$26.27	$23.71	$20.66	$18.82	$16.47

[a] Site preparation for a bedded citrus grove, including the cost of root removal, cultivation of the planting site with a rotovator, and leveling the site. Fumigation of the planting site costs approximately $2.50 per tree.

regions. Countries with the best statistical information will be able to manage their citrus production more efficiently and enable investors to identify the most profitable citrus regions. The discussion that follows provides an example comparing two orange-producing regions.

Everyone who works with citrus often receives requests for information on production and costs in different citrus-producing countries. These requests are usually from individuals who want to analyze comparative production or cost advantages for citrus investment. With current market information and knowledge of seasonal production, investors can then target "market window" opportunities which one production region may provide over another, such as early fresh-market grapefruit in Cuba and Honduras or complementary seasonal production in Northern and Southern Hemisphere production regions.

After identifying production and consumption trends along with potential market window advantages, an investor should analyze the comparative cost advantage of producing citrus. However, the cost comparison should be of production units or management units, such as cost per carton packed. For example, Table 15.8 shows cultural costs for oranges in two citrus-producing regions in 1995–1996. Total cultural costs in California ($2.29 per carton) were over 75% greater than costs in Argentina ($1.31 per carton). However, knowing the cultural

Table 15.8. Estimated freight-on-board (F.O.B.) cost per carton of fresh export oranges in California and the northeast region of Argentina, 1995–1996[a]

	California[b]	Argentina[c]
Cultural costs	$ 2.29	$1.31
Administrative costs	0.76	0.56
Harvesting costs	1.07	0.87
Packing and marketing costs	3.11	2.70
Freight to port, cold storage, harbor charges, and taxes	0.49	1.14
Total F.O.B. cost at port or harbor	$ 7.72	$6.60
Ocean freight cost	2.99	2.42
Total F.O.B. cost at destination	$10.71	$9.02

[a] Sources: U.S. data from California citrus growers and University of California Cooperative Extension Service; Argentinean data from Estación Experimental Agropecuaria, INTA, Concordia, Entre Ríos, Argentina. Figures are in U.S. dollars

[b] Costs are estimated for an irrigated cultural program for fresh fruit production, with an average yield of 12.7 tons per acre, producing 540 packed cartons per acre (a carton weighs 37.5 pounds), with 85% of total production packed for shipment to Japan.

[c] Costs are estimated for an unirrigated cultural program for fresh fruit production, with an average yield of 13.9 tons per acre, producing 483 packed cartons per acre (a carton weighs 37.5 pounds), with 65% of total production packed for shipment to Europe.

costs is only the first step. Other factors which may affect the total cost of producing and marketing citrus products must be considered when citrus production in different regions is evaluated, including packaging and handling of fresh fruit, juice processing and manufacturing, domestic taxes and regulations, transportation to markets, and import fees and tariffs.

For example, Table 15.8 compares the estimated 1995–1996 freight-on-board (F.O.B.) costs of fresh export oranges from California and the northeast region of Argentina, which include most of the other cost factors listed above. The California costs are for oranges exported to Japan or other East Asian countries, whereas the Argentinean costs are for oranges exported to Europe. Although the total F.O.B. cost per carton at the port or harbor are 17% higher in California than in Argentina, the individual cost items vary. Lower labor costs are the primary reason that cultural, harvesting, and packing costs are lower in Argentina. However, because of taxes on salaries, interest charges, infrastructure (lack of superhighways), and distance from ports, administrative costs and freight-to-port costs are higher in Argentina than California.

Since the F.O.B. cost at destination of Argentinean oranges is 19% less than that of California oranges, at first observation fresh orange production in Argentina would appear to be favored over production in California. However, each region has a production and marketing advantage. California's growing season and its proximity to the Asia allows California to export over 17 million cartons of fresh oranges to Asian markets. Argentina has a production and marketing advantage for export to Europe, because the Southern Hemisphere production season complements citrus production in the Mediterranean countries, and because of the competitive cost of shipping fresh packed oranges to Europe. Furthermore, because of the higher cost of ocean freight from Argentina to Asia, the total cost of Argentinean oranges at destination in Asia is estimated to be $1.33 per carton more than shipments from California. This cost advantage allows California to be competitive in shipping summer Valencia oranges to Asia. Argentina has a similar competitive advantage over California in shipping late-maturing orange varieties to Europe.

Conclusion

This chapter focuses on several management tools which can assist the citrus grower in making profitable economic decisions: market allocation decisions and decisions about grove operations and capital investments based on budgets. The examples of budgeting for pest management, nutrition, weed control, tree replacement, and pruning illustrate how to analyze the decision-making process in a citrus production program. This discussion should be helpful in several of the main areas of production management and provide growers with some ideas for a budget road map with which they will move into the future with some profitability.

PART SIX

Postharvest Considerations

G. Eldon Brown
Florida Department of Citrus
Citrus Research and Education Center, Lake Alfred

William R. Miller
U.S. Department of Agriculture, Agricultural Research Service
Horticultural Research Laboratory, Orlando, Florida

CHAPTER 16

Maintaining Fruit Health After Harvest

Maintaining the health of fresh citrus during postharvest preparation, storage, marketing, and utilization is essential to the success of the producer and packer-shipper and the satisfaction of the consumer. Losses during the postharvest period involve not only loss of investment in production, harvesting, packing, marketing, and transportation but also loss of consumer confidence and repeat sales.

Numerous pathological and physiological disorders, pests, and physical damage can cause significant losses during the marketing of fresh citrus. In this chapter, we describe practical methods of treatment and handling that will reduce postharvest losses to a minimal level, so that fresh citrus can move through the produce marketing chain in an effective and profitable manner.

Postharvest Disorders That Degrade Fruit Quality

Pathological Disorders

The major pathogens causing postharvest losses of citrus are listed in Table 16.1. Postharvest diseases can be separated into two categories, based on the time of infection. Some diseases develop from preharvest infection of immature fruit but are manifested only after a lengthy quiescent or inactive period. These diseases include stem-end rot (caused by *Diplodia natalensis, Phomopsis citri,* or *Alternaria citri*) and anthracnose (caused by *Colletotrichum gloeosporioides*). Other diseases are manifested shortly after infection, including green mold (caused by *Penicillium digitatum*) and sour rot (caused by *Geotrichum candidum*), which develop only at sites of injuries, and brown rot (caused by species of *Phytophthora*).

Physiological Disorders

Major peel disorders (Table 16.2) that occur shortly after packing or after storage are rind staining, stem-end rind breakdown, blossom-end clearing, zebra skin, oleocellosis (rind oil spot), and pitting. Creasing is a physiological disorder that develops before harvest but often does not become apparent until the fruit is harvested. Section drying and seed germination are internal conditions most commonly found in late-season grapefruit.

Insect Pests

Several species of fruit flies, in several genera, are endemic to certain citrus production regions and can cause economic postharvest losses by laying eggs beneath the epidermis of ripening fruit. When the eggs hatch, the larvae burrow into the rind and flesh of the harvested fruit, rendering it unfit for consumption.

Table 16.1. Major postharvest diseases of citrus fruit

Disease	Causal fungus	Type of infection	Site of infection
Preharvest infection			
Stem-end rot	*Diplodia natalensis*	Quiescent	Necrotic tissue (button)
Stem-end rot	*Phomopsis citri*	Quiescent	Necrotic tissue (button)
Stem-end rot (black rot)	*Alternaria citri*	Quiescent	Necrotic tissue (button)
Anthracnose	*Colletotrichum gloeosporioides*	Quiescent	Intact or injured rind
Brown rot	*Phytophthora* species	Active	Intact rind
Postharvest infection			
Green mold	*Penicillium digitatum*	Active	Injured rind
Blue mold	*Penicillium italicum*	Active	Injured rind
Sour rot	*Geotrichum candidum*	Active	Injured rind

Table 16.2. Major physiological disorders of citrus fruit

Disorder	Symptoms
Preharvest disorders	
Aging	Wrinkling, darkening, and collapse of the rind around the stem end of the fruit
Creasing	Irregular, collapsed grooves in the rind surface due to separations in the albedo
Water spot	Water soaking and swelling of the rind around the stylar end when water is absorbed through cracks in the cuticle, a disorder mostly of navel oranges under wet conditions
Postharvest disorders	
Blossom-end clearing	Water soaking, typically at the stylar end of the fruit, from juice released from ruptured juice vesicles
Chilling injury	Scattered brownish, sunken pits in the rind, particularly affecting grapefruit
Oleocellosis	Discoloration and collapse of epidermal cells surrounding injured oil glands
Pitting	Collapse of oil glands in clusters in small circular regions of the rind of waxed fruit, particularly white grapefruit
Puffiness	Separation of the rind from the pulp, sometimes occurring before harvest, almost exclusively affecting mandarins
Rind staining	Brownish discoloration of the rind surface
Section drying	Granulation and collapse of juice vesicles, commonly at the stem end of the fruit, occasionally occurring before harvest
Stem-end rind breakdown	Darkening and collapse of the rind around the stem end of the fruit, usually with a narrow ring of normal rind remaining near the button
Zebra skin	Striped pattern of darkened rind over the highest points of each fruit segment

Preharvest Practices to Maintain Fruit Health

Cultural Practices

Some cultural practices in the production of citrus may influence the incidence of certain diseases after harvest. The main effect of these practices is to lower inoculum production and thereby reduce the probability of infection.

The incidence of brown rot can be reduced by improving ventilation under the tree canopy to shorten the duration of high relative humidity and wetness during periods of heavy rainfall and cloudy weather. This is accomplished by pruning low-hanging branches and mowing or removing vegetation. Fruit close to the soil is more likely to become infected and subsequently involved in the spread of disease to healthy fruit higher in the tree canopy.

Proper irrigation, soil drainage management, and planting on well-drained soils also tend to prevent conditions conducive to inoculum production and subsequent disease. Good cultural practices that minimize the amount of deadwood in trees and removal of deadwood by pruning can aid in the reduction of Diplodia and Phomopsis stem-end rot and anthracnose. The fungi causing these diseases colonize dead limbs and twigs, and their spores are carried by irrigation and rainwater to the surface of immature fruit. Stem-end rot is usually less prevalent in healthy trees 10 years of age or younger, because they usually lack deadwood.

Observable reductions in green mold have been achieved by keeping the orchard floor free of fallen fruit, in which there is a high incidence of infection. The pathogen sporulates on infected fruit, and the airborne spores contaminate fruit in the tree canopy.

Certain growth regulators and fungicides reduce the incidence of some postharvest problems. Gibberellic acid sprayed on full-sized green fruit significantly reduces rind staining of California navel oranges, creasing and water spot of Australian oranges, and rind aging and natural color development of Florida seedless grapefruit, navel orange, and Minneola tangelo. Gibberellic acid may provide some control of sour rot in California lemons, because it decreases rind senescence.

2,4-D applied to control fruit drop may also reduce stem-end rot. Copper sprays to control melanose reduce Phomopsis stem-end rot. Copper or fosetyl-Al may be applied to fully expanded fruit to control brown rot. Applications of benomyl in the summer for greasy spot control also reduce the incidence of stem-end rot. However, an application of benomyl during the three weeks to the day before harvest is much more effective and also controls green mold during inadvertent delays after harvest or during degreening. Any field use of fungicides other than those for brown rot control must be followed with postharvest fungicide treatments to obtain commercially acceptable decay control.

Quarantine

Quarantine can be established to certify harvests from pest-free areas if the fruit is not subject to reinfestation during handling and transportation. An example of such a procedure is the harvest of Florida grapefruit certified free of the Caribbean fruit fly for export to Japan. Most Florida grapefruit shipped under quarantine is harvested from certified fly-free groves. A grove must be at least 40 acres to qualify for certification. In general, a 300-foot-wide buffer zone on the perimeter of the grove must be free of any preferred host plants of the pest. Fly traps must be set in the grove and buffer area (15 traps per square mile), and a 30-day period in which no fruit flies are trapped must pass before the grove is eligible for certification. Bait spray applied at seven- to 10-day intervals is optional but strongly recommended. The advantage of the fly-free protocol is that fruit from certified fly-free groves does not require postharvest quarantine treatments, which may be deleterious to fruit quality.

Harvesting Practices

Hand-harvesting is one of the most important factors affecting the postharvest health of citrus fruit. Injuries that occur in harvesting and transporting fruit to the packinghouse are sites of entry for the pathogens that cause green mold, blue mold, and sour rot. The probability of infection and decay is directly related to the number of injuries. Careful harvesting to minimize injuries has a significant impact on decay. Fruit should not be allowed to contact the soil, because soil particles cause abrasive injuries during fruit handling and harbor inoculum of *G. candidum*.

Fruit should be snapped or clipped off the trees properly to minimize plugging (tearing of a portion of the rind from the fruit at the stem end) and to avoid leaving long or short angled cut stems, which may puncture contiguous fruit in bags and pallet boxes. The correct technique for snap picking is to grasp the fruit with the fingers, lift it upward until it is in a horizontal position, and then break off the stem in a forward snapping motion of the wrist or by pulling vertically downward. Fruit is properly clipped with a 90° cut made across the stem as close to the button (calyx and disk) as possible. Softened or dehydrated fruit is more often plugged during picking than turgid fruit. Under dry conditions, trees can be irrigated prior to picking, but caution must be taken not to induce excessive fruit turgidity and susceptibility to oleocellosis.

The time of day when harvesting operations are conducted and the amount of time following irrigation or heavy rainfall before harvest can influence decay and peel injuries. Fruit harvested in the morning, with higher rind moisture, is more prone to sour rot than fruit harvested later in the day, when the rind is less turgid. However, blossom-end clearing of mature grapefruit is promoted by harvesting during the middle of hot days and by rough handling. Oleocellosis and zebra skin are more prevalent in fruit that is highly turgid as a result of irrigation or rainfall. Delaying the harvest for a few days after a significant rainfall or harvesting trees in the highest, driest, and most open area of the orchard will help reduce the incidence of these disorders.

To prevent oleocellosis, a measure of turgidity can be obtained from a pressure gauge that estimates the amount of force in pounds required to break oil glands of the fruit rind. Turgidity can be tested by placing a piece of tissue paper against the fruit, applying pressure until rind oil is exuded, and observing when the paper is wetted. A meaningful estimate for lemons tested with a ⅜-inch gauge cylindrical tip is that harvest should not be attempted at pressures of less than 3 pounds but can be conducted with extreme caution at pressures of 3 to 7 pounds and conducted normally at pressures of more than 7 pounds. Similar guidelines may apply to other cultivars, but each situation should be initially evaluated to determine the desired level of turgidity at harvest.

Degreening, a procedure to remove chlorophyll from the fruit rind to expose the orange and yellow pigments that enhance their appeal to consumers, significantly increases the occurrence of Diplodia stem-end rot and anthracnose. For early-season harvests that require extensive degreening, the trees that are harvested should have minimal deadwood, because they are less likely to contain inoculum of the stem-end rot pathogen. Harvesting fruit with better natural color, either by spot-picking for better color or delaying harvest to allow natural color development to shorten the time of degreening, will reduce the incidence of these two diseases.

Harvest can be delayed at least two weeks after infection with brown rot, in the absence of reinfection, to allow infected fruit to fall to the ground. Harvesting poorly drained groves during rainy periods should be avoided, and fruit from limbs in the lower part of the canopy, where the probability of fruit infection is greatest, should not be harvested.

When tangerines or late-season fruit, particularly grapefruit, are harvested, the internal condition of the fruit should be closely monitored for drying and seed germination.

Fruit should be carefully handled when it is placed in the picking bag, and the bag must be no larger than a picker can handle conveniently without tiring. If not, excessive injury of the fruit may occur from rough handling in picking and in dumping the fruit into a larger container for transport from the field to the packinghouse. These containers are usually boxes with a pallet base, holding enough fruit to require mechanical handling. The containers should be free of sand and should not have abrasive surfaces. The newer plastic pallet boxes are often smoother than wooden ones and tend to be less abrasive to the fruit. Containers should not be overfilled; losses due to physical damage to fruit in overfilled pallet boxes far exceed any benefits of overfilling. Pallet boxes do not require holes in the sides, but approximately 10% of the bottom surface should be open, usually with slots, to provide adequate ventilation during degreening or storage, when they may be stacked as many as six high.

Containers of harvested fruit should be removed carefully and immediately from the field to minimize physical injuries and damage due to heat or sunburn. Respiration and subsequent off-flavors can be significantly increased by physical damage and high temperature.

Packinghouse Operations

Proper design and operation of the packinghouse line are essential to the maintenance of fruit health. Citrus fruit is comparatively resistant to bruising, because of its spongy albedo, which absorbs much of the energy of impact. However, the fruit surface is very prone to scrapes, cuts, and punctures from sharp or abrasive objects. Mechanical damage most commonly occurs when fruit is transferred between pieces of equipment or at corners of the packing line. Properly designed packinghouse lines with limited turns will minimize handling damage to the fruit.

Packing and handling operations in the flow of citrus fruit through a packinghouse are listed in Figure 16.1. Procedures

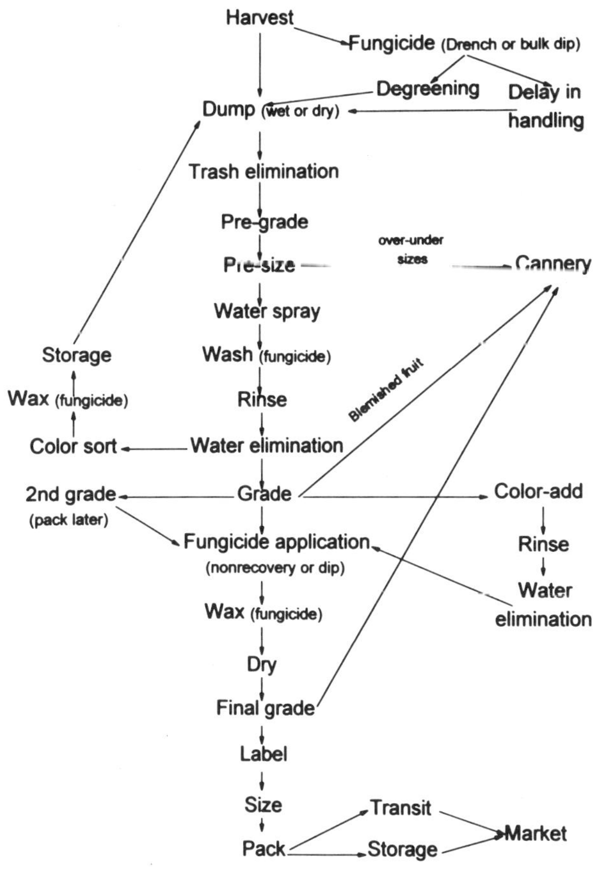

Fig. 16.1. Flow chart illustrating the normal steps in the handling of citrus fruit during packing. (Courtesy G. E. Brown)

may vary somewhat depending upon the type of fruit and the geographical location.

If fungicide treatment and packing are delayed more than 24 hours after harvesting (for example, because of degreening), the fruit should be treated with a fungicide drench or dip to protect against decay. Fruit may be dumped dry on roller or belt conveyors or into a dip tank with chlorinated water or fungicide. California citrus fruits, particularly lemons, are often dumped into a soda ash solution, washed, color-sorted, and then treated with a storage wax, which often contains a fungicide. The fruit is placed in storage for several months for color development before it is processed again for shipment (Fig. 16.1).

After the fruit is dumped, it receives preliminary grading and sizing, and wholesome fruit that is excluded at this stage is normally utilized in juice products. A fungicide is often applied during washing. The fruit is rinsed, and excess water is removed before major grading. Additional fruit unacceptable for the fresh market is then removed for processing into juice and by-products. The remaining fruit destined for packing may receive a color-add treatment, as early-season Florida fruit do, before fungicide application. The fungicide may be applied separately, before waxing, or combined with wax. If the fungicide is applied separately, the fruit may be dried again before the wax treatment. Fungicide- and wax-treated fruit is often subject to a final grading, and some may be removed before packing. Once packed, the fruit is transported to market or held in storage before marketing (Fig. 16.1).

More detailed discussion of many of these processes is presented in the remainder of this chapter.

Dip and Drench Treatments

The efficacy of postharvest fungicide treatments diminishes if the application is delayed more than 24 hours after harvest. For example, the efficacy of imazalil in controlling Diplodia stem-end rot in degreened oranges is significantly reduced when application is delayed by prolonged degreening (Fig. 16.2). After 24 hours, fungal pathogens have often reached a point of ingress into the fruit and can no longer be effectively controlled. The timing of fungicide application after harvest is therefore as important as the fungicide concentration. Delays in processing fruit on the packing line are commonly the result of logistical or marketing problems or, most often, because of ethylene degreening to improve rind color before processing.

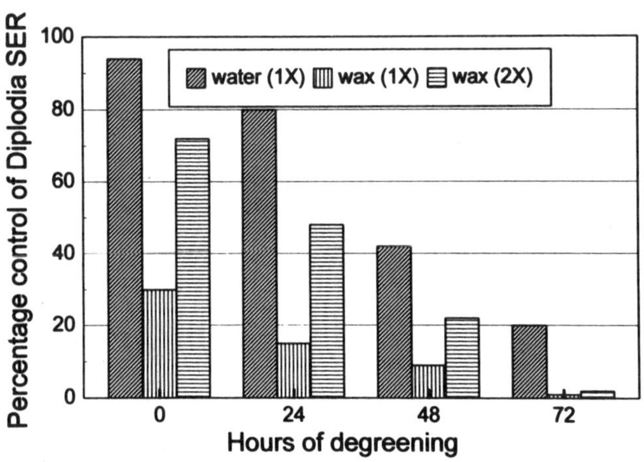

Fig. 16.2. Effect of degreening time, imazalil concentration (1× = 1,000 parts per million; 2× = 2,000 parts per million), and application of imazalil in water or resin solution water wax on the efficacy of control of Diplodia stem-end rot (SER) in degreened oranges. (Courtesy G. E. Brown)

When delays occur, the fruit in pallet boxes should be treated with a fungicide to prevent excessive decay, primarily sour rot, green and blue mold, and Diplodia stem-end rot. Maximum coverage of fruit is obtained when pallets are bulk-dipped in a fungicide suspension for 30 seconds, but effective coverage and decay control can also be achieved in properly

The efficacy of postharvest fungicide treatments diminishes if the application is delayed more than 24 hours after harvest.

operated drenching systems. Foaming of the suspension must be controlled with antifoaming agents, to prevent encapsulation of the fungicide by foam and excessive removal of fungicide on the fruit, which would dilute the concentration. Fungicide mixes should be constantly agitated after initial preparation to prevent wettable powders from settling out or materials in solution from becoming adsorbed by soil on the bottom of the treating tank.

Recovery drench applications can be applied to as many as 50 pallet boxes of fruit stacked two or three layers high on parked trailer trucks (Plate 60). The bottoms of the pallet boxes must be perforated or slotted to allow adequate drainage of the liquid fungicide preparation through the stacks. Optimum coverage can be achieved by flooding large volumes of liquid over the entire trailer load for 5 minutes to thoroughly wet the fruit throughout the pallets. Systems may contain as much as 1,000–1,600 gallons of liquid material, which is pumped through spray nozzles over the trailer load by two pumps with capacities of at least 450 gallons/minute. The liquid is distributed to the nozzles through a PVC pipe 4 inches in diameter. Fixed or rolling conveyor screens should be utilized to remove large pieces of debris from the recirculated liquid and prevent clogging of the spray nozzles. Smaller particles are usually macerated by recirculating pumps, which continually agitate the treatment solution to prevent settling. The orifices of the spray nozzles should be at least 0.5 inch in diameter to minimize clogging. Fruit in the pallet boxes in a stack should be inspected immediately after treatment in the initial evaluation of the drench application to confirm adequate coverage. About 1–1.5 gallons of liquid will be removed by a pallet box that contains approximately 850 pounds of fruit. The concentration of fungicide in the drench tank should be monitored, and the desired strength should be maintained by adding additional fungicide.

After treatment, another 5-minute period is required for excess liquid to drain back into the main holding tank. A timing system with red and green lights to indicate truck movement to the driver can be utilized to ensure adequate treatment exposure and drainage time. The old treatment mixture must be removed and discarded periodically, and the tank should be cleaned to eliminate accumulated dirt and debris before new material is added. To avoid discarding the entire contents of a full tank, the recharge add-back can be discontinued until the amount of liquid remaining in the system is insufficient for treatment. The waste material remaining in the drench system must then be discarded in a environmentally safe manner, by certified disposal methods.

Chlorine must constantly be added to the dip or drench if the treatment material does not control common contaminating organisms, such as *Geotrichum, Phytophthora,* and fungicide-resistant strains of *Penicillium. Geotrichum* is the major decay pathogen contaminating treatment material, because it is present in soil that accumulates in the tank during treatment. Chlorine should not be added to imazalil, because of incompatibility, and heat sterilization may be necessary to keep an imazalil drench solution relatively sanitary. Chlorine may not be needed with guazatine, which is active against *Geotrichum.*

Chlorine can be added in either gaseous or liquid form. Care is required in handling chlorine; in liquid form, it is caustic and can harm the eyes and skin in the event of accidental contact, and the gaseous form is toxic. Sodium hypochlorite, the form of liquid chlorine normally used, is buffered at a high pH to maintain its stability. In fresh solutions the available chlorine content can be as much as 15% by weight, but the concentration drops to half this level after three to four months of storage. It should be stored in a cool, dark area. A suitable acid, such as muriatic, must be metered into the fungicide mixture to counteract the alkaline hypochlorite and maintain the pH near neutral.

Gaseous chlorine hydrolyzes in water to form hypochlorous acid and hydrochloric acid, and the hypochlorous acid then partly dissociates to yield hydrogen and hypochlorite ions. The three forms of available chlorine resulting from these reactions—molecular chlorine, hypochlorous acid, and the hypochlorite ion—exist together in equilibrium. Their relative proportions are determined primarily by pH (Fig. 16.3). The proportions are the same whether the chlorine is introduced as chlorine gas or as sodium hypochlorite. Hypochlorous acid is the primary active sporicidal component. The addition of chlorine gas to water lowers the pH, because of the formation of acids, and therefore an alkali, such as caustic soda, must be added to raise the pH to an optimum of 6.5–7.5. In this range, hypochlorous acid predominates. Chlorine gas is evolved from solution at low pH, especially below pH 4, and the odor can

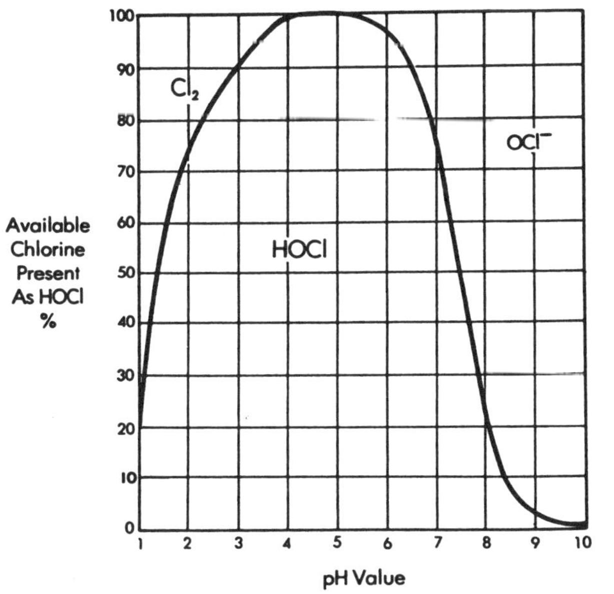

Fig. 16.3. Effect of pH on the forms of free available chlorine in water: molecular chlorine (Cl_2), hypochlorous acid (HOCl), and hypochlorite ion (OCl^-). (Reprinted, by permission, from A. T. Palin, 1983, Chemistry and Control of Modern Chlorination, 2nd ed., LaMotte Chemical Products Co., Chestertown, Md.)

cause irritation in workers. Above pH 7.5, more of the chlorine is in form of the hypochlorite ion, which is a weak sporicide. Chlorine which is available in water in any of these three forms is known as free available or unreacted chlorine.

In addition to the odor generated by the loss of chlorine gas at low pH, the odor of chloramines is produced when chlorine reacts with the ammonium ion or amino groups to form di- or trichloramines, which bubble from the water. A strong chloramine odor may indicate that a large amount of organic matter has been introduced into the drench or that the makeup water contains an excessive amounts of ammonium. It may also indicate that not enough chlorine is being added to the water to maintain the presence of unreacted chlorine.

Free available chlorine should be maintained at a concentration of 50–150 parts per million (ppm) in drench or dipping solutions or suspensions. The concentration can be monitored with kits that are readily available. A proper test kit should measure free available chlorine rather than total chlorine. Chlorine rapidly oxidizes on contact with organic matter, and therefore the demand for it increases significantly during treatment, and the concentration can drop quickly. Continuous addition of chlorine during treatment is the best way to ensure the liquid contains a sufficient concentration of unreacted chlorine to kill fungal propagules. Some systems that add chlorine to water are based on continuous on-line measurement of the free chlorine concentration; electrodes of highly purified platinum are immersed in the water to measure the oxidation-reduction potential. Chlorine is added to the water as needed to maintain the desired concentration. The pH is also monitored with a sensor to meter the addition of acid or base as needed to maintain the proper pH. In lieu of automatic monitoring systems, drench or dipping treatments must be tested frequently during use to ensure that the proper pH and chlorine concentration are maintained.

Repeated exposure of wooden pallets to water and chlorine by dipping or drenching will significantly weaken them and shorten their life, particularly in subtropical regions. Wet wood is susceptible to wood-rotting fungi, and chlorine is corrosive to metal and an active delignifier. Wooden pallets should be treated with an approved preservative, such as copper-8-quinolinolate, and stored under dry, well-ventilated conditions. Plastic, plastic-lined, or epoxy-painted metal boxes may be more suitable than wooden pallet boxes in some instances.

Degreening

Degreening of citrus fruit with ethylene (Plate 61), which removes chlorophyll from the rind to expose the orange and yellow colors, is a common commercial practice to enhance the appeal of the fruit to consumers. It is conducted for cosmetic purposes only. Mature fruit with green rind is more common in warm, subtropical growing regions, such as Florida, than in more arid climates, where lower temperatures (below 50°F) at night during the growing season promote the degradation of chlorophyll. The conditions imposed during degreening, particularly temperature, vary depending upon the geographical region. In Florida, the temperature is maintained at 82–85°F in order to achieve degreening in the least amount of time, because chlorophyll degradation occurs most rapidly at this temperature. However, very little carotenoid synthesis occurs at this temperature, and not much orange color will develop. In other citrus regions, such as California, Israel, and South Australia, the treatment is more often conducted at 70–77°F to allow rapid degreening and some carotenoid synthesis along with the chlorophyll degradation.

The decision about when mature fruit can be harvested and degreened successfully should be based, in most instances, on the percentage of orange or yellow color in the rind. Sufficient color break should be allowed to develop before harvest so that mandarin cultivars can be degreened within 36 hours and oranges and grapefruit within 72 hours. Degreening times in excess of these often increase decay dramatically.

Ethylene for degreening should be maintained at a concentration of 5 ppm to no more than 10 ppm. Accurate control of the concentration is critical, since levels higher than recommended do not expedite degreening but will significantly enhance the development of Diplodia stem-end rot and anthracnose. A continuous trickle or flow system for administering ethylene should be used to obtain optimum degreening and minimum disease development. Flowmeters are available with flow ranges suitable for both large and small degreening rooms. Estimates of flow rates can be determined if the volume of the room and the desired ethylene concentration are known, and if the ventilation allows one air exchange per hour. A portable analyzer can be used to determine the ethylene concentration. The most common type draws a measured amount of air through a glass detector tube containing a color indicator that reacts with ethylene. Air samples should be drawn from well in the interior of the room, away from the influence of the outside atmosphere. Ethylene is explosive at concentrations of 3% or above, and care should be taken to prevent leaks and avoid ignition from any spark or flame.

An alternative to continuous exposure to ethylene during degreening has been developed for Israeli citrus, based on observations that biological processes triggered by ethylene will continue for an additional 12–24 hours after the removal of the gas. Intermittent exposures to ethylene for 12-hour periods have minimized increases in respiration, weight loss, and decay, which may occur in fruit degreened by the continuous process. Fruit may also be removed from the degreening room before attaining its full color, to allow final color development to occur during shipment.

The relative humidity in degreening rooms should be kept above 92%, maintained with humidifiers, to reduce dehydration, shrinkage, stem-end rind breakdown, and fruit deformation after packing. Peel injuries, such as oleocellosis and sand scratches, are accentuated by degreening when the humidity is too low. Minor injuries to the flavedo will heal at high humidity, and lignin and coumarin compounds inhibitory to *Penicillium* will accumulate at injury sites. Healing is more rapid at 85°F than at lower temperatures. The fungus also grows more slowly at higher temperatures, and thus healing can often occur before a successful infection can be established.

Air circulation in degreening rooms must be adequate to equalize the ethylene concentration, relative humidity, and temperature of the atmosphere surrounding the fruit and to remove excessive concentrations of carbon dioxide (CO_2), which may retard degreening. The rate of degreening is reduced at CO_2 concentrations of 0.1% and nearly stopped at 1.0% or higher. Air circulation is best obtained by arranging fruit between areas of comparatively high and low pressures. This is best achieved in rooms designed for horizontal air movement, in which air is pulled through the stacks of fruit by fans. The air is replenished with moisture, heat, and ethylene in a false ceiling before passing again through the stacks of fruit. Degreening rooms can also be used to store fruit during a delay in processing, especially if they are equipped with refrigeration, so that the fruit can be stored at a low temperature to maintain its quality. When degreening rooms are used for storage, the humidity should be controlled, but heat and ethylene are not required.

Dump and Wash Procedures

On modern packinghouse lines, a large amount of fruit in pallet boxes is mechanically dumped quite rapidly. It is imperative that this large mass of fruit be leveled out as rapidly as possible, both to avoid squeezing and scraping the fruit and to allow the removal of trash. An antisurge device is needed (such as two belts, one fast and one slow), which in turn delivers the fruit to a roller-spreader. The fruit should be graded immediately after the dump to remove long stems and rotten fruit. Removing rotten fruit at this point will reduce contamination with *Penicillium* and *Geotrichum* in the remainder of the packing line and the washer.

The fruit should receive an aqueous treatment as soon as possible after dumping, to remove surface debris, such as sand particles, which might damage the fruit during conveyance. Aqueous treatments can be applied by spraying the fruit or soaking it in a tank. Chlorine should be added to the water to enhance surface sanitation and control contamination. The concentration of free available chlorine should be near 200 ppm, and the aqueous treatment should be neutralized to pH 7. The lethal action of chlorine is time-dependent. The maximum lethal action against *Penicillium* and *Geotrichum* propagules occurs in a 2-minute exposure. Chlorine has no eradicant or residual activity and acts only as a surface disinfectant.

Clean, attractive fruit, free of surface dirt and fungal growth, is important in obtaining higher packouts, better wax application and fruit presentation, and more effective marketing.

Bath treatments with borax, sodium carbonate (soda ash), or sodium *o*-phenylphenate (SOPP) can be utilized in soak tank applications, primarily for control of green mold and sour rot. The use of borax is now limited, because of concerns about residue and disposal. Solutions of 4–6% soda ash are generally heated to about 110°F. At the lower concentration, exposure for a minimum of 2 minutes is required for best results. Care must be taken to ensure that the treatment material is replenished and maintained at the proper concentration, and the packing line must be designed to remove fruit from the bath in the event of an equipment breakdown. The bath system must be kept relatively free of dirt and debris, such as rotten or crushed fruit, and cleaned periodically. The fruit should not be excessively rinsed, or the remaining residues may be insufficient for effective disease control.

Solutions of SOPP must be maintained near pH 12.0 to prevent undissociated *o*-phenylphenol from exceeding phytotoxic concentrations of 200–400 ppm. Even if SOPP is diluted to minimize the risk of phytotoxicity, care must be taken to maintain the high pH, or enough *o*-phenylphenol will remain in solution to damage the fruit rind. SOPP is most commonly applied at the time of washing in the form of a nonrecovery foam or spray or, less often, in the form of a recovery drench or flood. In a recovery application, as in bath treatments, proper control of pH is essential. Nonrecovery application is more popular in commercial operations, because fresh, uncontaminated material of proper strength and pH is always consistently applied during washing, which normally lasts for 20–30 seconds and is followed by a rinse. The nonrecovery treatment leaves less residue on the fruit and is less effective than the bath treatment for decay control. However, it is easier to manage, and pH control is not a problem, because the treatment solution is not recycled. The potential for injury is lower in foam or spray applications than in bath treatments.

Clean, attractive fruit, free of surface dirt and fungal growth, is important in obtaining higher packouts, better wax application and fruit presentation, and more effective marketing. Washing is performed by passing prewetted fruit over washer brushes set at right angles to the flow of fruit. Brushes with spiral strips of long fibers improve the washing process, particularly if consecutive brushes spiral in opposite directions, so that the stem end of the fruit can be cleaned well. The last brush in the washer should be plain, to maintain an even flow of fruit to the next section of the line. Fruit should be washed for a minimum of 20–30 seconds, and the brush speed should not exceed 120 rpm. To prevent brush burn and minimize peel injuries, the washing time should be controlled with a continuous mechanical wipeout or one that is activated when the flow of fruit is interrupted. Chlorinated or nonchlorinated water can also be sprayed at high pressure during brushing to enhance the removal of sooty mold and scale. The pressure should not exceed 250 psi, to prevent fruit damage.

Thorough rinsing with at least two successive water nozzles is necessary after washing. Rinsing is best done over the last of the washer brushes or over a roller conveyor. All residues of washing detergents should be removed, to minimize dehydration and avoid deleterious effects on the wax applied later in processing. The freshwater rinse can be heated to 130–140°F to enhance drying. Excess water is removed mechanically with sponge rubber rollers, soft brushes, or an air blast to aid drying and to prevent dilution of the subsequent fungicide and wax treatments. Brushing the fruit with heated dry brushes to enhance drying and shine is not recommended, because it may cause rind injuries.

Color-Add Process

Color adding is a process in which a dye is used to intensify the color of pale oranges and tangelos produced in subtropical regions, such as Florida and Texas, to enhance their appeal to the consumer. Washed fruit is conveyed slowly through a tank where it is flooded with a recycled, heated solution of Citrus Red Dye No. 2 (1-[(2,5-dimethoxyphenyl)azo]-2-naphthalenol). The temperature of the solution should not exceed 120°F for oranges and 114°F for tangelos, and the treatment time should not exceed 4 and 2.5 minutes, respectively. Optimum results are obtained when the dye is applied at a low concentration for the maximum treatment time. After treatment, the fruit must be thoroughly rinsed over washer brushes or moving rollers, rather than on a mesh belt or a slat conveyor on which the fruit does not rotate.

The two major problems that occur in the color-add process are weak color intensity and uneven distribution of dye over the fruit surface. If the treatment time is short, the concentration of dye must be increased, and then the color is often not uniformly distributed. Temperature control is important; a drop of 6–10°F will have the apparent effect of reducing the dye strength by half. The optimum pH is 7–10, maintained by adding a base such as NaOH. The pH will drop if broken and decayed fruit enter the tank. Color is slow to develop if excessive dirt is allowed to accumulate in the tank. An excessive concentration of dye will cause dark spots to form on the fruit surface a minute or so after the treatment is completed; the concentration can be corrected by adding solvent.

Fungicide and Wax Applications

Postharvest fungicides (Table 16.3) can be applied in an aqueous formulation before waxing or in an aqueous emulsion with wax. The advantages of a separate aqueous application are that it allows better systemic movement of fungicides (particularly imazalil) and thus better eradicant activity, and it leaves less residue than application in combination with wax, in which the fungicide concentration is usually double to compensate for the deleterious effect of the wax on fungicide activity (Fig. 16.2). Imazalil provides optimum control of decay and sporulation when it is applied first in an aqueous treatment, for maximum eradicant activity, and again in combination with wax to enhance the control of sporulation of *Penicillium.* The residue of imazalil left on the fruit after either aqueous or wax application must be at least 2 ppm for good sporulation control. Some disadvantages of aqueous fungicide application are that the fruit has to be dewatered or dried before wax is applied, more equipment is required for separate fungicide and wax treatments, and mechanical failures that interfere with aqueous application are less easily detected than failures in wax application.

Waxes are applied primarily to retard moisture loss, which is accelerated by the washing process, and to improve the appearance and natural shine of the fruit surface. Waxing will also slow down the development of rind disorders such as stem-end rind breakdown. 2,4-D can be incorporated into the wax to help retard the senescence of the calyx, which controls the development of Alternaria stem-end rot in stored fruit. Most waxes used commercially are of the water emulsion type; solvent waxes containing coumarone-indene resins dissolved in napthalene have been discontinued, principally because of the cost of the wax, and because of atmospheric pollution from the solvent during application.

Aqueous fungicides and emulsion wax are most commonly applied by drip or pressure misting or sprays while the fruit is rotating on a bed of slowly rotating (100 rpm) brushes. For this application, a 50:50 mixture of horsehair and synthetic brushes is used, because horsehair has a rough texture that allows it to hold water or wax and wipe it on the fruit surface, and the synthetic brushes help prevent the horsehair from matting down. The applicator section should consist of a minimum of 12 brushes, with the fungicide or wax applicator positioned over the third or fourth brush. A gallon of material should treat about 4,000 pounds of fruit, though actual usage varies significantly in different geographical regions. Excessive amounts of wax will interfere with fruit respiration and lead to anaerobic conditions within the fruit, which result in off-flavors and rind pitting. The use of wax can be controlled by a microswitch that actuates the waxer delivery pumps only when fruit is being conveyed under the waxer. Wax applicator brushes should be thoroughly washed with a high-pressure water spray at the end of the day to remove wax deposits and thus prevent matting and hardening of the brushes, which could damage subsequently treated fruit.

Dryers for waxed fruit should consist of a short initial section and a longer final section. The accumulation of wax on the equipment is confined primarily to the first section, which simplifies cleanup with a suitable commercial cleaner and reduces fruit damage from hardened wax droplets. Waxed fruit should not roll during drying except once, halfway through the dryer, to expose the underside. Wet fruit can withstand higher air temperatures than dry fruit. Temperatures much above 130°F are not recommended, particularly at the end of the dryer where the fruit is dry or nearly so.

Biological Control

Interest in the development of living organisms for the control of disease has resulted in the recent registration of products containing a yeast or a bacterium for the control of citrus decay (Table 16.3). The development of alternatives to chemicals has been spurred by concerns over fungicide resistance in isolates of *P. digitatum* and *P. italicum,* pesticide residues in food products, and regulatory and economic issues affecting registration.

Biological control organisms are formulated in wettable powders, which are suspended in water, and the suspension is continuously agitated during application, much like fungicide formulations. Several important differences between chemical and biological products must be kept in mind. Biological products must be stored refrigerated, but they still have a much shorter shelf life than postharvest chemicals, because they are living organisms. Treatment suspensions must be fresh and generally should be prepared daily. They cannot be applied with pack waxes, soaps, or sanitizers, such as chlorine, because these products are lethal to biological cells. However, biological control organisms can survive in storage waxes, such as those applied to lemons before storage.

Biological control agents at the sites of injuries of the fruit surface apparently survive the normal drying and waxing operations that follow. They utilize nutrients released by injured fruit tissue and compete with wound pathogens, such as *P. digitatum,* to prevent them from establishing infections. Control by the production of chemicals toxic to decay fungi does not appear to be a major mode of action. Non-chemical-producing organisms are preferred for development, because chemicals would be affected by the same expense and regulatory issues as synthetic fungicides. Biological control agents are not effective eradicants, and they need to be applied within 6–10 hours of infection (in comparison, chemical fungicides should be applied within 24 hours). They do not suppress sporulation (as imazalil does), and they do not control decay arising from quiescent infections, such as stem-end rot. Biological control organisms may have the potential to be used in combination with chemicals, in applications to control fungicide-resistant pathogens, and in applications for postharvest chemical-free fruit.

Grading

Effective grading is crucial to good keeping quality. Overmature or diseased fruit removed in grading would be sources of inoculum for the spread of decay to healthy fruit during long periods of transit and storage. Injured fruit, such as fruit with plugs and punctures, which is more prone to infection, is also eliminated from the packed carton. The major part of the grading should be conducted when the fruit has been washed but before fungicide and wax treatments have been applied, to reduce packing costs and contamination of fruit delivered to the processing plant for juicing. Some final grading may be required on the final fruit presentation, but this should be kept to a minimum.

Grading stations should be properly lighted, with high-output, cool white fluorescent lamps producing an illuminance

Table 16.3. Treatments to control postharvest decays in various citrus production regions and diseases against which they are effective

	Application rate[a]	Diplodia stem-end rot	Phomopsis stem-end rot	Green mold	Blue mold	Sour rot	Brown rot	Alternaria stem-end rot
Chemical treatments								
Thiabendazole[b]	1,000 ppm[c]	×	×	×	×	. . .	. . .	×
Imazalil[b]	1,000 ppm[c]	×	×	×	×	. . .	. . .	. . .
Sodium *o*-phenylphenate tetrahydrate[b]	2%	. . .	. . .	×	. . .	×	. . .	. . .
Benomyl[b]	600 ppm,[c] 1–2 pounds/acre[d,e]	×	×	×	×	. . .	. . .	. . .
Guazatine	500 ppm	. . .	. . .	×	. . .	×	. . .	. . .
Sodium carbonate[b,f]	4%	. . .	. . .	×	. . .	×	×	. . .
Borax[b]	2%	. . .	. . .	×	. . .	. . .	. . .	. . .
Prochloraz	1,000 ppm[c]	×	×	×	×	. . .	. . .	. . .
Potassium sorbate[b]	2%	×	. . .	×	. . .	. . .	. . .	. . .
Carbendazim	500 ppm	×	×	×	×	. . .	. . .	. . .
Thiophanate-methyl	1,000 ppm	. . .	. . .	. . .	. . .	. . .	. . .	. . .
Copper fungicides[b]	4 pounds (metallic) per acre[e]	. . .	. . .	. . .	. . .	. . .	×	. . .
Fosetyl-Al[b]	5 pounds/acre[d,e]	. . .	. . .	. . .	. . .	. . .	×	. . .
Metalaxyl	1,000 ppm[c]	. . .	. . .	. . .	. . .	. . .	×	. . .
2,4-D[b]	250–500 ppm	. . .	. . .	. . .	. . .	. . .	. . .	×
Biological treatments								
Bio-Save 1000[b] (*Pseudomonas syringae*)	2 pounds/7 gallons[d]	. . .	. . .	×	×	. . .	. . .	. . .
Aspire[b] (*Candida oleophila*)	0.25 pound/10 gallons[d]	. . .	. . .	×	×	. . .	. . .	. . .

[a] Concentration of the active ingredient, unless stated otherwise; ppm = parts per million.
[b] Approved for use on citrus in the United States.
[c] Concentration is double or greater when incorporated in water emulsion wax.
[d] Amount of formulated product.
[e] Preharvest application.
[f] Heated to 109–118°F.

of about 200 foot-candles. The work area should be temperature-controlled, and the sound level should be kept below 85 decibels (decibel(A) scale). Cull chutes should be at the height of the grading table, and a center belt for second-grade fruit should not require more than a 2- to 3-inch lifting motion. Controlled rotation of the fruit, so that its entire surface is visible, facilitates grading. A forward rotation of 1.6 revolutions per foot of linear travel is adequate for fruit on grading conveyors. The rollers of the conveyor should be off-white to minimize glare.

The grading area should be designed so that graders can be added or removed, depending on the needs of the operation, with the number of graders determined primarily by the number of fruit to be inspected. When graders check for only one or two defects, their efficiency will be improved. The fruit flow should be divided among several lines, with no more than three or four graders per line. With four short grading lines as opposed to two long ones, the fruit passes before the grader at half the linear speed.

Ultraviolet light can be utilized to detect certain decay lesions and injuries that are difficult to discern under ordinary light. The initial lesions of *Penicillium* infections, a quarter of an inch or less in diameter, fluoresce yellow of various intensities, depending upon the type of fruit, against a dark purple background. Punctures and fissures tend to fluoresce dull green-yellow. The UV viewing booth should be constructed over the grading conveyor, and the walls and equipment should be painted black to prevent reflection into the eyes of the graders and to reduce the light level. UV lamps should be suspended at such a height that the graders' eyes are shielded from direct radiation, and metal reflectors should be used to reflect the light downward onto the fruit surface. At least two 40-watt tubular lamps emitting long-wavelength light (approximately 360 nanometers) should be used for each conveyor.

Sanitation

Sanitation of the packinghouse is one of the most important and least expensive decay control practices associated with citrus fruit handling and should receive close attention in packinghouse management. Effective packinghouse sanitation provides additional insurance that high-quality decay-free fruit will be consistently delivered to the consumer from one shipment to the next (Plate 62). Sanitation removes inoculum of *P. digitatum, P. italicum, G. candidum,* and *Phytophthora* species, which can proliferate and accumulate in the packinghouse environment. Complete eradication usually cannot be achieved and is not necessary for a significant reduction in decay, since a certain level of inoculum is required at an infection site to induce disease. Sanitation practices that keep inoculum at low levels will greatly reduce the chances of infection.

Inoculum of these four fungi can accumulate in recirculated or containerized water, and such sources of water should be chlorinated, by the procedures previously discussed. Inoculum of *Penicillium* is primarily airborne spores produced on moldy fruit. Arthrospores and mycelium of *Geotrichum* are borne in soil and produced during fruit infection. Pallets and packinghouse lines are commonly contaminated by *Geotrichum,* and *Penicillium* contaminates these areas as well as degreening and storage rooms. *Penicillium* species are the only decay pathogens disseminated by air currents in the packinghouse, and large populations of spores can be present. The highest spore population is usually present at the dump, where spores are released from infected fruit in the dumping process. A significant reduction in decay has been noted when the dump and primary grade areas, where decayed fruit are first encountered, are spatially separated from the rest of the packing line and loading facilities. Exhaust fans that remove air from the dump area can be used to exclude spores from the packinghouse, and the packing line can be designed so that prevailing wind currents do not spread spores from the degreening and dump areas to the rest of the line.

Sanitizers, such as hot water (160°F), chlorine, quaternary ammonium, and formaldehyde, can be used to kill inoculum. In a steam-pressurized system, the temperature of the water must be verified, since steam can be formed at lower than lethal temperatures. Chlorine (200 ppm) is approved for application to equipment and fruit, and quaternary ammonium (200 ppm) is approved only for equipment. Chlorine is less expensive than quaternary ammonium, but it is corrosive to metal and deleterious to wood. Quaternary ammonium is a wetting agent with detergent properties and an effective penetrant of porous surfaces. It is relatively noncorrosive, except to aluminum and exposed electrical equipment, and is more stable in the presence of organic matter than chlorine.

> **Sanitation of the packinghouse is one of the most important and least expensive decay control practices associated with citrus fruit handling.**

Dirt and debris should be removed from the pallets, and they should be sanitized before they are returned to the field for more fruit. The packing line from the dump through the waxer should be cleaned daily. Ambient water and detergent will remove a significant amount of inoculum, but additional sanitation can be achieved by following the wash with a spray of sanitizer. All fruit and trash should be removed daily to eliminate substrates for *Penicillium* sporulation, and a flume system with recirculated water for the removal of rotten fruit from the dump and grading areas will help reduce spore dispersal. Flume water should be sanitized to eliminate inoculum and odor. Degreening and storage rooms should be washed and sanitized by spraying or misting to leave a wet deposit on all wall, ceiling, and floor surfaces. Packed fruit with green mold should never be repacked in the packinghouse; it should be removed and handled in a remote area to prevent contamination of the packinghouse facilities with spores, many of which are likely to be resistant to the fungicides used on the fruit.

A 1–3% formaldehyde solution (weight/volume [w/v]) may be used to fumigate the packinghouse, applied through a permanently installed spray manifold system in closed rooms. Formaldehyde is readily absorbed on moist surfaces, and therefore high humidity should be introduced, if possible, for 24 hours before fumigation, to increase the amount of moisture on empty boxes and other surfaces in the room. The area should not be wetted down with water hoses. Recirculating fans should be turned on during fogging and run for 2 hours after it is completed, and the room should remain closed for at least 24 hours. The application rate should not exceed 1 pound of 37%

formaldehyde solution per 1,000 cubic feet of treated space. Formaldehyde is a hazardous material, and all entrances and exits should be posted with warning signs. The fumigated area should be thoroughly ventilated before reentry. Fruit should not be treated or exposed to treated areas until the formaldehyde odor has dissipated.

Assessment of Fruit Damage

An objective evaluation of harvesting and packing practices that may promote disease can be achieved with the use of chemical dyes and pH indicators that change color when they contact citrus juice or peel oil, indicating where surface injuries to the rind have occurred. The entire packing line or portions of it can be evaluated to determine if fruit are being excessively damaged during packing. Trouble spots are identified so that corrective action can be immediately taken to prevent problems that appear later during distribution and marketing.

Several stains accumulate in injured tissue, so that damage is more easily detected by visual observation. When fruit is submerged for 5 minutes in a solution of indigo carmine (0.1%, w/v) containing a few drops of wetting agent and 0.5% hydrochloric acid, injured areas turn blue-green. In another dye procedure, fruit is submerged in a ferric chloride solution (5%, w/v) containing a few drops of wetting agent for 5 minutes; it is then rinsed or wiped dry and placed in a tannic acid solution (5%, w/v) for 2–3 minutes, and injuries on the fruit surface are stained blue-black. The tannic acid solution should be discarded after one use.

Another procedure that is quite easy to use in the packinghouse to detect mechanical injury is to soak fruit in a solution containing triphenyltetrazolium chloride (TTC). This is a water-soluble powder which becomes insoluble upon contact with living tissue, causing a bright red discoloration. Bruises and minute scratches in the fruit surface can be detected with this procedure. Fruit is lightly hand- or machine-washed and then submerged for about 30 minutes in a solution of TTC (0.5%, w/v). It is then soaked in fresh water for 5 minutes, removed, and held in brown paper bags for at least 8 hours to permit color development. The degree of damage can be visually assessed.

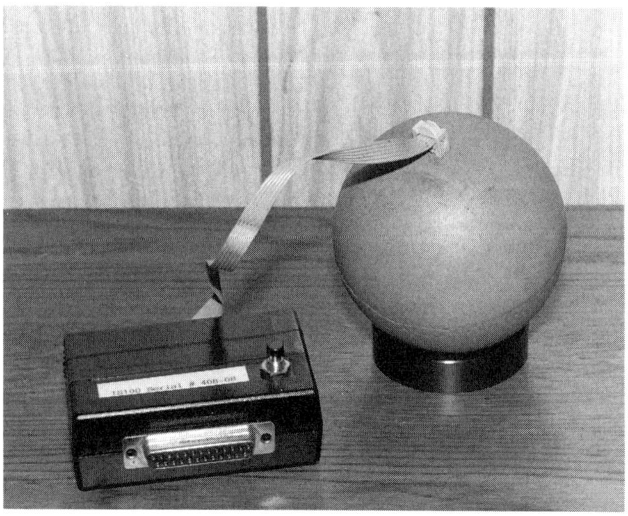

Fig. 16.4. Instrumented sphere with interface to measure forces affecting fruit on the packing line. (Courtesy G. E. Brown)

Extensive damage, such as crushing or splitting that may occur during harvesting and hauling, can be evaluated with filter paper saturated with an alkaline solution of methyl red. The paper turns from brilliant yellow to red at low pH when moistened with juice. Filter paper treated with an alkaline alcoholic solution of phenolphthalein can detect less severe injuries that cause only the release of peel oil. The paper also reacts to juice, and the contrast between the peel oil reaction and the juice reaction may be difficult to discern. If the paper is also treated with an oil-soluble blue dye, peel oil is indicated by a lemon yellow color, and juice is indicated by pale blue (solution A: phenolphthalein [1%, w/v] and sodium hydroxide [1%, w/v] in isopropyl alcohol; solution B: Victoria Blue [0.25%, w/v] in acetone; solution C: 150 milliliters of solution A, 100 milliliters of solution B, and 250 milliliters of water). Sheets of these papers can be placed at various impact points in harvesting and handling processes to estimate the degree of damage.

A relatively new technique, first used in the apple industry, has been developed to accurately quantitate information on dynamic conditions that initiate fruit damage. An instrumented sphere approximately the size of a citrus fruit, containing a triaxial accelerometer, signal conditioning unit, and computer memory, is placed on the packing line to detect movement and identify forces strong enough to cause fruit damage (Fig. 16.4). Measurements are transferred through an RS-232 interface to a portable data acquisition system. Impact data are reported in terms of the acceleration of gravity (g), where $g = 9.8$ m/s^2 (meters per square second). For example, a 12-inch drop to a steel surface will produce a force corresponding to an acceleration of approximately $300g$. Handling or harvesting practices that produce forces corresponding to an acceleration in excess of $150g$ definitely must be modified to reduce damage, while those below $100g$ will probably not lead to any significant injury to citrus fruit. Impact at values between $100g$ and $150g$ may or may not cause damage. High impact is most likely to occur during dumping and sizing operations.

The production of CO_2 can also be used as a measure of fruit damage. A sample of fruit is removed at any point on the packing line, weighed, and stored in sealed plastic containers at approximately 68°F. CO_2 production is measured over a period of 10 hours with a portable infrared analyzer and expressed in terms of milligrams per kilogram of fruit per hour. An increase in CO_2 production over that of the control fruit is an indication of increased mechanical injury.

Specific weight loss is also an indication of mechanical injury, since damaged fruit dehydrate more rapidly than sound fruit. A sample of fruit is collected from the packing line and stored for 24 hours at 70°F and 70% relative humidity to equilibrate. Since washing and waxing affect weight loss, control and treated fruit should be subjected to similar handling in these two processes. The surface area, weight, and the ratio of weight to surface area are calculated for each fruit in a sample of about eight fruits. After 24 hours, the fruit is weighed again, and the difference between the weight–to–surface area ratio of the control fruit and that of the treated fruit is a measure of damage.

Fungicide Resistance

Spontaneous mutation in fungal populations results in the emergence of fungicide-resistant biotypes of *P. digitatum* and *P. italicum* in the field and packinghouse. These biotypes are not controlled by highly selective fungicides such as SOPP, benomyl, thiabendazole, and imazalil. This resistance has caused a major obstacle to the effective use of postharvest fungicides to control green mold and blue mold on citrus fruit. Fungicide resistance has not been a problem in other decay fungi.

Type of Resistance

Because of their ability to produce large numbers of airborne spores, the mold fungi can rapidly produce large numbers of resistant spores with relative ease. Potentially, a single spore of *Penicillium* can produce about 100 million spores on an infected fruit in seven days under optimum environmental conditions. In a natural population of 100 million *Penicillium* spores, one to 10 resistant individuals can be produced. These resistant strains constitute a minor component of the population, because they have no ecological advantage over the wild types in the same population. However, once the population is subjected to selection pressure by continuous application of a selective fungicide, such as thiabendazole, the fungicide-resistant mutants are able to proliferate free of competition.

The shift of the natural population to one that is resistant under the selection pressure of the fungicide is reached through either a disruptive or a directional scheme. In disruptive selection, the fungicide is essentially inactive against resistant strains, which suddenly proliferate and cause a rapid loss of disease control. This is the kind of selection that can happen with resistance to thiabendazole or benomyl. In directional selection, the loss of fungicide activity is much more gradual, beginning with a few less sensitive strains, which increase in numbers and resistance slowly over time. An example of this is the development of *Penicillium* strains resistant to imazalil.

Factors Favoring the Development of Resistance

Repeated and continuous use of the same fungicide increases the possibility that resistance will develop. The selection pressure exerted by postharvest fungicide treatment of citrus fruit is considerably greater than that of the same fungicide sprayed in the field. Postharvest fungicides are applied in a manner that usually results in complete coverage, and the fungicide deposit deteriorates only slightly during the storage life of the fruit. Treated fruit is stored under conditions that permit the growth and sporulation of fungicide-resistant strains of the pathogen. The resistant spores are readily dispersed by air currents in the packinghouse to other treated fruit, continuing the process of selection for resistance.

Field applications of the same fungicide used in the packinghouse or a similar fungicide will enhance selection for resistance.

Field applications of the same fungicide used in the packinghouse or a similar fungicide will enhance selection for resistance. Preharvest sprays of benomyl to control postharvest decay may lead to greater resistance of the *Penicillium* population to thiabendazole applied in the packinghouse.

Storage of treated fruit for extended periods gives the resistant biotypes ample time to infect, sporulate, reinfect, and increase in numbers in the storage facility. Thus, resistance can be a problem in long-term storage of lemons in California, in storage of Florida Valencia oranges for summer markets, or in storage of export fruit at docks and in transit containers.

Assessing Resistance

Resistance to fungicides is evaluated by collecting mold spores from the atmosphere or from surfaces of walls, floors, and equipment in the packinghouse. These are then dispersed on a culture medium, usually containing potato dextrose agar, yeast extract, and peptone. The medium is amended with the fungicide of interest and other amendments, such as dichloran (3 ppm) or *o*-phenylanisole (100 ppm) and PCNB (500 ppm), to suppress contaminating fungi and to cause colonies of *Penicillium* to be compact and nondiffuse for easier counting. The concentrations of fungicides used in resistance assays have varied, and attempts have been made to standardize these levels so that surveys can be compared. Recommended levels of fungicides for use in culture media are as follows: thiabendazole, 10 ppm; benomyl, 2 ppm; *o*-phenylphenol, 15–20 ppm; and imazalil, 0.05–0.1 ppm. Spores are also cultured at the same time on unamended media, so that the total population can be measured to determine the percentage of resistant spores. When the resistant portion of a relatively high *Penicillium* population in the packinghouse approaches 10%, the performance of the fungicide treatment is usually significantly reduced.

Several methods have been used to collect spores. The most common procedure is to remove the lids from petri dishes containing a medium, usually for 1 minute, and allow spores to settle on the agar surface. The exposure time is reduced to 5–15 seconds if the spore concentration is high (for example, at dump sites). The dishes are returned to the lab and incubated at 75–80°F for several days, and colonies formed by each germinating spore are counted. Another version of this assay is to expose filter papers in sterile, uncovered petri dishes for several days at various locations in the packinghouse. The dishes are covered and returned to the lab, spores on the papers are removed by washing with sterile water, and measured aliquots of the water are spread over the surface of agar plates. The plates are then dried, covered, and handled in a manner similar to the previous procedure. These two methods are used to sample airborne spores that happen to settle on the exposed test surfaces. These techniques provide an indication of the airborne contamination level. In another, more quantitative method, measured quantities of air are drawn across the assay medium with an air sampler unit. The air flora impinge directly on the agar surface, and the number of spores per cubic foot of air can be determined.

Combating Resistance

Strict sanitary practices and proper design of the packinghouse, as previously discussed, are required if resistance is to be kept in check. The use of fungicides with different chemistries will help to minimize resistance, since two genes must mutate simultaneously to unrelated fungicides before resistance to both can be expressed. A fungicide with antisporulant properties, such as imazalil, will help suppress sporulation during storage to reduce spore populations. However, single biotypes with resistance to all fungicides with specific modes of activity, such as SOPP, thiabendazole, and imazalil, do occur in commercial practice. Continued use of any of these three fungicides may promote the persistence of these multiresistant strains in the packinghouse. The only known defense against multiresistant biotypes is proper sanitary practices along with broad-spectrum biocides, such as borax, chlorine, quaternary ammonium, formaldehyde, or soda ash. Fungicides with specific modes of activity are best utilized on the packing line for final treatment immediately before shipment.

Packaging

The package must protect the fruit from mechanical damage and provide a suitable microenvironment if quality is to be maintained from harvest to consumption. Corrugated full-telescoping fiberboard cartons with vent holes are the primary container used for the packaging and transport of citrus fruit. In addition, a significant amount of fruit is packaged in polyethylene mesh or net or transparent perforated polyethylene consumer packs containerized in fiberboard "master cartons." The design of the carton and the arrangement of cartons in unitized stacks are very important physical considerations in the maintenance of proper fruit temperatures and ventilation.

The constant demand from buyers for a "good, full pack" leads to overfilling of cartons, but overfilling or bulge-packing of fiberboard containers causes fruit deformation, particularly in instances of lengthy transport, such as export shipments. Moisture evaporates more readily from compressed areas of the deformed fruit, causing hollow areas to form in the pulp. The fruit is puffy and dry and unsalable. Packers and buyers should be aware of this problem and avoid overfilling and packing fruit too tightly in cartons.

Harvesting and handling conditions also have a significant influence on fruit deformation during packing. Gently handled grapefruit kept under humid, cool, shaded conditions until waxed and packed is far more resistant to deformation than fruit handled roughly and subjected to conditions that favor desiccation. Also, fruit of advanced maturity loses elasticity and is more prone to deformation. Fruit packed in cartons in the flat orientation is more resistant to deformation than fruit packed on its equatorial plane. Tray-pack containers and honeycomb cell-pack cartons provide protection against deformation, soilage, and the spread of decay to healthy fruit.

> **The package must protect the fruit from mechanical damage and provide a suitable micro-environment if quality is to be maintained from harvest to consumption.**

Humidity within the container is influenced by transpiration, respiration, and changes in ambient conditions (such as a change from cold to warmer conditions, which causes water to condense on fruit surfaces. "Chalking" often occurs in waxed fruit exposed to free water, when the wax partially dissolves and forms white deposits upon drying. High humidity and poor ventilation in polyethylene bags promote decay, and therefore film with perforations is required, to prevent the accumulation of excess moisture and high CO_2 levels. In contrast, fruit packed in mesh bags must be properly maintained under high relative humidity to prevent excessive moisture loss and shriveling.

With proper packing, most of the weight in stacks of packed cartons is supported by the cartons and not by the fruit. Cartons should be arranged to provide stability and adequate ventilation for cooling and gas exchange. They should be stacked on wooden pallets, wrapped with mesh, tape, or plastic, and protected with corner boards to provide maximum protection and stability during long transit times typical of export (Plate 63). In Florida, cartons are unitized on slip sheets rather than wooden pallets for domestic markets. On arrival at terminal markets, the slip sheets may be placed on wooden pallets. Fruit can be shipped packaged in bulk form, particularly for domestic markets. Fiberboard ventilated bulk containers or bins holding as much as 945 pounds of loose or bagged oranges or grapefruit can be displayed in the produce sections of large retail stores (Plate 64). The use of these bulk containers reduces the cost of packaging and labor, and colorful advertising can be displayed directly on the bin.

Shrink-wrapping of individual fruit in high-density polyethylene or low-density polymers film has been studied to evaluate keeping quality. Compared to waxing, film wrapping has provided better weight loss control, resistance to deformation, internal gas levels, soilage control, control of stem-end rind breakdown, and extended shelf life. However, this method has not developed commercially, because of the unavailability of effective high-speed wrapping machines, and because the significant additional expense of the process cannot be recovered under present marketing conditions.

Transit and Storage Conditions

The optimal time to harvest citrus fruit for prolonged storage is dependent primarily upon judgment and not any clearly defined physiological parameter. Fruit harvested when it is at its best quality will retain its quality during storage much better than fruit left on the tree until a later harvest. Generally, fruit acidity decreases and other characteristics of senescence increase more rapidly in fruit attached to the tree than in comparable fruit held in storage for the same period of time.

Loss of weight after packing is a significant factor in postharvest deterioration of citrus fruit. Transpiration leads to desiccation, accelerated softening, shriveling, loss of attractiveness, and accelerated senescence. For these reasons, fruit should be held at near 100% relative humidity when stored in wooden or plastic containers. For storage in fiberboard cartons, humidity should be as high as possible without damaging the container, which is usually around 85–90% relative humidity. Water should not be allowed to settle on the fruit surface. This can be prevented by using efficient humidifiers and eliminating temperature fluctuations at very high relative humidity. The relative humidity in transit and storage facilities should be established before they are filled, to ensure that the humidification system can achieve the desired humidity without the contribution of moisture from the fruit.

Temperature management is another important practice in maintaining fruit quality. It complements other postharvest practices that minimize decay and peel injury—proper handling, sanitation, and fungicide application. Temperatures near ambient conditions favor fungal decay, rapid moisture loss, ascorbic acid loss, and shriveling, while excessive cold may cause chilling injury. The development of decay, particularly sour rot and Diplodia stem-end rot, is significantly retarded by low temperature, and metabolic processes that lead to loss in quality and disease resistance are also delayed.

Precooling before shipment is an effective method of removing field heat from fruit before it is placed in transit containers, which often lack adequate refrigeration for rapidly low-

Table 16.4. Typical storage conditions for citrus fruit

	Temperature (°F)	Relative humidity (%)
Oranges		
Arizona, California	40–44	85–90
Florida, Texas	32–34	85–90
Australia, Israel, South Africa	45–50	85–90
Grapefruit[a]	50–60	90–95
Lemons, limes[a]	50–60	90–95
Mandarins[a]	38–42	90–95

[a] Typical storage conditions in all citrus-growing regions.

ering temperature. Determining the right temperature depends on various factors, including susceptibility to chilling injury, length of storage, decay potential, preharvest treatments, prestorage cooling, type of container, or cold sterilization for fruit fly eradication. Care must be taken to avoid damage to certain types of citrus that are sensitive to chilling. Grapefruit, lemons, and limes are especially prone to chilling injury and should be stored at higher temperatures than oranges or specialty fruits (Table 16.4). However, mid- and late-season Florida grapefruit has been successfully handled commercially at temperatures of 38–40°F with little chilling injury. The minimum storage temperature for oranges is dependent upon the geographical region of production.

Accurate records of relative humidity and temperature should be maintained during transit and storage to verify proper handling and to relate these conditions to any problems with keeping quality that may arise.

Adequate ventilation is necessary to prevent the buildup of CO_2 and the accumulation of volatiles, primarily ethylene. The level of CO_2 should not exceed 0.1%, and the ethylene level should be below 1.0 ppm. Air purification may also be achieved by the use of activated carbon, potassium permanganate, or a shielded ultraviolet radiation source.

Controlled or modified atmosphere has not shown any significant benefits in extending storage of citrus fruit. Since there are no intensive ripening changes to suppress, efforts with controlled atmosphere have been directed toward the control of color changes and chilling injury, but these efforts have not been successful.

Storage life is difficult to predict accurately, so provisions should be made for a weekly sampling of fruit of each harvest, size, and planting as the end of the storage life approaches. Small fruit should be adequately represented, since indications of storage problems usually appear in the smaller sizes. Samples should be divided into two lots. One is placed at room temperature and observed for disorders for at least a week, and the other should be examined for visible and internal decay, peel injuries, and internal dryness and granulation.

Quarantine Treatments

Quarantine treatments are commonly applied to fruit before or during transit to ensure that pests such as fruit flies or scale are not introduced into the importing region. A cold treatment is approved for Florida grapefruit exported to some countries, including Japan. The protocol allows for either a standard treatment or a short-term cold treatment, depending on the results of insect trapping conducted earlier in the grove. The standard cold treatment requires that fruit be subjected to 32, 33, 34, or 35°F for 14, 17, 20, or 24 days, respectively, whereas short-term treatment requires 11, 13, 15, or 17 days, respectively. The treatment may be applied prior to shipment or in approved transport vehicles (seavans, break bulk ships) during transit. Prior to cold treatment, a seven-day conditioning pretreatment at 60°F and 85–90% relative humidity is recommended for grapefruit; this conditioning reduces physiological stress on grapefruit by alleviating or reducing chilling injury of peel tissue.

Cold treatment has several disadvantages. It delays shipments, unless it is applied aboard ship, as in the case of Florida grapefruit exported to Japan. It imposes additional costs in certifying treatment facilities and monitoring treatment procedures. The treatment also increases the potential for chilling injury, and lesions on the peel due to chilling injury as a result of cold treatment are susceptible to infection by decay organisms. Cold treatment was a common practice during the 1980s after ethylene dibromide was banned from use as a fumigation agent in quarantine treatments, but it has since given way to the fly-free zone protocol.

Fumigation with methyl bromide (MB) is an approved quarantine treatment used mostly against the Caribbean fruit fly on citrus requiring pest-free certification in shipments from Florida to California, Arizona, and Texas and, when required, against the Mediterranean fruit fly in shipments out of California. It is also approved for use against the citrus blackfly and the Fuller rose beetle, which occurs predominantly in California. Fruit should be free of moisture during treatment, and the fumigation chamber must not be filled to more than 50% capacity in treatments for the Mexican fruit fly and 80% capacity in treatments for the Caribbean fruit fly. In general, the MB dosage varies, depending on the fruit temperature, the air temperature, and the targeted pest. A posttreatment aeration period of at least 2 hours is required, and fruit cannot be released from the treatment facility until the MB residue in the air has diminished to 5 ppm. Importing countries usually have maximum allowed limits on MB residue. Grapefruit usually will not tolerate the dosage required for certification against the Caribbean fruit fly and develops a surface discoloration of the peel, but MB is commonly used against the same pest in oranges. Growers must determine the level of tolerance to MB stress in specific citrus fruits prior to treatment.

California has an approved combination treatment of MB fumigation plus cold storage for all types of citrus fruit to control the Mediterranean fruit fly and the Oriental fruit fly. For example, this combination treatment involves MB fumigation for 2, 2.5, or 3 hours followed by cold storage for a period ranging from 11 days at 38–47°F to 3 days at 43–47°F. Users must determine the level of treatment stress for various types of citrus fruit prior to treatment, because the potential surface injury will likely reduce the commercial market grade of fruit, thus lowering the profit to the grower.

The advantages of MB are that the procedure is relatively fast and that it eliminates other undesirable pests, especially surface scale insects. However, MB is scheduled to be phased out of use in the United States and other developed countries by the year 2005, because it has been identified as a major depleter of atmospheric ozone.

Fumigation with hydrogen cyanide is also a quarantine treatment used to combat surface pests, such as mites, scales, mealybugs, and whitefly nymphs on citrus fruit. This treatment is required by certain states and countries. It is authorized for use in California under a special local need registration. Hydrogen cyanide may be phytotoxic to plant tissue with high moisture content, and therefore this treatment must be closely monitored by experienced workers.

High temperature may be used as a quarantine treatment against certain fruit flies, such as the Caribbean fruit fly and Mexican fruit fly. The high-temperature treatment regime for the Mexican fruit fly is more harsh than that for the Caribbean fruit fly and is approved for use with grapefruit and oranges. The heat in this treatment may be provided by forced air, which does not allow moisture to accumulate on the fruit surface, or by vapor heat at high humidity, which causes moisture to condense on the fruit during treatment. The forced-air treatment requires the pulp temperature to be increased in stages, to 104°F for 120 minutes, 122°F for 90 minutes, and then 126°F until the center pulp reaches 118°F. The vapor heat treatment is applied at 110°F until the center pulp reaches 110°F within 8 hours, and then the fruit is held at this temperature for an additional 6 hours. Heat treatment may cause injury to the peel of some types of citrus. For example, grapefruit harvested early in the season is very susceptible to peel damage from heat treatment. Tolerance for heat treatment may vary in different production seasons.

Conclusion

Successful delivery of high-quality fresh citrus fruit to the consumer is dependent upon close attention to the many details outlined in this discussion. All aspects of fruit preparation, transit, and storage are critical. If any one process in the preparation and distribution chain is neglected, the quality of the final product will suffer. Production of high-quality fresh citrus is only part of the task; the postharvest efforts must also be consistent and reliable to preserve quality until the product is consumed.

Sources of Additional Information

General Pest Management

Altieri, M. A. 1994. Biodiversity and Pest Management in Agroecosystems. Food Products Press (Haworth Press), New York.

Browning, H. W., McGovern, R. J., Jackson, L. K., Calvert, D. V., and Wardowski, W. F. 1995. Florida Citrus Diagnostic Guide. Florida Science Source, Lake Alfred.

Dent, D. 1991. Insect Pest Management. CAB International, Wallingford, UK.

Dent, D. 1995. Integrated Pest Management. Chapman and Hall, New York.

Knapp, J. L., ed. 1983. Florida citrus integrated pest and crop management handbook. Fla. Coop. Ext. Serv., Univ. Fla., Inst. Food Agric. Sci. Publ. SP-14.

Knapp, J. L., ed. 1998. 1998 Florida citrus pest management guide. Fla. Coop. Ext. Serv., Univ. Fla., Inst. Food Agric. Sci. Publ. SP-43.

Metcalf, R. L., and Luckmann, W. H., eds. 1975. Introduction to Insect Pest Management. 1st ed. John Wiley and Sons, New York.

Price, P. 1997. Insect Ecology. 3rd ed. John Wiley and Sons, New York.

Rosen, D., Bennett, F. D., and Capinera, J. L., eds. 1994. Pest Management in the Subtropics: Biological Control—A Florida Perspective. Intercept, Andover, Hants., UK.

Smith, D., Beattie, G. A. C., and Broadley, R., eds. 1997. Citrus Pests and Their Natural Enemies. Queensland Department of Primary Industries and Horticultural Research and Development Corp., Brisbane, Australia.

University of California Statewide Integrated Pest Management Project. 1991. Integrated Pest Management for Citrus. 2nd ed. Publ. 3303. Division of Agriculture and Natural Resources, Oakland.

Rootstock and Scion Varieties

Castle, W. S. 1987. Citrus rootstocks. Pages 361–399 in: Rootstocks for Fruit Plants. R. C. Rom and R. E. Carlson, eds. John Wiley and Sons, New York.

Castle, W. S., Tucker, D. P. H., Krezdorn, A. H., and Youtsey, C. O. 1993. Rootstocks for Florida citrus. Univ. Fla. Publ. SP-42.

Ferguson, L., Sakovich, N., and Roose, M. 1990. California citrus rootstocks. Univ. Calif. Publ. 21477.

Hodgson, R. W. 1967. Horticultural varieties of citrus. Pages 431–591 in: The Citrus Industry. W. Reuther, H. G. Webber, and L. D. Batchelor, eds. University of California Press, Berkeley.

Saunt, J., ed. 1990. Citrus Varieties of the World. Sinclair International, Norwich, UK.

Wutscher, H. K. 1979. Citrus rootstocks. Pages 237–269 in: Hort Reviews. J. Janick, ed. AVI Publishing, Westport, Conn.

Soils and Fertilization

Alva, A. K., and Graham, J. H. 1991. The role of copper in citriculture. Adv. Agron. 1:145–170.

Alva, A. K., Graham, J. H., and Tucker, D. P. H. 1993. Role of calcium in amelioration of copper phytotoxicity for citrus. Soil Sci. 155:211–218.

Alva, A. K., Graham, W. D., and Wheaton, T. A. 1996. Citrus nitrogen management and groundwater nitrate in Florida. Pages 14–18 in: Proc. Annu. Fertil. Res. Educ. Prog. Conf., 4th.

Alva, A. K., and Obreza, T. A. 1993. Variation in soil pH and calcium status influenced by microsprinkler wetting pattern for young citrus trees. Hortic. Sci. 28:1166–1167.

Bach, B. W. 1976. The measurement of calcium exchange capacity of soils. J. Sci. Food Agric. 27:273–800.

Bradfield, R. 1950. Soil structure. Page 9 in: Int. Congr. Soil Sci., 4th., vol. 2.

Buckman, H. O., and Brady, N. C. 1969. The Nature and Properties of Soils. Macmillan, New York. p. 653.

Carlisle, V. W., Sodek, F., III, Collins, M. E., Hammond, L. C., and Harris, W. G. 1989. Characterization data for selected Florida soils. Univ. Fla., Inst. Food Agric. Sci., Soil Sci. Dep. Res. Rep. 89-1.

Chapman, H. D. 1966. Diagnostic criteria for plants and soils. Univ. Calif., Div. Agric. Sci.

Hanlon, E. A., and DeVore, J. M. 1989. IFAS extension soil testing laboratory chemical procedures and training manual. Univ. Fla. Coop. Ext. Serv. Circ. 812.

Jones, G. W., Dehaven, E. C., Clark, L. F., Rauch, J. T., Rasmussen, J. R., and Guillen, G. C. 1990. Groundwater quality sampling results from wells in the southwest Florida water management district: Northern, central, and southern region. Joint Rep. Southwest Florida Water Management District, Brooksville, and Florida Department of Environmental Regulation, Tallahassee.

Koo, R. C. J. 1979. The influence of N, K and irrigation on tree size and fruit production of 'Valencia' orange. Proc. Fla. State Hortic. Soc. 92:10–13.

Koo, R. C. J. 1988. Fertilization and irrigation effects on fruit quality. Pages 32-45 in: Factors Affecting Fruit Quality—Short Course Proceedings. J. J. Ferguson and W. F. Wardowski, eds. University of Florida and Florida Cooperative Extension Service.

Obreza, T. A., Alva, A. K., Hanlon, E. A., and Rouse R. E. 1993. Citrus grove leaf-tissue and soil testing: Sampling, analysis, and interpretation. Soil Science Fact Sheet, Univ. Fla., Inst. Food Agric. Sci. Publ. SL115.

Smith, P. F. 1966. Citrus nutrition. Pages 174–207 in: Nutrition of Fruit Crops: Tropical, Sub-Tropical, Temperate Tree and Small Fruits. N. F. Childers, ed. Somerset Press, N.J.

Smith, P. F. 1966. Leaf analysis of citrus. Pages 208–228 in: Nutrition of Fruit Crops: Tropical, Sub-Tropical, Temperate Tree and Small Fruits. N. F. Childers, ed. Somerset Press, N.J.

Topp, G. C. 1993. Soil water content. Pages 541–557 in: Soil Sampling and Methods of Analysis. M. R. Carter, ed. Canadian Society of Soil Science.

Tucker, D. P. H., Alva, A. K., Jackson, L. K., and Wheaton, T. A. 1995. Nutrition of Florida citrus trees. Univ. Fla. Coop. Ext. Serv. Publ. SP 169.

Irrigation and Drainage

Boman, B. J. 1994. Microirrigation management considerations for flatwoods citrus. Proc. Soil Crop Sci. Soc. Fla. 53:3–9.

Levy, Y. 1983. Acclimation of citrus to water stress. Sci. Hortic. 20: 267–273.

Levy, Y., and Syvertsen, J. 1981. Water relations of citrus in climates with different evaporative demands. Pages 501–503 in: Proc. Int. Soc. Citric., vol. 2.

Maas, E. V. 1992. Salinity and citriculture. Pages 1290–1301 in: Proc. Int. Soc. Citric., vol. 3.

Marsh, A. W. 1973. Irrigation. Pages 230–279 in: The Citrus Industry. Vol. 3. W. Reuther, ed. University of California, Division of Agricultural Sciences, Berkeley.

Parsons, L. R., and Wheaton, T. A. 1987. Microsprinkler irrigation for freeze protection: Evaporative cooling and extent of protection in an advective freeze. J. Am. Soc. Hortic. Sci. 112:897–902.

Parsons, L. R., Wheaton, T. A., and Tucker, D. P. H. 1986. Florida freezes and the role of water in citrus cold protection. HortScience 21:1–2, 175–176.

Shalhevet, J., and Levy, Y. 1990. Citrus trees. Pages 951–986 in: Irrigation of Agricultural Crops. B. A. Stewart and D. R. Nielsen, eds. Agron. Monogr. 30. Agronomy Society of America, Crop Science Society of America, and Soil Science Society of America, Madison, Wisc.

Cultural Practices

Cary, P. R. 1981. Citrus tree density and pruning practices for the 21st century. Pages 165–168 in: Proc. Int. Soc. Citric., vol. 1.

Koo, R. C. J., and Muraro, R. P., 1982. Effects of tree spacing on fruit production and net returns of 'Pineapple' oranges. Proc. Fla. State Hortic. Soc. 95:29–33.

Wheaton, T. A. 1981. Fruit thinning of Florida mandarins using plant growth regulators. Pages 263–268 in: Proc. Int. Soc. Citric., vol. 1.

Wheaton, T. A., Castle, W. S., Tucker, D. P. H., and Whitney, J. D. 1978. Higher density plantings for Florida citrus—Concepts. Proc. Fla. State Hortic. Soc. 91:27–33.

Wheaton, T. A., Whitney, J. D., Castle, W. S., Muraro, R. P., Browning, H. W., and Tucker, D. P. H. 1995. Citrus scion and rootstock, topping height, and tree spacing affect tree size, yield, fruit quality, and economic return. J. Am. Soc. Hortic. Sci. 120:861–870.

Wheaton, T. A., Whitney, J. D., Tucker, D. P. H., and Castle, W. S. 1984. Cross hedging, tree removal, and topping affect fruit yield and quality of citrus hedgerows. Pages 109–114 in: Proc. Int. Soc. Citric., vol. 1.

Whitney, J. D., and Wheaton, T. A. 1984. Tree spacing affects citrus fruit distribution and yield. Proc. Fla. State Hortic. Soc. 97:44–47.

Weeds

Alva, A. K., and Singh, M. 1991. Use of adjuvants to minimize leaching of herbicides in soil. Environ. Manage. 15:263–267.

Gómez de Barreda, D. 1994. Sistemas de manejo del suelo en citricultura. Generalitat Valenciana, Conselleria de Agricultura, Pesca y Alimentación, Valencia, Spain.

Ingles, C. A. 1994. Management strategies for protecting groundwater quality. Pages 15–32 in: Protecting groundwater quality in citrus production. UC Sustainable Agric. Res. Educ. Program, Univ. Calif., Div. Agric. Nat. Res. Leafl. 21521.

Jordan, L. S. 1978. Benefits and problems of herbicide use in citriculture. Pages 209–214 in: Proc. Int. Soc. Citric.

Jordan, L. S., and Day, B. E. 1973. Weed control in citrus. Pages 82–96 in: The Citrus Industry. Vol. 3. W. Reuther, ed. University of California, Division of Agricultural Sciences, Berkeley.

Knapp J. L., Fasulo, R. T., Tucker D. P. H., and Parsons, L. R. 1982. The effects of different irrigation and weed management practices on mite population in citrus groves. Proc. Fla. State Hortic. Soc. 95:47–50.

Prichard, T. L., Sills, W. M., Asai, W. K., Hendricks, L. C., and Elmore, C. L. 1989. Orchard water use and soil characteristics. Calif. Agric., July–Aug., pp. 23–24.

Singh, M., and Achhireddy, N. R. 1984. Tolerance of citrus rootstocks to preemergence herbicides. J. Environ. Hortic. 2:73–76.

Singh, M., and Siyuam, T. 1994. Weed management in citrus. Pages 316–323 in: Weed Management for Developing Countries. R. Labrada, J. C. Caseley, and C. Parker, eds. FAO Plant Production and Protection Paper. Food and Agriculture Organization, Rome.

Singh, M., and Tucker, D. P. H. 1988. Herbicide effects on tree growth and weed populations in citrus. Proc. Fla. State Hortic. Soc. 101:59–61.

Singh, M., Tucker, D. P. H., and Futch, S. H. 1990. Multiple applications of preemergence herbicide tank mixtures in young citrus groves. Proc. Fla. State Hortic. Soc. 103:16–21.

Tucker, D. P. H., Muraro, R. P., and Abbitt, B. 1980. Two weed control systems for Florida citrus. Proc. Fla. State Hortic. Soc. 93: 30–33.

Diseases

Bar-Joseph, M., Marcus, R., and Lee, R. F. 1989. The continuous challenge of citrus tristeza virus control. Annu. Rev. Phytopathol. 27:291–316.

Beattie, B., Wild, B., Tugwell, B., Bagshaw, J., and Keenan, P. 1992. Postharvest management. Pages 122–184 in: Guide to Quality Management in the Citrus Industry. B. Beattie and L. Revelant, eds. Australian Horticultural Corp., Sydney.

Bové, J. M. 1995. Virus and Virus-Like Diseases of Citrus in the Near East Region. Food and Agriculture Organization, Rome.

Bové, J. M., and Vogel, R. 1980. Descriptions and Illustrations of Virus and Virus-Like Diseases of Citrus. Vols. 1–4. SETCO-IFAC, Paris.

Brown, G. E. 1980. Fruit handling and decay control techniques affecting quality. Pages 193–224 in: Citrus Nutrition and Quality. S. Nagy and J. J. Attaway, eds. ACS Symp. Ser., no. 143. American Chemical Society, Washington, D.C.

Duncan, L. W., Graham, J. H., and Timmer, L. W. 1993. Seasonal patterns associated with *Tylenchulus semipenetrans* and *Phytophthora parasitica* in the citrus rhizosphere. Phytopathology 83:573–581.

Eckert, J. W. 1990. Impact of fungicide resistance on citrus fruit decay control. Pages 286–302 in: Managing Resistance to Agrochemicals: From Fundamental Research to Practical Strategies. M. B. Green, H. M. Lebaron, and W. K. Moberg, eds. ACS Symp. Ser., no. 421. American Chemical Society, Washington, D.C.

Eckert, J. W., and Eaks, I. L. 1989. Postharvest disorders and diseases of citrus fruits. Pages 179–260 in: The Citrus Industry. Vol. 5. W. Reuther, E. C. Calavan, and G. E. Carman, eds. University of California, Division of Agriculture and Natural Resources, Oakland.

Frison, E. A., and Taher M. M., eds. 1991. FAO/IBPGR Technical Guidelines for the Safe Movement of Citrus Germplasm. Food and Agriculture Organization and International Board for Plant Genetic Resources, Rome.

Garnier, M., and Bové, J. M. 1993. Citrus greening disease and the greening bacterium. Pages 212–219 in: Proc. Conf. Int. Org. Citrus Virol., 12th. P. Moreno, J. V. da Graça, and L. W. Timmer, eds. IOCV, Riverside, Calif.

Graham, J. H. 1995. Root regeneration and tolerance of citrus rootstocks to root rot caused by *Phytophthora nicotianae*. Phytopathology 85:111–117.

Graham, J. H., and Gottwald, T. R. 1991. Research perspectives on eradication of citrus bacterial diseases in Florida. Plant Dis. 75: 1193–1200.

Graham, J. H., and Timmer, N. H. 1991. Peat-based media as a source of *Thielaviopsis basicola* causing black root rot on citrus seedlings. Plant Dis. 75:1246–1249.

Graham, J. H., and Timmer, L. W. 1992. *Phytophthora* diseases of citrus. Pages 250–269 in: Plant Diseases of International Importance. Vol. 3, Diseases of Fruit Crops. J. Kumar, H. S. Chaube, U. S. Singh, and A. N. Mukhopadhyay, eds. Prentice-Hall, Englewood Cliffs, N.J.

Graham, J. H., Timmer, L. W., Drouillard, D. L., and Peever, T. L. 1998. Characterization of *Phytophthora* spp. causing outbreaks of citrus brown rot in Florida. Phytopathology 88:724–729.

Hadidi, A., Khetarpal, R. K., and Koganezawa, H., eds. 1998. Plant Virus Disease Control. American Phytopathological Society, St. Paul, Minn.

International Board for Plant Genetic Resources. 1991. Technical Guidelines for the Safe Movement of Citrus Germplasm. Food and Agriculture Organization and IBPGR, Rome.

Kader, A. A., ed. 1992. Postharvest Technology of Horticultural Crops. 2nd ed. University of California, Division of Agriculture and Natural Resources, Berkeley.

Klotz, L. J. 1978. Fungal, bacterial, and nonparasitic diseases and injuries originating in the seedbed nursery and orchard. Pages 1–66 in: The Citrus Industry. Vol. 4. W. Reuther, E. C. Calavan, and G. E. Carman, eds. University of California, Division of Agricultural Sciences, Richmond.

Kotzé, J. M. 1981. Epidemiology and control of citrus black spot in South Africa. Plant Dis. 65:945–950.

Lee, A. T. C., and Roxburgh, K. 1993. Guideline for the Production of Container-Grown Citrus Nursery Trees in South Africa. Outspan International, Port Elizabeth, South Africa.

Lee, R. F., Baker, P. S., and Rocha-Peña, M. A. 1994. The Citrus Tristeza Virus (CTV). International Institute of Biological Control, Ascot, Berks., UK.

Lutz, A., and Menge, J. A. 1986. Citrus root health. II. Phytophthora root rot. Citrograph 72:33–39.

Navarro, L. 1993. Citrus sanitation, quarantine and certification programs. Pages 383–391 in: Proc. Conf. Int. Org. Citrus Virol., 12th. P. Moreno, J. V. da Graça, and L. W. Timmer, eds. IOCV, Riverside, Calif.

Pelser, P. du T. 1988. Recommendations for the Control of Post-Harvest Decay of Citrus Fruits. South African Co-operative Citrus Exchange, Pretoria.

Rocha-Peña, M. A., Lee, R. F., Lastra, R., Niblett, C. L., Ochoa-Corona, F. M., Garnsey, S. M., and Yokomi, R. K. 1995. Citrus tristeza virus and its aphid vector *Toxoptera citricida*. Plant Dis. 79: 437–445.

Roistacher, C. N. 1992. Graft-Transmissible Diseases of Citrus: Handbook for Detection and Diagnosis. Food and Agriculture Organization, Rome.

Roistacher, C. N. 1993. Psorosis—A review. Pages 139–154 in: Proc. Conf. Int. Org. Citrus Virol., 12th. P. Moreno, J. V. da Graça, and L.W. Timmer, eds. IOCV, Riverside, Calif.

Stall, R. E., and Civerolo, E. L. 1991. Research relating to the recent outbreak of citrus canker in Florida. Annu. Rev. Phytopathol. 29: 399–420.

Semancik, J. S., and Duran-Vila, N. 1991. The grouping of citrus viroids: Additional physical and biological determinants and relationships with diseases of citrus. Pages 178–188 in: Proc. Conf. Int. Org. Citrus Virol., 11th. R. H. Brlansky, R. F. Lee, and L. W. Timmer, eds. IOCV, Riverside, Calif.

Sharp, J. L., and Hallman, G. L., eds. 1994. Quarantine Treatments for Pests and Food Plants. Westview Press, Boulder, Colo.

Smoot, J. J., Houck, L. G., and Johnson, H. B. 1971. Market diseases of citrus and other subtropical fruits. U.S. Dep. Agric. Agric. Handb. 398.

Timmer, L. W., Agostini, J. P., Zitko, S. E., and Zulfiqar, M. 1994. Postbloom fruit drop, an increasingly prevalent disease of citrus in the Americas. Plant Dis. 78:329–334.

Timmer, L. W., Priest, M., Broadbent, P., and Tan, M.-K. 1996. Morphological and pathological characterization of species of *Elsinoe* causing scab diseases of citrus. Phytopathology 86:1032–1038.

Timmer, L. W., Sandler, H. A., Graham, J. H., and Zitko, S. E. 1988. Sampling citrus orchards in Florida to estimate populations of *Phyophthora parasitica.* Phytopathology 78:940–944.

Timmer, L. W., Sandler, H. A., Graham, J. H., and Zitko, S. E. 1989. Phytophthora feeder root rot of bearing citrus: Fungicide effects on soil populations of *Phytophthora parasitica* and citrus tree productivity. Proc. Fla. State Hortic. Soc. 102:5–9.

Tugwell, B. L., and Gillespie, K. J., eds. 1989. A Citrus Handling Guide. Department of Agriculture, Adelaide, South Australia.

Wallace, J. M. 1978. Virus and virus-like diseases. Pages 67–184 in: The Citrus Industry. Vol. 4. W. Reuther, E. C. Calavan, and G. E. Carman, eds. University of California, Division of Agricultural Sciences, Richmond.

Wardowski, W. F., Miller, W. M., and Grierson, W. 1987. Packingline machinery for Florida citrus packinghouses. Univ. Fla. Bull. 239.

Wardowski, W. F., Nagy, S., and Grierson, W., eds. 1986. Fresh Citrus Fruits. AVI Publishing, Westport, Conn.

Whiteside, J. O. 1977. Behavior and control of greasy spot in Florida citrus groves. Pages 981–986. Proc. Int. Soc. Citric., vol. 3.

Whiteside, J. O. 1977. Sites of action of fungicides in the control of citrus melanose. Phytopathology 67:1067–1072.

Whiteside, J. O., Garnsey, S. M., and Timmer, L. W., eds. 1988. Compendium of Citrus Diseases. American Phytopathological Society, St. Paul, Minn.

Nematodes

Baines, R. C., Van Gundy, S. D., and DuCharme, E. P. 1978. Nematodes attacking citrus. Pages 321–345 in: The Citrus Industry. Vol. 4. W. Reuther, E. C. Calavan, and G. Carman, eds. University of California, Division of Agricultural Sciences, Richmond.

Cohn, E. 1972. Nematode diseases of citrus. Pages 125–244 in: Economic Nematology. J. E. Webster, ed. Academic Press, London.

Duncan, L. W., and Cohn, E. 1990. Nematode parasites of citrus. Pages 321-346 in: Plant Parastic Nematodes in Subtropical and Tropical Agriculture. M. Luc, R. A. Sikora, and J. Bridge, eds. CAB International, Wallingford, UK.

Esser, R. P., O'Bannon, J. H., and Riherd, C. C. 1988. The citrus nursery site approval program for burrowing nematode and its beneficial effect on the citrus industry of Florida. EPPO Bull. 18:579–586.

O'Bannon, J. W., and Tarjan, A. J. 1984. Nematodes in citrus. Pages 395–433 in: Plant and Insect Nematodes. W. R. Nickle, ed. Marcel Dekker, New York.

Arthropod Pests

Dekle, G. W., ed. 1976. Arthropods of Florida and Neighboring Land Areas. Vol. 3, Florida Armored Scale Insects. Florida Department of Agriculture and Consumer Services, Division of Plant Industry, Tallahassee.

Duncan, L. W., and McCoy, C. W. 1996. Vertical distribution in soil, persistence and efficacy against citrus root weevil of two species of entomogenous nematodes. Environ. Entomol. 25:174–178.

Gerson, U. 1996. Arthropod root pests. Pages 797–809 in: Plant Roots, The Hidden Half. Y. Waisel, A. Eshel, and U. Kafkafi, eds. Marcel Dekker, New York.

Hamon, A. B., and Williams, M. L., eds. 1984. Arthropods of Florida and Neighboring Land Areas. Vol. 11, The Soft Scale Insects of Florida (Homoptera: Coccoidea: Coccidae). Florida Department of Agriculture and Consumer Services, Division of Plant Industry, Tallahassee.

Haney, P. B. 1988. Identification, ecology, and control of ants in citrus: A world survey. Pages 1227–1251 in: Proc. Int. Soc. Citric., vol. 3.

McCoy, C. W. 1994. Citrus root weevils: Biology and current IPM strategies in Florida. Pages 145–170 in: Citrus Integrated Management of Insects and Mite Pests: A World Overview. S. Gravena and L. C. Donadio, eds. Estação Experimental de Citricultura de Bebedouro, São Paulo, Brazil.

O'Brien, C. W., and Wibmer, G. J. 1982. Annotated check list of the weevils (Curculionidae sensu lato) of North America, Central America, and the West Indies (Coleoptera: Curculionoidae). Pages 28–31 in: Memoirs of the American Entomological Institute. AEI, Ann Arbor, Mich.

Schroeder, W. J., and Beavers, J. B. 1977. Citrus root weevils in Florida: Identification, biology and control. Pages 498–500 in: Proc. Int. Soc. Citric., vol. 2.

Stansly, P. A., and Rouse, R. E. 1993. Causes and prevention of injury to young citrus by termites. Citrus Ind. 74(2):10–13.

Van Whervin, L. W. 1968. The citrus weevils of Jamaica and some of their parasites. Univ. West Indies Citrus Res. Bull. 1:1–23.

Whitwell, A. C. 1990. *Diaprepes* spp. problems in Dominica and some possible solutions. Pages 529–541 in: Rencontres Caraibes en Lutte Biologique, Guadeloupe.

Woodruff, R. E. 1985. Citrus weevils in Florida and the West Indies: Preliminary report on systematics, biology and distribution (Coleoptera: Curculionidae). Fla. Entomol. 68:370–379.

Economics

Browning, H. W., and Muraro, R. P. 1993. Approaches to integrated pest management on citrus fruit destined for processing. Citrus Ind. 74(3):49–56.

Futch, S. H., and Muraro, R. P. 1993. Budgeting citrus production costs to maintain profitability. Citrus Ind. 74:48–51.

Larocca, L., and Mika, R. 1995. Proyecto de diversificacion productiva proyecto regional procitrus. Bol. Inf. Citricola No. 1. INTA, Concordia, Entre Ríos, Argentina.

Muraro, R. P. 1997. 1996–97 Comparative citrus budgets. Citrus Ind. 78(7):25–36.

Muraro, R. P., Fairchild, G. F., and Behr, R. M. 1991. Citrus allocation: Decisions between fresh and processed markets. Proc. Fla. State Hortic. Soc. 104:214–217.

Muraro, R. P., Roka, F. M., and Rouse, R. E. 1997. Budgeting costs and returns for southwest Florida citrus production, 1996–97. Univ. Fla., Inst. Food Agric. Sci., Food Res. Econ. Dep. Econ. Inf. Rep. EI 97-6.

O'Connell, N., Klnonsky, K., Freeman, M., Kallsen, C., and Livingston, P. 1995. Sample costs to establish an orange orchard and produce oranges. Univ. Calif. Coop. Ext. Serv., Davis, Bull. 7.

Stricker, J. A., and Muraro, R. P. 1981. Determining most profitable level of nitrogen fertilization for citrus. Proc. Fla. State Hortic. Soc. 94:8–11.

Index